GÖDEL'S DISJUNCTION

# Gödel's Disjunction

*The scope and limits of mathematical knowledge*

LEON HORSTEN

*Professor of Philosophy, University of Bristol*

PHILIP WELCH

*Professor of Mathematical Logic, University of Bristol*

OXFORD
UNIVERSITY PRESS

OXFORD
UNIVERSITY PRESS

Great Clarendon Street, Oxford, OX2 6DP,
United Kingdom

Oxford University Press is a department of the University of Oxford.
It furthers the University's objective of excellence in research, scholarship,
and education by publishing worldwide. Oxford is a registered trade mark of
Oxford University Press in the UK and in certain other countries



First Edition published in 2016



Published in the United States of America by Oxford University Press
198 Madison Avenue, New York, NY 10016, United States of America

British Library Cataloguing in Publication Data
Data available

Library of Congress Control Number: 2016930237

ISBN 978–0–19–875959–1

# CONTENTS

# 1

# Introduction

LEON HORSTEN AND PHILIP WELCH

**Background**—In the month of March, in 2012 and 2013, a small group of invited participants gathered at the University of Bristol to discuss a particular subject within the philosophy of mathematics. This centred around, but was not limited to, some consequences that they and others drew from Gödel's incompleteness theorems in Mathematical Logic. One view of the incompleteness theorems (drawn by people such as Roger Penrose and John Lucas) is that these theorems indicate, or even prove, that the human mind surpasses any machine. Gödel did not hold this view, but rather, encapsulated his view in a two-limbed disjunction, which is known as *Gödel's disjunction* [Gödel 1951]:

> Either the human mathematical mind cannot be captured by an algorithm, or there are absolutely undecidable problems.

In the literature on the subject there is a consensus that Gödel's arguments for his disjunctive conclusion are definitive. At the same time there is a widespread feeling that the existing arguments for or against either of the disjuncts are unconvincing as they stand. The intention of the two workshops was to gain, collectively and through interaction, deeper insight into the content of both disjuncts, and to come closer to deciding the truth value of either of them. The present volume is the outcome of this process.

## 1.1 Gödel's Disjunction and Beyond

In the early 1930s, the celebrated incompleteness and undecidability theorems appeared ([Gödel 1931], [Church 1936], [Turing 1936]). These theorems have profound implications for the limits of certain classes of abstract computing devices (Turing machines) and their physical embodiments (electronic computers). They also appear to point towards conclusions about the limits of human mathematical knowledge. However, informative results about such scope and limits turn out to be very hard to establish conclusively.

*Gödel's Disjunction*. First Edition. Leon Horsten & Philip Welch (eds).

### 1.1.1 *The Disjunctive Thesis*

In [Gödel 1951], Gödel argued for the disjunctive thesis that either the human mathematical mind cannot be captured by an algorithm, or there are absolutely undecidable problems. It would appear that there is not, and cannot be, a mathematical proof of Gödel's disjunction. The reason for this is that the disjunctive thesis involves informal concepts (human mathematical mind, algorithm, absolute undecidability), whereas mathematical proofs only involve formal concepts. Nonetheless, there is agreement in the literature that Gödel's *a priori* argument establishes his disjunctive thesis as conclusively as any philosophical argument ever can.

Gödel's argument goes roughly as follows. Suppose that the human mathematical mind is ultimately algorithmic, i.e. that there is an algorithm that produces all the mathematical theorems that the human mathematical mind is capable of producing. The Church–Turing thesis says that every algorithmically computable function is computable by a Turing machine. This is mathematically equivalent to saying that the collection of humanly knowable theorems can be recursively axiomatised in some formal theory $T$. This theory will evidently be consistent. Gödel's second incompleteness theorem says that consistent theories (that contain the elementary laws of arithmetic) cannot prove their own consistency, even though they can express their own consistency (via numerical coding). So there will be mathematical propositions $\varphi$ that cannot be decided by $T$. But since $T$ is supposed to capture what is humanly provable, this means that these propositions $\varphi$ are humanly, i.e. absolutely, undecidable.

Gödel observed out that his disjunction is not exclusive: it is possible that both disjuncts are true. Nonetheless, the disjuncts seem to point in opposite directions. The first disjunct expresses an aspect of the *power* of the human mind. The second disjunct expresses an aspect of its *limitations*. The negation of the first disjunct says that, in some sense, the mathematical human mind is an algorithmic device. This thesis lies at the heart of the most influential theories in contemporary philosophy of mind. The negation of the second disjunct says that the human mind is capable of complete mathematical knowledge. Weaker theses can be obtained by restricting the mathematical realm about which knowledge is gained, for instance to arithmetic or to analysis.

Even though Gödel believed that the first disjunct holds and that the second disjunct is false, he did not find a way conclusively to establish either disjunct. However other mathematicians, scientists, and philosophers attempted to strengthen Gödel's result by arguing for or against one of the disjuncts. The aim is for such arguments to be as incontrovertible as Gödel's argument for his disjunctive thesis. Preferably, these arguments should be of a purely *a priori* nature.

### 1.1.2 *The First Disjunct*

J.R. Lucas formulated an argument for the first disjunct ('the human mind is not algorithmic') based on Gödel's incompleteness results [Lucas 61]. Lucas' original argument goes roughly as follows. Suppose, for a reductio, that the human mathematical mind is algorithmic. Then by the Church–Turing thesis there is a Turing machine that enumerates all humanly knowable mathematical theorems. As before, this is mathematically equivalent to saying that the collection of humanly knowably theorems can be recursively axiomatised in some formal theory $T$. Then, by the second incompleteness theorem, the (consistent) theory $T$ cannot prove its own consistency. Since $T$ is supposed to encapsulate what is humanly knowable, this means that the consistency of $T$, i.e. the consistency of what is mathematically knowable, is not itself humanly knowable. But we *do* know that what is humanly knowable is consistent. So, under the supposition that $T$ is the extension of what is humanly knowable, we do know that $T$ is consistent. But then $T$ does not capture the extension of humanly knowable mathematical facts after all. In sum, we have reached a contradiction, and have to reject the supposition that the human mind is algorithmic.

Lucas' argument and variations on it were discussed (albeit not widely) in the philosophical community, where they did not meet with much enthusiasm. Penrose later revived Lucas' argument, and produced, in two phases ([Penrose 89], [Penrose 94]), essentially two variations of it. His arguments have since then been widely discussed, not only in the philosophical community, but also in the physical and the logical communities. A critical discussion of Penrose's arguments can be found in issues 13.4(1990) and 16.3(1993) of the journal *Behavioural and Brain Sciences*.

Most philosophers and logicians believe that variants of the arguments of Lucas and Penrose are not fully convincing. For instance, Benacerraf objected that Lucas' argument does not exclude the possibility that the human mathematical mind is indeed a Turing machine (or, equivalently, a formal mathematical theory), *but it is not humanly knowable which one it is* [Benacerraf 67].

Lucas' argument makes use of a deep theorem in mathematical logic (the second incompleteness theorem) and a deep thesis which is almost universally accepted (the Church–Turing thesis). Nonetheless, his argument, and his and Penrose's variations on it, are formulated in informal terms. Indeed, the reactions to Lucas' and Penrose's informal arguments are also formulated in informal terms, even though they too appeal to mathematical results.

Gödel's argument for his disjunctive thesis was exactly of this nature (informal but making use of formal results) and it is, as we have seen, highly compelling. But the informal arguments for or against the first or the second disjuncts have remained contentious. To this day it is not completely clear exactly what the status of the central arguments in this debate is.

### *1.1.3 The Second Disjunct*

There have been attempts to establish variants of the second disjunct, which says that there are absolutely undecidable mathematical propositions.

Fitch produced an argument for the thesis that there are true but absolutely unknowable propositions of the form '$p$ and it is not known that $p$' [Fitch 63]. Such propositions are only semi-mathematical (because the concept of knowledge enters into them). However, if some of them are true but unknowable, then these must indeed be absolutely undecidable. Fitch's argument uses the extremely weak empirical premise that there are in fact unknown truths. Fitch's argument was not much discussed until it was revisited by Williamson

[Williamson 00], who concludes that it is simply incontrovertible. A consensus seems to be emerging that Fitch's argument is convincing. But one could hope for a stronger conclusion. Fitch's argument establishes the existence of semi-mathematical, absolutely undecidable propositions without producing an instance of an absolutely undecidable proposition. So one might hope for more detailed information: it would be desirable if we could either give an instance of an absolutely undecidable semi-mathematical proposition, or could show that one cannot give an *a priori* argument that establishes of a proposition that it is absolutely undecidable.

Some have argued that there are certain purely mathematical propositions, such as the continuum hypothesis, that are truth-valueless. If there indeed are such sentences, then they must be absolutely undecidable. However, no consensus has emerged as to the existence of mathematical propositions that have no truth value. Another question is whether there are absolutely undecidable sentences that, in addition to the arithmetical vocabulary, contain a primitive notion of provability. Reinhardt has observed that absolutely undecidable sentences can be shown to exist if we have a provably sound absolute provability *predicate* [Reinhardt 86, p. 439]. Reinhardt's absolutely undecidable sentence is produced by diagonalisation (the fixed-point lemma). Intuitively, it is a sentence which says of itself that it is absolutely unprovable. It is not *purely* arithmetical because it contains an occurrence of the absolute provability predicate. Like the liar sentence, which says of itself that it is true—and perhaps like the continuum hypothesis—it seems that Reinhardt's self-referential sentence has no truth value.

So there seem to be reasons to believe that there are propositions that are absolutely undecidable because they are truth-valueless. But when attention is restricted to mathematical propositions that do have a truth value, it seems very difficult to argue on purely *a priori* grounds that some of them must be absolutely undecidable.

Hilbert expressed a strong form of rational optimism according to which there are no purely mathematical propositions that are absolutely undecidable [Hilbert 1970, p. 387]. Wang reported that Gödel shared Hilbert's optimism [Wang 1974, pp. 324–326]. But it seems fair to say that both Hilbert and Gödel realised that they did not possess decisive arguments to back up their optimism. Indeed, some authors argue that it is likely that there are absolutely undecidable propositions [Feferman & Solovay 1992, p. 292].

So again we are in a situation where going beyond Gödel's disjunction proves difficult.

## 1.2 Formal Frameworks

### *1.2.1 Mathematical Philosophy to the Rescue?*

Lucas' argument makes use of a deep theorem of mathematical logic (the second incompleteness theorem) and a deep thesis which is almost universally accepted (the Church–Turing thesis). Nonetheless, we have seen that his argument, and his and Penrose's variations on it, are formulated in informal terms. Indeed, the reactions to Lucas' and Penrose's informal arguments are also formulated in informal terms, even though they too appeal to mathematical results. One reason why the soundness of the Lucas–Penrose arguments has remained unclear for so long, while the question of the soundness of the Fitch–Williamson argument can be settled more easily, or at least discussed more transparently, is that the Fitch–Williamson argument has been formulated and discussed within a formal setting: the framework of propositional modal-epistemic logic.

If we want sharper results than Gödel's disjunction to be established decisively, then we must go beyond conducting the debate in informal terms. We must try to bring logico-mathematical methods to bear on our research questions. We must engage in what Kreisel called an exercise of informal rigour [Kreisel 1972], or what today is called an exercise in mathematical philosophy [Leitgeb 2013], [Horsten 2013]. In particular, two things are presently needed on the logical side. On the one hand, we must try to *formalise* some of the central arguments in the debate, so that we are clear about their exact logical form. This involves formally connecting informal notions (such as that of human knowability, and that of algorithm) with mathematical notions (such as that of formal derivability, and the notion of a Turing machine). On the other hand, we need to produce a class of *models* (in the logician's sense of the word) for mathematical knowability. For instance, we want to know whether it is indeed a possibility that the human mind is a Turing machine (or a formal system), but we don't and cannot know which one, as Benacerraf surmises. To show that it is, we would seem to need to produce a somewhat realistic (but schematic) model that witnesses this possibility.

Of course further conceptual analysis is also badly needed. First and foremost, an analysis of the level of idealisation involved in the notion of human knowability is needed, as was already observed in [Kripke 1980, pp. 34–35]. Furthermore, a deeper conceptual analysis of the notion of algorithm is wanted. But even the extent to and ways in which the notion of consistency of a formal system can be expressed in formal systems is not fully clear.

Even after all this, the truth value of the disjuncts in Gödel's disjunctive thesis will not turn out to be decided. But we will see that deeper insight into their content will have been gained.

### 1.2.2 *Epistemic Mathematics*

One attempt to bring logico-mathematical methods to bear on questions regarding the disjunctive thesis emerged in the early 1980s. Following a suggestion of Myhill [Myhill 1960], Stewart Shapiro and a group of mathematical logicians (H. Friedman, Flagg, N. Goodman, Shapiro, and Scedrov) started investigating mathematical theories (arithmetical, type-theoretical, set theoretical) to which a sentential absolute provability operator ($K$, for knowability) is added [Shapiro 85]. The notion of absolute provability is intended to be intensionally different to the notion of provability in a formal system, in that absolute provability is not conceptually connected to any given formal system. The common view is that whereas the notion of provability in a formal system is well understood, the notion of absolute provability is presently unclear at best. The worry regarding the unclarity of the notion of absolute provability is addressed in the contributions of Shapiro, Williamson, and Koellner. Nonetheless, Visser shows in his contribution that there are also conceptual questions relating to provability in a formal system that are presently not completely understood. In particular, in his article he addresses the question of under what conditions it is reasonable to say that a given formal notion expresses provability in a particular formal system?

It was noted by the early investigators in epistemic mathematics that this epistemic framework has a surprisingly large expressive power. First, intuitionistic mathematical theories can quite generally (by Gödel's modal translation) be seen as subtheories of classical epistemic theories [Flagg & Friedman 86]. Thus, the epistemic framework has the expressive power at least of intuitionistic theories. Secondly, in the framework of epistemic arithmetic one can come close to expressing the Church–Turing thesis (CT) as follows ([Shapiro 85a], [Flagg 85], [Horsten 98]):

> If it is (absolutely) provable that for every number $x$ there is a number $y$ which can be shown to stand in the functional relation $\varphi$ to $x$, then the function $\varphi$ is Turing-computable.

This thesis is known as ECT (epistemic Church's thesis"). So in this framework one can come closer to expressing CT than the so-called intuitionistic versions of CT: the intuitionistic version of Church's thesis has the form of a constructive entailment, whereas the Church–Turing thesis and ECT have the form of a non-constructive implication. In other words, while it is arguable that though the negative translation intuitionistic mathematics has at least the power of classical mathematics, epistemic mathematics has a greater expressive power than both.

A problem thus this research programme was that quantification into the context of an absolute provability operator was never well understood. In the context of the expressively rather weak language of Peano arithmetic this did not really present a problem (for all its terms are sufficiently 'transparent'). But when a substitutional interpretation of the quantifiers is no longer at hand (as in the case of second-order epistemic arithmetic), problems arise that are well known from the philosophy of language and that have not been satisfactorily resolved [Kaplan 68]. This is why the authors in this volume are restricting their attention to formal frameworks, such as the language of first-order arithmetic, in which only 'transparent' terms can be formulated.

### *1.2.3 Computation and the Nature of Algorithm*

In the framework of epistemic arithmetic, an approximation to the informal notion of algorithmic computability can be expressed, and the formal relation between this approximation and the notion of Turing-computability can be systematically investigated. But one can also just model the notion of algorithm directly. General frameworks have in recent years been developed for investigating the notion of algorithm ([Moschovakis 98], [Blass & Gurevich 06]). Dean, in his chapter for this volume, seeks to analyse the ontological status of algorithm, and in particular whether algorithms can be considered as mathematical objects, which one might call *algorithmic realism*. The main problem is that of identifying algorithms with mathematical objects in a non-arbitrary way. Two solutions have been proposed: algorithms are to be represented by either (a) equivalence classes of machines or (b) equivalence classes of programs in some suitably large and general programming language.

The notion of algorithmic computability is a form of 'lawlike' behaviour of a function. The opposite of the notion of algorithmic computability is that of complete randomness. This can be seen as a form of lawlessness. Traditionally, this notion of lawlessness has been investigated (by mathematicians such as Brouwer) in an intuitionistic setting. But just as epistemic arithmetic in a sense subsumes constructivist considerations in a classical setting, one can investigate randomness (lawlessness) in a classical setting. This is done by Joan Rand-Moschovakis, who develops a model-theoretic perspective in which both lawlike and lawless sequences can be integrated in one classical model.

## 1.3 Organisation of the Contributions

We have attempted to group the articles around the task to which they contribute. This is not straightforward, because most of the articles contribute to more than one task.

Part I contains the contributions of Dean, Visser, Moschovakis and Achourioti. The principal aim of these articles is to contribute to the formal understanding of key notions involved in Gödel's disjunction: the notions of algorithm, formal consistency, and human knowability.

Part II contains the contributions of Carlson, Koellner, and Shapiro. They are all mainly concerned with the first disjunct. The articles of Koellner and Shapiro evaluate arguments for the first disjunct. The aim of the article of Carlson is to gain deeper insight into Benacerrafian scenarios, according to which the human mathematical mind is a Turing machine, but it does not know which one.

Part III contains the contributions of Leach-Krouse, Williamson, and of Antonutti & Horsten. All these articles are mainly concerned with the second disjunct, i.e., the question whether there are absolutely undecidable propositions.

### *1.3.1 Algorithm, Consistency, and Epistemic Randomness*

#### 1.3.1.1 Dean

Dean addresses the problem of the status of *algorithm* both in mathematics and in computer science. He considers two questions. The first concerns the ontological status of algorithm: are algorithms objects, and if so are they abstract or concrete, or somewhere in between? The second question is more epistemological: how can we justify mathematical results by means of computational methods? (The second has been much discussed since the advent of computational methods in the first proof of the four colour theorem.) One answer to the first could be that algorithms are *mathematical* objects. This Dean dubs *algorithmic realism,* and he claims that it is a sort of conventional wisdom adopted implicitly by researchers such as Moschovakis and Gurevich. A variant which takes into account intensional features of algorithm he denotes by *direct algorithmic realism,* as this then bears affinity with constructive or intuitionistic mathematics.

One aim of the chapter is to discuss algorithmic realism; ultimately Dean argues that it cannot be sustained as a thesis. Although initially plausible, this seems less so when one considers the methodologies of complexity theory and algorithmic analysis (he argues). The main problem is that of identifying the algorithms with mathematical objects in a non-arbitrary way. Two solutions have been proposed: algorithms are to be represented by either (a) equivalence classes of machines or (b) equivalences classes of programs in some suitably large and general programming language. Both approaches have their adherents and their difficulties, and Dean alludes to the corresponding bad company and Julius Caesar difficulties in the neo-logicist program of using abstracta for, e.g. numbers.

Nevertheless this view is in the author's judgement the most popular and the most promising proposal for studying the above two questions. In Section 1.4 he discusses the notion of *simulation* that has been suggested as a substitute for either of the equivalence relations. He argues that it is nevertheless impossible to use this notion as an adequate basis for algorithmic realism. In a final section, he discusses the implications of this for Moschovakis's and Gurevich's theories of algorithms.

#### 1.3.1.2 Visser

In this section the author considers two problems related to the celebrated *second incompleteness theorem* (G2) of Gödel. A first question is whether the G2 applies to very weak theories or even unsound theories $S$. It is well known that weak theories also cannot prove their own consistency statements (e.g. [Bezboruah & Shepherdson 1976]), but that the weakness of the theories undermines any attribution of true consistency to the formalised counterpart $Con_S$. We may therefore ask: when can we reasonably say that a consistency statement such as $Con_T$ really does express the consistency of $T$? Visser introduces variants of G2 most of which deal, not with outright non-provability of a consistency statement from an axiomatic theory $T$, but the *non-interpretability* of a weak arithmetic plus the consistency statement back into $T$. This gives us a range of subtly differing relations.

Visser discusses what it means for a sentence to be a *consistency statement* for a theory $T$. This is nearer to the 'classical' approach of Gödel: what makes something a consistency statement is that it is a statement of the form 'falsum is not provable in $T$', where we have a provability predicate satisfying a number of general global properties. Thus the predicate has a conceptual role as well as some internal properties which characterise that role. These have distilled out in the three Hilbert–Bernays derivability conditions, which have been widely discussed by Feferman. The author considers several generalisations in the setting of interpretability of theories.

His overriding conclusion is that for weak theories it is not the theory $T$ that is the 'locus classicus' where $Con_T$ gets its meaning, but it should be seen rather as being about the interpretability relation between the theory $T$ and the theory $S^1_2 + Con_T$.

Towards the end of the chapter, Visser addresses the second problem of formulating G2 in a general way so that it is 'coordinate-free' (that is, independent of coding), and considers two strategies for dealing with the large number of choices we are forced to make concerning coding, the representation of syntax, and the like. This section summarises previous work of the author.

An appendix gives a basic outline of the fundamental definitions and basic results.

#### 1.3.1.3 Moschovakis

Moschovakis gives a unified presentation of a series of results that yield a model for intuitionistic analysis that accommodates the lawlike sequences of Brouwer in part of the model where classical logic holds, while at the same time in another part of the model, a constructive segment is constituted by the lawless sequences.

Kreisel had proposed axioms for numbers, constructive functions, and lawless sequences [Kreisel 1968]. A strong form of continuous choice was pointed out by Troelstra to be problematic. While Kreisel had identified independence with inequality for lawless sequences, most of his axioms can be made true (with a more natural notion of independence) by defining 'lawless' relative to a well-ordered collection of lawlike sequences whose intended interpretation is countable.

Ultimately Brouwer's continuum can be thought of as an expansion of a completely determined well-ordered classical continuum by what the author describes as a 'relatively chaotic expansion' consisting of the lawless sequences. The method used is to iterate definability in a manner not un-reminiscent of ramified analysis to build a model.

Moschovakis shows that there are parallels with Gödel's wish to move away from constructibility alone as a mechanism from building sets, to a universe defined using so-called 'ordinal definability' as somehow maximally definitional. The method is to define 'lawlike sequence' by iterating definability, with quantification over numbers and lawlike sequences and (restricted) quantification over arbitrary choice sequences, to build a classical model of a three-sorted theory.

#### 1.3.1.4 Achourioti

It was noticed early on by both Reinhardt and Shapiro that in-principle provability exhibits a number of intensional features, which Achourioti believes separate it from standard quantified **S4** that would otherwise be a good candidate logic for a notion of absolute provability (which she calls *in-principle provability* in the chapter). Her aim has been to provide a semantics for in-principle provability that takes into account these intensional features, while keeping the relation between in-principle provability and actual provability. She thus proposes a model for epistemic mathematics together with a modal operator to express this.

The basic principle of the semantics is that mathematics grows both with respect to the entities it constructs and the theorems it proves, where growth is understood in a non-monotonic manner. This means that the modal structures that provide an interpretation to the in-principle provability operator cannot be of the standard quantified **S4** kind, but are given by more general inverse systems. These structures model geometric theories of a restricted logical kind. While there is thus significant difference between this approach and that of Shapiro and Carlson, there is also significant agreement in that the epistemic operator is also meant to define what Carlson calls a theory of knowledge, i.e. mathematical knowledge. Central to this formalisation is an interpretation of an intensional notion of truth to match the intensional notion of absolute provability.

## 1.3.2 *Minds and Machines*

### 1.3.2.1 Carlson

Carlson produces a new theorem in *epistemic arithmetic, EA*. This theorem establishes the existence of *knowing machines*, which know the axioms of *EA* together with *epistemic Church's thesis, ECT*:

> For any function I can compute, I can find a Turing machine which computes the same function.

Clearly this is a strong version of Church's thesis put into an epistemic setting. As Carlson mentions, *EA* + *ECT* was known to be consistent from the results of Flagg [Flagg 85]. What is here added is the idea of a *machine* knowing *EA* [Carlson 00] together with *ECT*. To the language $\mathcal{L}_{PA}$ is added a modal 'knowledge' operator *K*. We may then also express it more formally as follows.

*ECT* is the schema consisting of the universal closures of formulae of the form

$$K\forall x\exists yK\varphi \rightarrow \exists eK\forall x\exists y(\{e\}(x) = y \wedge \varphi),$$

where $e$ does not occur free in $\varphi$ and $\{e\}$ is the $e$th Turing machine (where in turn we should more formally express the clause $\{e\}(x) = y$ by a suitable representing formula). Interpreting this over $\mathbb{N}$, we see the antecedent in this scheme as expressing that it is known that there is some passage from an $x$ to a $y$ that is known to satisfy a formula $\varphi$; the conclusion then states that there is a Turing machine where it is known or can be recognised as (i) total, and (ii) for every $x$ the machine produces a requisite $y$. Carlson identifies *knowing machines* with computably enumerable theories $T$ that encode structures satisfying the axioms of *EA*, together with the knowledge assertions about those axioms: $K\varphi$ for $\varphi \in EA$.

The argument is ingenious and involves building a stratified hierarchy of knowledge languages indexed by ordinals up to $\epsilon_0 \cdot \omega$. Using proof theoretic techniques of his own devising, which consider collapses of structures along such ordinals, he defines a computable collapsing relation on such ordinals which involve $\Sigma_1$-elementarity; by means of this he can use the recursion theorem to build up a knowing entity containing the right theory. The method is subtle and has been used by the author to find a knowing machine that knows *EA* + *SMT*, where *SMT* is Reinhardt's strong mechanistic thesis which says that the human mathematical mind extensionally coincides with the output of a Turing machine. He claims that further elaborations can also establish the independence of *SMT* and *ECT* as well as the existence of a knowing machine of the theories together simultaneously: *EA* + *ECT* + *SMT*.

### 1.3.2.2 Koellner

This chapter provides a deep analysis of Gödel's disjunction and in particular of those arguments of Lucas, Penrose, and others ([Lucas 61], [Penrose 89], [Penrose 94]), purporting to show that the incompleteness theorems demonstrate that the *anti-mechanist thesis* (in short, that the idealised human mind outstrips in capabilities any Turing machine) holds. This is the first disjunct of the two in Gödel's disjunction. If the thesis were true, then it would show something fundamental about human reasoning and its (non-)reducibility to the computational aspect of the physical world.

There is an extensive discussion among logicians concerning Penrose's arguments, with most coming out heavily against Penrose. Indeed they intend to show that Penrose's first arguments cannot really stand, resting as they do upon an unsubstantiated conditional antecedent (as many others have remarked). One of the difficulties, as Koellner says, is that background assumptions on the underlying notions—most importantly, truth, absolute provability, and the related notion of idealised human knowability—are never fully articulated by their proponents. Part of the chapter is taken up by making precise some of these underlying assumptions.

Koellner is ultimately sceptical of Penrose's original arguments and shows how to get some definitive results once the notions have been sufficiently formalised. He discusses earlier results of Reinhardt that show that Gödel's disjunction is provable in $EA_T$, that is epistemic arithmetic with a truth predicate, but that the first conjunct is not so provable.

However, he argues that Penrose's later arguments in [Penrose 94] are more subtle and cannot be so easily dismissed. He devises a theory, DTK, which extends one of Feferman's involving a determinateness predicate D, a truth predicate T, and an absolute provability or 'knowledge' predicate [Feferman 2008]. In a new twist he shows that these must be formalised in a type-free theory of absolute provability and simultaneously a type-free theory of truth. Both notions are treated as *predicates*, and while one might at first expect that Montague's results on the impossibility of modal operators being predicates would stymie this treatment, he manages to sidestep them.

However, using type-free versions of truth and determinateness, one runs the danger of encountering indeterminate sentences during the course of an argument. This is precisely what happens: Penrose's later argument cannot be formulated in DTK since the premises Penrose needs are provably indeterminate in DTK!

Indeed, DTK proves that both Gödel's disjunction and its disjuncts are indeterminate. However, if we restrict the Gödel's disjunction etc. to *determinate* sublanguages (such as PA), which is what we are really after, then we do have that DTK proves the determinateness of Gödel's disjunction; but now DTK neither proves nor refutes either disjunct. Again: Gödel's disjunction is provable and determinate, but the component disjuncts are neither determinate, nor provable, nor refutable. The disproof of the latter two possibilities is extensive and requires an analysis of the axioms system, and is indeed the major technical result of the chapter.

The possibility is left open that one day we shall find the right type-free theory of truth and use it to show that the first conjunct is provable. Thus, the case against any kind of Penrosian argument is not yet water-tight in Koellner's opinion. However he has shown that none of the systems we currently have, allow for such an argument, and in fact the generality of the independence results allows us reason to think that no such system will be discovered.

Koellner takes mostly an uncritical stand towards the notion of absolute provability (since this is in fact what the proponents do as regards the first conjunct—and indeed they do not make precise their notions) for most of the chapter. As he states, he wishes to be as generous as possible to the Penrose side of the argument, so that his demonstration of their impossibility will have the most force. However in the final section he reveals his criticisms of this notion. The main difficulty he argues is with the notion of 'idealising the mind' in such discussions, as we currently do not have a good model of the mind to idealise. Indeed he thinks it is suspect to talk of 'mechanising the mind' at all. He thus concludes with his own dichotomy: either the notions of absolute provability and knowability by an idealised human being are problematic (for the reasons he gives), or his results place serious limitations on any argument for the first disjunct.

This chapter arguably amounts to one of the most thorough-going and penetrating analyses of Gödel's disjunction to date.

#### 1.3.2.3 Shapiro

Shapiro also takes up the argument that the mechanist thesis concerning the first clause of the Gödel disjunct is nowhere near precise and plausible enough to be undermined by the incompleteness theorems. He asks if the collection of unassailable, mathematically certain, and so knowable sentences, has crisp boundaries; and points out that the Lucas–Penrose side of the argument requires this. In a second section he outlines the point again that Gödel held a rational optimist's view that the second conjunct did not hold, but that what puts Gödel outside the Penrose–Lucas camp is that Gödel did not regard rational optimism as established.

The problem again is *idealisation*. This discussion is the major part of the chapter. In short, while it is clear (or clearer) what an idealised Turing machine may be, the idealised human entity, even when medical limitations are ignored, somehow has to have the attribute of some sort of insight. We recognise that *if* a formal theory is consistent, then the Gödel sentence for that theory is true. This is not just for some straightforward axiomatisations such as Peano Arithmetic (PA), but for all future devised formal theories. Shapiro is saying that as we embed PA and other theories in further theories such as those of a 'higher type', we may lose our unassailable confidence that the theories are consistent. We thus run into epistemological questions concerning this idealised human or human community.

A final section considers ordinal analysis of the kind initiated by Turing in his ordinal logics [Turing 1939]. Shapiro runs through the steps of Turing's analysis and sketches Feferman's study of various reflection principles that advanced Turing's work beyond the adding of simple consistency statements [Feferman 1962]. As Shapiro says, no machine can iterate the Lucas–Penrose procedure of somehow just adding Gödel sentences to a theory to obtain more knowledge, but neither can Lucas (or Penrose, but it is principally Lucas who has advanced these kinds of argument), since they also have to 'recognise' notations for recursive ordinals. Rather like Penrose, Lucas is stuck with conditionals: *if* $e$ is a code of a Turing machine that enumerates an increasing sequence of ordinals, *then* $3.5^e$ denotes an ordinal greater than any enumerated by machine $e$. How is he to recognise the antecedent as true?

### 1.3.3 *Absolute Undecidability*

#### 1.3.3.1 Leach-Krouse

The author approaches the workshop's theme by discussing the possibility of defining a notion of *absolute provability* via two differing conceptions of the notion through the work of

Gödel ([Gödel 1951]) and previous work of Post [Post 2004]. We have seen that Gödel formulated the dichotomy that either (a) there exist absolutely undecidable propositions, or (b) the process of human reasoning cannot be imitated by a machine. Both Post and Gödel arrived at (equivalent) versions of this dichotomy and both held it to be valid. But whereas Gödel hoped to refute (a) and establish (b), for Post it was the other way around. The chapter traces their respective paths to these dichotomies. The first question Leach-Krouse asks is: why did these two come to take opposing sides of the dichotomy? The second question is: how do we know that they were talking about the same sense of absolute provability? The first part of the chapter lays out this historical groundwork and seeks to explain Post's reasons for believing there would be absolutely undecidable propositions, and Gödel's reasons for the contrary.

Leach-Krouse, towards an answer to the second question, describes a notion of *sub-groundedness,* which is an adequacy condition that both Gödel and Post seem to have accepted on account of absolute provability. In one section he describes the tentative steps both researchers took to producing formal frameworks for studying absolute provability—although both would ultimately reject their own definitions. He then gives a problem (a *diagonalisation problem*) that a set of sensible desiderata of an absolute provability predicate—which contains Hilbert-Bernays like conditions on provability predicates—cannot all be satisfied by the set $P$ of absolutely provable sentences. He argues that this problem confronts any attempt to produce such formal frameworks that satisfy the sub-groundedness requirement, and argues that both the earlier accounts are unlikely to overcome this.

#### 1.3.3.2 Williamson

Williamson discusses the Gödelian disjunction, arguing for the non-analyticity of first principles of mathematics. Creatures might evolve (indeed may just be future evolved humans) to be hard-wired to accept the truth of a mathematical proposition $\varphi$ and come to accept it as an axiom, because they find it primitively compelling, even if we (current day) humans do not. As mathematicians continue their work they accumulate mathematical knowledge, not only by the purely mathematically formal methods of proof, but also in other ways (as we have accumulated more mathematical knowledge than our forbears), and we now find certain propositions compelling, which, e.g. the Greeks would not.

Williamson argues that this process leads to the conclusion that such a proposition $\varphi$ is absolutely provable, not because there are such creatures (somewhere in the universe) but that there *could* be such in principle. Absolute provability of a proposition is then equivalent to saying that it can be known by a normal mathematical process. Absolute provability is not contingent on evolution. Therefore, Williamson argues, every truth of mathematics is absolutely provable in that an evolved creature may find it primitively compelling. It is overwhelmingly laudable that every sentence of arithmetic is either true or false; hence all are absolutely decidable.

#### 1.3.3.3 Antonutti and Horsten

Antonutti and Horsten investigate an analogue of Gödel's disjunction, which says that either *ECT* is true, or there are absolutely undecidable propositions (involving the notion of absolute provability). Using the framework of epistemic arithmetic they establish that this new disjunctive thesis holds. More in particular, they show that if *ECT* is true, then there are absolutely undecidable propositions that are of low arithmetical complexity.

As with Gödel's disjunction, there then is also for this new disjunctive thesis the question of what is the truth value of each of the disjuncts? And as with Gödel's disjunction, we have no conclusive results. All that they are presently able to show is that in a class of models for absolute provability that are based on Feferman's transfinite progressions of formal theories, *ECT* holds.

## References


[Benacerraf 67] Benacerraf, P. God, the devil and Gödel. *The Monist* **51**(1967), pp. 9–32.

[Blass & Gurevich 06] Blass, A. & Gurevich, Y. Algorithms: A quest for absolute definitions. In: A. Olszewski et al. (eds.) *Church's Thesis after 70 Years.* Ontos Verlag (2006), pp. 24–57.

[Bezboruah & Shepherdson 1976] Bezboruah, A. & Shepherdson, J. Gödel's second incompleteness theorem for Q. *Journal of Symbolic Logic* **41**(1976), pp. 503–512.

[Carlson 00] Carlson, T. Knowledge, machines, and the consistency of Reinhardt's strong mechanistic thesis. *Journal of Pure and Applied Logic* **105**(2000), pp. 51–82.

[Church 1936] Church, A. An unsolvable problem of elementary number theory. *American Journal of Mathematics* **58**(1936), pp. 345–363.

[Feferman 1962] Feferman, S. Transfinite recursive progressions of formal theories. *Journal of Symbolic Logic* **27**(1962), pp. 259–316.

[Feferman & Solovay 1992] Feferman, S. & Solovay, R. Remark 2 of the Introductory Note to 1972a. In: Gödel, K. *Collected Works,* Volume 2, Oxford University Press (1992), pp. 287–292.

[Feferman 2008] Axioms for determinateness and truth. *Review of Symbolic Logic,* **1**(2008), pp. 204–217.

[Fitch 63] Fitch, F. A logical analysis of some value concepts. *Journal of Symbolic Logic* **28**(1963), pp. 135–142.

[Flagg 85] Flagg, R.C. Church's Thesis is consistent with Epistemic Arithmetic. In: [Shapiro 85, pp. 121–172].

[Flagg & Friedman 86] Flagg, R. & Friedman, H. Epistemic and intuitionistic formal systems. *Annals of Pure and Applied Logic* **34**(1986), pp. 53–60.

[Gödel 1931] Gödel, K. On formally undecidable propositions in *Principia Mathematica* and related systems I. In: S. Feferman et al. (eds.) *Kurt Gödel. Collected Works.* Volume I: *Publications 1929–1936.* Oxford University Press (1986), pp. 144–195.

[Gödel 1951] Gödel, K. Some basic theorems on the foundations of mathematics and their implications. In: S. Feferman et al. (eds.) *Kurt Gödel. Collected Works.* Volume III: *Unpublished Essays and Lectures.* Oxford University Press (1995), pp. 304–323.

[Hilbert 1970] Hilbert, D. *Gesammelte Abhandlungen*, Volume 3, Second Edition, Springer (1970).

[Horsten 98] Horsten, L. In defense of epistemic arithmetic. *Synthese* **116**(1998), pp. 1–25.

[Horsten 2013] Horsten, L. Mathematical philosophy? In: H. Andersen et al. (eds.) *The Philosophy of Science in a European Perspective.* Volume 4, Springer (2013), pp. 73–86.

[Kaplan 68] Kaplan, D. Quantifying in. *Synthese* **19**(1968), pp. 178–214.

[Kreisel 1968] Kreisel, G. Lawless sequences of natural numbers. *Compositio Mathematica* **20**(1968), pp. 222–248.

[Kreisel 1972] Kreisel, G. Informal rigour and completeness proofs. In: I. Lakatos (ed.), *Problems in the Philosophy of Mathematics.* North-Holland (1972), pp. 138–186.

[Kripke 1980] Kripke, S. *Naming and Necessity.* Harvard University Press (1980).

[Leitgeb 2013] Leitgeb, H. Scientific philosophy, mathematical philosophy, and all that. *Metaphilosophy* **44**(2013), pp. 267–275.

[Lucas 61] Lucas, J. Minds, machines, and Gödel. *Philosophy* **36**(1961), pp. 120–124.

[Moschovakis 98] Moschovakis, Y. On founding the theory of algorithms. In: H. Dales & G. Oliveri (eds.) *Truth in Mathematics.* Clarendon Press (1998), pp. 71–104.

[Myhill 1960] Myhill, J. Some remarks on the notion of proof. *Journal of Philosophy* **57**(1960), pp. 461–471.

[Penrose 89] Penrose, R. *The Emperor's New Mind.* Oxford University Press (1989).

[Penrose 94] Penrose, R. *Shadows of the Mind.* Oxford University Press (1994).

[Post 2004] Post, E. *The Undecidable. Absolutely Unsolvable Problems and Relatively Undecidable Propositions.* Dover Publications (2004).

[Reinhardt 86] Reinhardt, W. Epistemic theories and the interpretation of Gödel's incompleteness theorems. *Journal of Philosophical Logic* **15**(1986), pp. 427–474.

[Shapiro 85] Shapiro, S. (ed.) *Intensional Mathematics.* North-Holland (1985).

[Shapiro 85a] Shapiro, S. Epistemic and intuitionistic arithmetic. In: [Shapiro 85, pp. 11–46].

[Turing 1936] Turing, A. On computable numbers, with an application to the Entscheidungsproblem. *Proceedings of the London Mathematical Society* **42**(1936), pp. 230–265.

[Turing 1939] Turing, A. Systems of logic based on ordinals. *Proceedings of the London Mathematical Society* **45**(1939), pp. 161–228.

[Wang 1974] Wang, H. *From Mathematics to Philosophy.* Routledge and Kegan Paul (1974).

[Williamson 00] Williamson, T. *Knowledge and its Limits.* Oxford University Press (2000).

# PART I

## Algorithm, Consistency, and Epistemic Randomness

# 2

# Algorithms and the Mathematical Foundations of Computer Science

WALTER DEAN

## 2.1 Introduction

The goal of this chapter is to bring to the attention of philosophers of mathematics the concept of *algorithm* as it is studied in contemporary theoretical computer science, and at the same time to address several foundational questions about the role this notion plays in our practices. In the most general sense, an algorithm is simply a procedure for achieving a particular mathematical end—paradigmatically computing the values of a function, or deciding whether a given mathematical object has a particular property. Most readers will be familiar with a variety of such procedures—e.g. carry addition, long division, Euclid's greatest common divisor algorithm—and will also be able to provide a characterization of the features in virtue of which we traditionally classify such methods as practical aids to calculation.

Methods such as these form an important part of the intellectual heritage of mathematics in several respects. For instance, methods for calculating quantities such as areas, inverses, powers, and roots appeared early in the ancient Babylonian, Sumerian, and Egyptian mathematical traditions. In the early middle ages, Arabic mathematicians developed a variety of procedures for performing arithmetic and algebraic operations which exploit the features of positional notation systems. The greater efficiency of these algorithms—some of which are the ancestors of procedures still taught to school children today—over earlier methods for calculating with Roman numerals is often cited as having been the determining factor in the acceptance of Hindu–Arabic numerals in medieval Europe.

In the modern era, the discovery of algorithms of practical import has often gone hand in hand with significant mathematical discoveries. For instance, Gauss first noted that both long division and Euclid's algorithm can be generalized to the ring $K[X]$ of polynomials over a field $K$. Together with the procedure now known as Gaussian elimination for solving systems of linear equations, these algorithms figured prominently in the proofs of a variety of theorems of nineteenth-century algebra—e.g. Gordan's demonstration

*Gödel's Disjunction*. First Edition. Leon Horsten & Philip Welch (eds).

that the ring of invariants of binary forms of fixed degree is finitely generated, Hilbert's Nullstensatz, and Strum's theorem. An important contemporary example in the same tradition is Buchberger's algorithm [9] for computing Gröbner bases.[1] Many similar examples can be cited in linear algebra (e.g. the Gram–Schmidt algorithm), analysis (e.g. the Newton–Raphson method), and graph theory (e.g. Krusal's algorithm).

In the face of such examples, it seems reasonable to suggest that the notion of algorithm deserves a place alongside concepts such as number, set, and function as an important substrate of contemporary mathematics. This role is partially accounted for in light of well-known analyses of computability undertaken during the 1930s. For note that all of the methods mentioned thus far are paradigmatically *effective*, i.e. they are finitely specifiable in terms of operations (e.g. adding or subtracting natural numbers, comparing the order of leading terms of polynomials, etc.) which may be carried out using finite resources by a mechanical computing agent. Such characteristics inform the conception of an *effectively computable function*—i.e. one computable by an effective procedure—which was analyzed in distinct but extensionally equivalent ways by Church, Turing, Post, Kleene, and Gödel.

Reflection on these analyses led via a well-known route to the framing of Church's thesis:

(CT) $f : \mathbb{N}^k \rightarrow \mathbb{N}$ is effectively computable if and only if $f(\vec{x})$ is recursive.

Note, however, that an effectively computable function is standardly understood to be one whose values are computable by an algorithm. Church's thesis may thus also be formulated as follows:

($\mathrm{CT}_a$) There exists an algorithm for computing $f : \mathbb{N}^k \rightarrow \mathbb{N}$ if and only if $f(\vec{x})$ is recursive.

$\mathrm{CT}_a$ is indeed a significant contribution to our understanding of the notion of algorithm. In particular, it enables us to analyze the truth conditions of statements such as

(1) There is no algorithm for determining whether an arbitrary formula of first-order logic is valid.

in terms of a proposition which quantifies, not over algorithms, but rather over the members of some mathematically well-defined class (e.g. that of Turing machines). $\mathrm{CT}_a$ thus provides a means of assigning significance to the formal results which Church and Turing demonstrated in the course of answering Hilbert's *Entscheidungsproblem* in the negative (which is what is reported by (1)). And it has a similar effect for other formal undecidability results for problems such as the word problem for semigroups or Hilbert's tenth problem.

Taken on its own, however, $\mathrm{CT}_a$ provides little direct insight into how the notion of algorithm is understood in contemporary computer science. For note that while several of the examples mentioned previously suggest that mathematical advances have often been tied to the discovery of new algorithms, the statements of the associated mathematical *theorems* rarely mention algorithms explicitly.[2] Thus it might appear that the task of formalizing classical mathematics (say within axiomatic set theory) does not require that we accord algorithms the status of freestanding objects, or that we provide a rigorous reconstruction of the means by which we apply or analyze them in practice.

In computer science this order of inquiry is often inverted. For instance, in fields like complexity theory and algorithmic analysis, individual algorithms are treated as the presumptive objects of study and mathematical methods are then used to investigate their properties. This emphasis is reflected, for instance, in the practice of referring to individual algorithms by what appear to be grammatically proper names (e.g. Euclid's algorithm, MERGESORT, etc.). Moreover, results in algorithmic analysis are often reported by predicating computational properties directly of individual procedures by using these names (e.g. "The AKS primality algorithm has polynomial running time") or by quantifying over classes of procedures (e.g. "There exists a polynomial time primality algorithm," "There is no linear time comparison sorting algorithm").

As the use of such language bears many of the hallmarks which philosophers have traditionally associated with ontological commitment, the practice of computer science raises a variety of questions about the status of algorithms which have been overlooked by philosophers of mathematics. Perhaps most prominent among these is the following:

> (Q1) Are individual algorithms properly regarded as objects? If so, are they abstract, concrete, or somehow intermediate between these possibilities? How can we account for our apparent ability to make reference to and prove both singular and general propositions about them?

Questions of type (Q1)—which are of an overtly ontological nature—can be contrasted with epistemological concerns about the use of algorithms in mathematical practice, such as the following:

> (Q2) How can we justify the use of computational methods in the derivation of mathematical results? In particular, why ought we to accept a calculation carried out by using an algorithm as an adequate demonstration of a mathematical proposition?

Questions in the vicinity of (Q2) have attracted more attention within philosophy of mathematics than (Q1) largely because of the difficulties which are thought to arise in accounting for the status of proofs—most famously of the Four Color Theorem [78]—which have substantial computational components.

I will discuss such examples further in §2.2. Therein I will also argue that they lead naturally to the formulation of the following thesis about how questions of type (Q1) should be answered:

> (A) Algorithms are mathematical objects.

(A) answers (Q1) by proposing that algorithms are no different in kind than the sorts of objects (e.g. numbers, sets, groups, matrices, graphs, etc.) studied in recognized branches of classical mathematics. Such a view also provides a convenient answer to questions of type (Q2): if algorithms are just another type of mathematical object, then the use of an algorithm to perform a lengthy calculation which may be required in the course of a proof can be justified by a supplemental mathematical proof that the algorithm is *correct* (i.e. that it computes the function that it is claimed to).

I will refer to the view expressed by (A) as *algorithmic realism*.[3] Although it is often not identified as a thesis requiring an explicit statement and defense, I will argue in §2.3 that (A) is presupposed throughout much of contemporary theoretical computer science.[4] Variants of (A) are also implicit in the writings of most theorists who have considered the notion of algorithm in a foundational setting. This most prominently includes Yiannis Moschovakis and Yuri Gurevich, both of whom have set out theories wherein algorithms are explicitly identified with certain classes of mathematical objects (respectively in [54], [56], [57] and [36], [37], [35]). Other formulations of algorithmic realism can also be found in the writing of many other theorists who have addressed the nature of algorithms in relation to other concepts studied in computer science—e.g. Gödel [32], Rogers [66], Kreisel [48], Milner [51], Knuth [41], Odifreddi [59], Foster [23], and Yanofsky [82]—many of whose views will be discussed below (at least in passing).

The primary goal of this chapter will be to address the question of whether algorithmic realism can be sustained as a foundational thesis about the nature of algorithms in light of the details of our mathematical and computational practices involving them. In spite of the realistic character of much of our discourse about algorithms, I will ultimately argue that it *cannot be so sustained*. The central reason I will offer for this is that our practices simply do not exert sufficient pressure on the use of the expressions we employ to refer to algorithms, to allow us to assign them mathematical denotations in a non-arbitrary manner. In fact, I will suggest that the fundamental problem confronting someone wishing to defend algorithmic realism is the familiar one of showing how to associate algorithms with mathematical objects such that the identified objects represent all and only the computational properties which we associate with them directly in the course of our practices.

I will refer to variants of the view on which the class of mathematical objects with which algorithms are identified are members of one of the conventional models of computation studied in computability theory—e.g. Turing machines, partial recursive function definitions, lambda terms—as *Strong Church's Thesis*.[5] When seen simultaneously through the lens of computer science and philosophy of mathematics, this view faces a strong *prima facie* challenge. For, note that if we propose to identify, say, Euclid's algorithm with a *particular* mathematical object $M$ (say a specific Turing machine), we thereby equip the former with all of the properties—e.g. having a particular number of states, an exact as opposed to asymptotic running time, etc.—which are possessed by the latter. As we will see, however, it is out of keeping with the methodologies of algorithmic analysis and complexity theory to think that such properties are appropriately attributed to individual algorithms, as opposed, e.g., to their implementations with respect to a given model of computation or their expression in a given programming language.

The adoption of Strong Church's Thesis thus faces a potential objection on the basis of what we might call the *what-numbers-could-not-be problem*. In its original form, this is the charge famously leveled by Benacerraf [5] that a successful reduction of number theory to set theory must provide not only a means of identifying numbers with sets, which preserves the truth of our chosen arithmetical axioms, but also an account of what makes a particular means of making such an identification correct relative to other identifications, which also preserve the truth of the axioms. For, to take the familiar example, were we to propose to identify the natural numbers with the finite von Neumann ordinals—as opposed to, say, the finite Zermelo ordinals—we would be faced with the seemingly unanswerable question of why 2 has the properties of $\{\emptyset, \{\emptyset\}\}$ (e.g. having two members) as opposed to those of $\{\{\emptyset\}\}$ (e.g. having a single member).

Perhaps for this reason, the contemporary consensus appears to be that individual algorithms should be identified not with particular instances of models of computation, but rather with equivalence classes of such items determined by an appropriate notion of "computational equivalence" defined over an appropriate class $\mathfrak{C}$ of mathematical objects. I will refer to this as the *algorithms-as-abstracts* view.[6] The most popular versions of this view take $\mathfrak{C}$ to coincide with the class of machine $\mathfrak{M}$ of some (sufficiently generalized) model of computation or the class of programs $\mathfrak{P}$ over a (sufficiently generalized) programming language. These proposals respectively lead to what I will refer to as the *machine-based* and *program-based* variants of the algorithms-as-abstracts view. In order to develop this view it is also necessary to propose an equivalence relation $\sim$ by which these classes can be factored into the classes which its proponents hold to be algorithms. As I will discuss further, standard choices for $\sim$ on the machine-based and program-based variants are respectively *mutual simulation* (which I will denote by $\approx$) and various technical definitions of *program synonymy* (which I will denote by $\simeq$).

Once appropriate definitions for these relations have been fixed, the proponents of the algorithms-as-abstracts view can then be understood to propose that the following implicit definitions provide an analysis of statements about algorithmic identity and non-identity:

(MP) the algorithm implemented by machine $M_1$ = the algorithm implemented by machine $M_2$ if and only if $M_1 \simeq M_2$

(PP) the algorithm expressed by program $\Pi_1$ = the algorithm expressed by program $\Pi_2$ if and only if $\Pi_1 \approx \Pi_2$

Putting aside for the moment how a proponent of the algorithms-as-abstracts view might go about defining $\mathfrak{M}, \mathfrak{P}, \simeq$, and $\approx$, it is also evident that the schema (MP) and (PP) have the form of *abstraction principles* of the sort which have been widely discussed in relation to Frege's [24] analysis of natural numbers as equivalence classes of finite sets with respect to the relation of equi-cardinality. It might thus at first appear that it is open to such theorists to propose that these schemas provide a means by which we can understand the status of algorithms as abstract objects in something like the manner in which neo-logicist philosophers of mathematics, such as Wright and Hale [81], [38], propose that the adoption of Hume's principle (i.e. the abstraction principle *the number of Fs = the number of Gs* just in case the *Fs* and *Gs* are equinumerous) can be understood to introduce the natural numbers as a new class of "logical objects."

But in spite of the outward affinity between the two proposals, the problems which must be initially confronted to develop the algorithms-as-abstract view are of a different character than those which have traditionally been taken to confront neo-logicism. In particular, the neo-logicist project is traditionally understood to be premised on the claim that Hume's principle should be understood as a conceptual or analytic truth about our understanding of the concept *natural number*. In order to defend this view, it is usually understood to be incumbent on the neo-logicist not only to defend this claim, but also to respond to traditional challenges such as the so-called Bad Company objection and the Julius Caesar problem, which might appear to stand in the way of its technical or philosophical development.[7]

As we will see, however, a significant problem which the algorithms-as-abstracts view must face is that it is difficult to find non-trivial instances in which the schema (MP) or (PP) are applied in practice against which we can test our intuitions about their potential analyticity. In fact, the practice of fields like algorithmic analysis and complexity theory appear to leave not only the choice of the classes $\mathfrak{M}$ and $\mathfrak{P}$, but also the choice of the equivalence relations $\simeq$ and $\approx$ highly unconstrained. Proponents of the algorithms-as-abstracts view thus face the initial challenge not only of defining what is meant by "computational equivalence," but also of characterizing the class of objects over which it is defined. In particular, before a proposal in this family could plausibly be taken to serve as an *analysis* of what we mean when we speak of algorithms in the course of our practices, it would have to be informed by the same sort of linguistic, conceptual, and technical investigation to which Frege [24], [25] famously subjected the notion of *natural number*. There is thus much work to be done even before analogues of the traditional challenges to neo-logicism can be raised for the algorithms-as-abstracts view.

These points notwithstanding, the algorithms-as-abstracts view is currently the best developed strategy for addressing the status of algorithmic realism and questions of type (Q1) more generally. In order to understand the challenges facing this view it will be useful to first consider in greater detail the factors which motivate algorithmic realism and also the practices of the subfields of theoretical computer science which bear on the details of our contemporary understanding of algorithms. These will be the respective purposes of §2.2 and §2.3. In §2.4, I will further develop several of the strategies for developing algorithmic realism just surveyed, so as to provide a more systematic case that the machine based variant of the algorithms-as-abstracts view is currently the most promising approach for defending this view. In §2.5, I will then consider the relation of mutual simulation which several theorists have proposed can play the part of $\simeq$. Upon concluding that it is not possible to provide a definition of this notion which serves the needs of algorithmic realism, in §2.6 I will then discuss this conclusion in light of Moschovakis's and Gurevich's theories of algorithms, as well as offering some final thoughts on the significance of algorithms and identity statements among them in mathematical practice.

## 2.2 Motivating Algorithmic Realism

The purpose of this section will be to set out a series of considerations which serve to motivate the consensus view that algorithmic realism is a correct means of responding to questions of type (Q1). As mentioned, questions of this sort have received considerably less attention in the philosophy of mathematics than those of type (Q2). For although there has been relatively little philosophical engagement with theoretical computer science to date, what interaction there has been has largely focused around evaluating the epistemic questions raised by mathematical arguments which involve lengthy algorithmic calculations. The best-known example of this sort is, of course, the so-called "computer proof" which was employed by Appel and Haken ([2], [3]) in their original demonstration of the four color theorem.[8] However, in order to illustrate how attempts to answer questions of type (Q2) lead inevitably to consideration of questions of type (Q1), it will be useful to examine not this example but rather the use of algorithmic methods to solve a more commonplace problem: given a natural number $n$, is $n$ prime?[9]

Until the late nineteenth century, all methods for solving this problem appear to have been based on variants of the "naive" method of trial division—i.e. in order to determine if $n$ is prime, show $p_1 \nmid n, p_2 \nmid n, \ldots$ for all primes $p_i \leq \sqrt{n}$. But it has long been recognized that this method is both laborious and unreliable due to the large number of divisions it required. The first significant advance in primality testing was the development of a procedure by the French mathematician Édouard Lucas which allows a Mersenne number (i.e. a number of the form $m_p = 2^p - 1$ for $p$ prime) to be tested for primality without checking whether it is divisible by a large number of smaller primes.

The basis of Lucas's method is a number theoretic lemma whose general form is now known as the *Lucas–Lehmer primality test*:

(2) For $p$ an odd prime, $2^p - 1$ is prime if and only if $r_{p-2} \equiv 0 \mod 2^p - 1$, where $r_0 = 4$ and $r_{i+1} = r_i^2 - 2$.

This statement implies that in order to determine whether $2^p - 1$ is prime it suffices to carry out the following procedure (which I will refer to as *Lucas's algorithm*): compute the sequence of numbers $r_0, r_1, \ldots, r_{p-2}$ and then check if $2^p - 1$ divides $r_{p-2}$.

In 1876 Lucas employed this procedure to show that the 39 digit number $m_{127}$ was prime. This required him to compute the values of 125 products on numbers up to 39 decimal digits in length. While this is still an arduous task to perform without mechanical computing equipment, it should be compared with the fact that more than $1.9 \times 10^{36}$ divisions would be required to test the primality of $m_{127}$ using the trial division algorithm. The contrast between trial division and Lucas's algorithm thus illustrates the reason why we are often concerned with the existence of an algorithm for solving a given mathematical problem—i.e. by exploiting a "clever" or indirect method, algorithms often allow us to solve instances of problems of antecedent interest which would be infeasible for us to solve by direct or "naive" methods alone.[10]

The question arises, however, why we ought to accept Lucas's calculation as an admissible mathematical demonstration of the proposition expressed by "$m_{127}$ is prime." For even if we put aside the possibility that Lucas made a mistake during his calculation, we must presumably also demonstrate that Lucas's algorithm accurately decides the primality of $2^p - 1$ before we are justified in accepting that a computation carried out by this algorithm is sufficient to demonstrate that $m_p$ is prime. In order to understand the significance of such a requirement, note that the foregoing specification of the algorithm does not bear a transparent relationship to the traditional definition of primality (i.e. $n$ is prime just in case $n$ is divisible by only 1 and itself). A computation carried out by the algorithm will thus bear little resemblance to an execution of the trial division algorithm previously described, and even less resemblance to a "canonical" deductive proof that $m_{127}$ is prime—i.e. one whose structure mirrors the logical form of the proposition $\forall x[x|m_{127} \rightarrow (x = 1 \vee x = m_{127})]$). When viewed in isolation, we would thus have little reason to accept such a calculation as a demonstration that $m_{127}$ is prime.

The desired connection between this calculation and the definition of primality is, of course, provided by the Lucas–Lehmer test. This result allows us to see informally why repeatedly carrying out the operation $s \mapsto (s^2 - 2)$, by which the value associated with the variable $s$ is replaced by that of $s^2 - 2$, results in a final value of $s$ such that $s \bmod 2^p - 1 = 0$ just in case $2^p - 1$ is prime. Note, however, that in order to turn this observation into what we normally regard as a mathematical proof, some explanation must presumably be given of the meaning of expressions such as *repeatedly carrying out* or *replacing* one value of a variable with another.

A related question is how we are to understand such language if we wish to give a purely mathematical proof that a given algorithm computes a function or decides a predicate which has an antecedently given mathematical definition. For note that while it is standard to think of mathematical objects as static, the use of operational terms in the specification of algorithms reflects our understanding that to execute an algorithm is to carry out a sequence of operations which are *ordered in time.*[11]

In the current case, however, a resolution to this apparent incongruity still appears near at hand: in order to prove that Lucas's algorithm is correct, it suffices to construct a formal representation of the procedure—call it $\phi_{\text{LUCAS}}$—which relates the value associated with the variable $s$ at the $i + 1$st stage in its operation to the value of the term $r_i$ occurring in (2). This may be accomplished by specifying a mathematical structure, $\phi_{\text{LUCAS}}$, sometimes known as an *iterator.*[12] $\phi_{\text{LUCAS}}$ may be taken to consist of a set of *computational states, $St \subseteq \mathbb{N} \times \mathbb{N}$*, whose first member keeps track of the index $i$ and whose second member keeps track of the value of $s_i$ together with a *transition function*, $\sigma : St \rightarrow St$, such that $\sigma(\langle x, y\rangle) = \langle x - 1, y^2 - 2\rangle$. We may then define $App(\phi_{\text{LUCAS}}, p)$—i.e. the result of applying LUCAS to $p$—to be 1 if upon iterating $\sigma$ we obtain a sequence of states of the form $s_0 \equiv \langle p, 4\rangle, s_1 = \sigma(s_0), \ldots, s_{i+1} = \sigma(s_i), \ldots, \sigma_{p-1} = \sigma(s_{p-1}) \equiv \langle 2, y\rangle$ and $y = 0$ and 0 otherwise.[13]

Using (2) it is now straightforward to prove a statement expressing the correctness of LUCAS in the following form:

**Proposition 2.2.1** *For all odd primes $p$, $App(\phi_{\text{LUCAS}}, p) = 1$ if and only if $2^p - 1$ is prime.*

Unlike the observations about the operation of Lucas's algorithm previously recorded, Proposition 2.2.1 is a result involving only mathematical definitions and structures (i.e. natural numbers, finite sequences of ordered pairs, etc.). As such, it admits to a standard proof by mathematical induction which does not require us to reason informally about notions like "repeatedly carrying out" an operation.

Proposition 2.2.1 reports what computer scientists call the *correctness* of Lucas's algorithm—i.e. that for all odd primes $p$, the algorithm outputs 1 or 0 according to whether

$2^p - 1$ is prime. The derivation of Proposition 2.2.1 from the Lucas–Lehmer test is sufficiently straightforward that this result might at first appear trivial. In the general case, however, correctness proofs for more complex algorithms may require sophisticated mathematical arguments themselves. But what is most significant in the current case is not the difficulty of demonstrating correctness, but rather the proposal that algorithms admit to mathematical proofs of correctness in the first place.

For as we have just seen, the availability of such a proof for an algorithm $A$ appears to presuppose the existence of a mathematical structure, such as $\phi_A$, which provides a precise representation of what we might informally describe as $A$'s "mode of operation." Once we have specified $A$ informally, it is often a routine exercise in formalization to construct $\phi_A$. But even after we have provided such a representation, the question remains of what relationship such a structure bears to the algorithm $A$, in the sense that we originally understood it relative to an informal specification similar to that by which Lucas's algorithm was introduced previously.

One obvious proposal is that the relationship between an algorithm and its formal representation as a mathematical structure, such as an iterator, is sufficiently direct and systematic that the former may simply be *identified* with the latter. A consequence of this is that the procedure we started out calling Lucas's algorithm simply *is* the structure $\varphi_{\text{LUCAS}}$, and similarly, that all of the other algorithms we have previously identified in our mathematical practices may be identified with such structures. This is not to deny that our initial apprehension of how such procedures operate is often grounded in the sort of informal description provided for Lucas's algorithm, or that, such descriptions are sufficiently precise to guide us in their execution. Rather, the import of both the current example and several others to be considered below, is to suggest that nothing of mathematical import is lost when we move from such informal descriptions to the sort of formal representation of a procedure which is required to support a correctness proof.

Our apparent ability to make such identifications in a manner which preserves the details of how we apply and reason about algorithms in practice provides the basis of the consensus that algorithmic realism is correct. In particular, the need to provide correctness proofs before we can assimilate algorithmic calculations to traditional mathematical proofs, coupled with the apparent consensus that we are indeed justified in believing the outcomes of algorithmic calculations on the basis of something akin to proof, points towards a strong *prima facie* case in favor of this view. The burden of §2.3–§2.5 will be to show that the practice of contemporary computer science exerts sufficiently many other pressures on the form which an adequate foundational theory of algorithms must take so as to render algorithmic realism untenable. The goal of §2.6 will then in part be to provide an alternative reconstruction of the justification of the use of computational methods in mathematics.

## 2.3 Algorithms in Theoretical Computer Science

The goal of this section is to provide a concise account of how algorithms are treated within theoretical computer science. This will inform the detailed considerations which are required to evaluate algorithmic realism relative to our contemporary practices. In so doing, we face the challenge that despite its relative youth, theoretical computer science has spawned several distinct subfields in which algorithms are treated, each with their own motivating questions and methodologies. Of these, considerations originating in complexity theory and algorithmic analysis (which will be considered in this Section), and programming language semantics (which will be considered in §2.4) will be most relevant in the following.[14]

An elementary (but deceptively simple) class of examples is provided by the study of procedures for sorting a finite array $B$ of numbers or other items linearly ordered by a relation $\leq$.[15] Consider, for instance, the following passage from a popular algorithmic analysis textbook:

> In this chapter, we introduce another sorting algorithm: HEAPSORT. Like MERGESORT, but unlike insertion sort, HEAPSORT's running time is $O(n \log_2(n))$. Like INSERTIONSORT, but unlike MERGESORT, HEAPSORT sorts in place: only a constant number of array elements are stored outside the input array at any time. Thus, HEAPSORT combines the better attributes of the two sorting algorithms we have already discussed.
>
> [14], p. 151

This passage illustrates how expressions like HEAPSORT, MERGESORT, and INSERTIONSORT are used to denote individual methods for computing the sorting function—i.e. the function sort($x$) which on input array $B$ returns a permutation $B'$ of $B$ such that $B'[0] \leq B'[1] \leq \ldots \leq B'[n]$. A notable feature of the discourse of algorithmic analysis and related subjects is that such terms play the apparent grammatical role of *proper names*—i.e. linguistically primitive expressions which function as singular terms for the purpose of predication and formulation of statements of identity and non-identity. I will refer to such expressions as *algorithmic names*.

A question close to the heart of algorithmic realism is how we should understand the reference of expressions like HEAPSORT or MERGESORT. As a first step toward addressing this, note that algorithmic names are often introduced along with informal mathematical descriptions, as exemplified by the following specification of INSERTIONSORT:

> [C]onsider the elements one at a time, inserting each in its proper place among those already considered (keeping them sorted). The element being considered is inserted merely by moving larger elements one position to the right, then inserting the element into the vacated position.
>
> [71], pp. 95–96

Although such informal descriptions are sometimes taken to be sufficient for the purpose of proving that an algorithm is correct or analyzing its computational complexity (a task I will discuss further in a moment), it is also conventional to supplement them with more regimented formulations using an idiom known as *pseudocode*. A specification of INSERTIONSORT in this manner might take the following form:[16]

```
InsertionSort(B)
for j ← 2 to |B|
    do m ← B[j]
       i ← j − 1
       while i > 0 and m ≤ B[j]
          do B[i + 1] ← B[i]
             i ← i − 1
       B[i + 1] ← m
```

Expressions like LUCAS, INSERTIONSORT, or EUCLID (i.e. Euclid's algorithm) are introduced in much the same way that names for theorems and definitions are used in conventional mathematical texts (e.g. Desargues's Theorem, the Snake Lemma, the Riemann integral, etc.)—i.e. as means of facilitating reference to prior complex definitions or results. Within computer science, however, algorithmic names also commonly appear as the grammatical subjects of sentences expressing that individual algorithms have various computational properties—e.g.

(3) a) EUCLID computes the function $gcd(x, y)$.
b) INSERTIONSORT has running time complexity $O(n^2)$.
c) MERGESORT has smaller running time complexity than INSERTSORT.

The frequency with which such statements appear in computer science textbooks and other sources underscores the fact that what is often of interest in fields like algorithmic analysis and complexity theory is not merely the discovery of novel algorithms, but rather the use of mathematical methods to show that certain properties and relations hold of procedures which have already been introduced.

In addition to the property of correctness with respect to a given mathematically defined function (e.g. as reported by (3a)), of paramount concern in computer science are the sorts of complexity theoretic properties reported by (3b,c). For it is the attribution of such properties by which the efficiency of algorithms for solving particular mathematical problems is measured and compared. The manner in which such attributions are treated in algorithmic analysis and complexity theory will play a significant role in the evaluation of algorithmic realism developed in the sequel. It will thus be useful to offer the following brief account of how such statements are demonstrated.

Defined informally, the *running time complexity* of an algorithm $A$ on an input $x$ is the number of basic computational steps which are required for $A$ to halt and return an output when applied to the input $x$ measured as a function of some standard measure $|x| = n$ of the size of $x$ (e.g. the length of its binary representation in the case where $x$ is a natural number). This property of $A$ is typically reported as a function, $time_A(n)$ of type $\mathbb{N} \rightarrow \mathbb{N}$, which returns the maximum number of steps consumed by $A$ for all inputs $x$ of length equal to $n$.[17] As much of theoretical computer science is aimed at finding and analyzing efficient methods for solving mathematical problems, this property is often taken as the single most significant feature of an algorithm. But before $time_A(n)$ can be precisely defined, some prior stipulation must be made as to what should be counted as a "basic computational step" of $A$.

Complexity theory and algorithmic analysis can be understood to offer distinct but complementary answers to this question. On the one hand, complexity theory is typically developed so that measures of *time* (and also *space* or *memory*) *complexity* are reported relative to a particular fixed model of computation $\mathfrak{M}$.[18] Although most familiar models are sufficient for this purpose, complexity theory does impose additional constraints beyond those imposed by computability theory on the choice of $\mathfrak{M}$ (wherein $\mathfrak{M}$ is typically only required to be Turing complete and provide a suitable means of indexing machines so that results like the $s$–$n$–$m$ Theorem are satisfied). In particular, $\mathfrak{M}$ must be powerful enough to allow for the representation of numbers in binary form, but not so powerful as to allow for arbitrarily branching parallel computations.[19] Such models—which include the familiar single- or $k$-tape Turing machine model $\mathfrak{T}_k$ with a binary tape alphabet, or the standard random access [RAM] model $\mathfrak{R}$ with unit time addition and subtraction—are often referred to as *reasonable* and are taken to comprise a natural category known as the *first machine class* [79].

Although the reliance of complexity theory on the existence of a well-defined class of models of the sort just described is often not stressed explicitly, it is difficult to understate its overall significance for theoretical computer science. For complexity theory is standardly taken to provide our most complete account of which mathematical problems are *feasibly decidable*—i.e. decidable by a method which can be carried out in practice for inputs of a reasonable size, as opposed to merely in the "in principle" sense of computability theory. Thus if there were no means of precisely delimiting the basic operations which can be carried out in a single step by a concretely embodied computing device of the sort we can employ in practice, there would be little hope of giving a mathematical analysis of feasible computability in the manner which complexity theory aims to provide.[20]

Once a model $\mathfrak{M}$ in the first machine class has been fixed as a benchmark, complexity theory suggests that the running time, $time_A(x)$, of an algorithm $A$ should be defined by first constructing a machine, $M \in \mathfrak{M}$, which is said to *implement* $A$. In order to state in precise terms what it means for $M$ to implement $A$ requires solving what I will refer to as the *implementation problem*—i.e. that of providing necessary and sufficient conditions for when a given $M \in \mathfrak{M}$ is an adequate mathematical representation of $A$. As should already be evident from the example of §2.2, providing a non-stipulative answer to this question is of central importance to assessing the status of algorithmic realism itself. In broad terms, however, the manner in which $M$ is often constructed from the specification of $A$ will be familiar to anyone who has attempted (for example) to construct a Turing machine for performing addition or multiplication on natural numbers represented in binary notation—i.e. it must be shown how the basic operations and data structures of $\mathfrak{M}$ may be used to mimic the step-by-step operation of $A$ in terms of the (possibly "higher-level") primitive operations in terms of which $A$ is informally specified. Upon so doing, the function $time_A(n)$ can then be defined as $time_M(n)$—i.e. the mathematically defined running time complexity of the machine $M$ constructed in this manner.

After a choice of machine model has been fixed, complexity theory then goes on to study the minimal time complexity of solving various problems of mathematical interest—e.g.

primality testing, determining whether a propositional formula is satisfiable or is a tautology, determining whether a given graph has a clique of a certain size, or whether two graphs are isomorphic. This gives rise to the notion of a *complexity class*, $\mathbf{Time}_{\mathfrak{M}}(f(n))$, which is defined to consist of those computational problems $X$ which can be solved by a machine, $M \in \mathfrak{M}$, such that $time_M(n) \leq O(f(n))$.[21] Such classes form the familiar hierarchy, $\mathbf{LogTime} \subseteq \mathbf{P} \subseteq \mathbf{NP} \subseteq \mathbf{PSpace}$, respectively defined as the classes of computational problems which can be solved in time $k \log_2(n)$ and $k_1 \cdot n^{k_2}$ using (for example) the deterministic Turing machine model $\mathfrak{T}_k$ (respectively described as *logarithmic time* and *polynomial time*), the class of problems which can be solved in time $k_1 \cdot n^{k_2}$ using the non-deterministic Turing machine model $\mathfrak{N}\mathfrak{T}_k$ (*non-deterministic polynomial time*), and the class of problems which can be solved in space $k_1 \cdot n^{k_2}$ using $\mathfrak{T}_k$ (*polynomial space*)—see, e.g., [60] for details.

As should be evident from the foregoing discussion, complexity theory is primarily concerned with the classification of computational problems understood extensionally as classes of their instances. The most fundamental distinction in this regard is between those problems which can be shown to be members of **P**—which is commonly thought to provide an upper bound on the class of problems whose instances can be uniformly solved in practice—and those which can be shown to be *hard* for a class such as **NP**, which is strongly believed (but not known) to properly contain **P**. A problem $X$ is shown to be hard in this sense by showing that there is a polynomial time reduction which efficiently transforms any instance of $X$ into an instance of some problem which is known to be *complete* for **NP**. Such a problem can thus be understood to embody the underlying computational difficulty of solving all problems in this latter class.[22]

One well-known example of such a problem is the set SAT consisting of binary strings encoding formulas of propositional logic for which there exists a satisfying assignment—e.g. $P_0 \vee P_1 \in$ SAT, but $P_0 \wedge \neg P_0 \notin$ SAT. Since SAT is known to be **NP**-complete, it is very unlikely that SAT is feasibly decidable. Such a fact would typically be reported as follows:

(4) If $\mathbf{P} \neq \mathbf{NP}$, then there does not exist an algorithm with polynomial running time complexity which solves the problem SAT of determining whether a formula of propositional logic is satisfiable.

Note that like many other limitative results in computability theory, such as (1), statements like (4) are often reported informally using quantifiers over algorithms. But it should be clear on the basis of the foregoing that just as $\mathrm{CT}_a$ allows us to paraphrase away the quantification over algorithms in "There is no algorithm for deciding first-order validity" in favor of quantification over the members of any Turing complete model of computation, it is possible to paraphrase away the quantifier in (4) in favor of quantification over the members of an appropriate model of computation in the first machine class.[23]

Matters are somewhat different in algorithmic analysis wherein the focus of study is not on computational problems, but rather on individual algorithms themselves. Note, however, that to justify a statement such as (3b) requires that we are able to somehow count the number of steps which are required by an algorithm like INSERTIONSORT or MERGESORT to sort a list of length $n$. This is typically accomplished by performing a so-called *running time analysis* of the algorithm—i.e. a combinatorial argument which counts the maximum number of distinct operations that must be performed when the algorithm is carried out for an input of length $n$. For instance, a routine calculation shows that as described by the previous pseudocode specification, INSERTIONSSORT has *exact* running time complexity $(3/2)n^2 + 5/2(n) - 4$.

It should be borne in mind, however, that such calculations are conducted under the assumption that it is admissible to determine the running time complexity of an algorithm by counting each instruction in its pseudocode specification as a single primitive step. This convention raises two questions: (1) should the exact running time complexity derived in this manner be understood as an intrinsic property of an algorithm itself?; (2) given that a pseudocode specification can in principle be stated in terms of arbitrary mathematical operations (inclusive, e.g., of non-effective ones), does it make sense to ascribe a running time complexity to an algorithm directly, or should such ascriptions be understood modulo a further specification of what operations are to be taken as primitive?

A comparison of different textbooks is often sufficient to confirm that conventional usage of common algorithmic names allows for slight variations in the way in which an algorithm is specified—e.g. INSERTIONSORT might be specified by using a "for" or "repeat-until" loop instead of a "while" loop, or it might increment its counter variables in a different order. Experience bears out, however, that such small variations in how an algorithm is specified typically affect the computation of its exact running time by at most a scalar factor. By virtue of this, the running time complexity of individual algorithms is conventionally reported not as an exact function, but rather using so-called *asymptotic* notation.

Recall that the notation $O(f(n))$ is used to denote the class of functions whose rate of growth is dominated by a scalar multiple of $f(n)$ for sufficiently large $n$[24]—i.e.

$$(5)\quad O(f(n)) =_{df} \{g : \mathbb{N} \rightarrow \mathbb{N} \mid \exists c, n_0 \forall n \geq n_0 [g(n) \leq cf(n)]\}$$

Based on the previous calculation, INSERTIONSORT would thus typically be reported as having running time $O(n^2)$. And based on similar (but more complex) calculations, the algorithms MERGESORT and HEAPSORT would both be reported as having running time $O(n \log_2(n))$. It is easy to see that any function in the latter class is eventually dominated by one in the former. And it is such facts that are typically understood to provide the mathematical content of comparisons of the efficiency of algorithms as reported by (3c).

The use of asymptotic notation goes some distance toward justifying the claim (which appears implicit in the methodology of algorithmic analysis) that a running time complexity can be associated directly with an algorithm. However, a yet more serious challenge to the basis of this practice arises in regard to the second question flagged above—i.e. is it licit to use arbitrary effective operations in a pseudocode specification of an algorithm?

In order to address this question, it is useful to keep in mind that the development of algorithmic analysis has been largely guided by the goal of developing methods for comparing the efficiency of procedures for solving problems which are of practical interest. But at the same time, it is also easy to see that by selecting an appropriate choice of primitive operations, it is often possible to trivialize the comparison of the relative efficiency of algorithms $A_1$ and $A_2$, by simply defining another algorithm $A_3$ which performs the same task as $A_1$ and $A_2$ in fewer (effective, but intuitively complex) steps.

For instance, although one might think that (3c) reports that a genuine gain in practical efficiency can be had by sorting a list using MERGESORT instead of INSERTIONSORT, even the $O(n \log_2(n))$ running time complexity of the latter is expensive in comparison with that of the $O(1)$ (i.e. constant time) algorithm

TRIVIALSORT($B$),

1 **return** sort($B$)

TRIVIALSORT sorts lists of arbitrary length in a single step by simply calling the function *sort*($B$)—an operation whose effectiveness is in turn attested to by the existence of algorithms such as MERGESORT and INSERTIONSORT.

The conventional manner in which this problem is addressed in algorithmic analysis is to assume that while it is legitimate to specify and reason about algorithms informally via their pseudocode specifications, such specifications should be understood relative to a choice of a fixed model of computation, $\mathfrak{M}$, for which it is assumed that they can be implemented. Although this model is often chosen for convenience, it is still important that it be among those in the first machine class. One hallmark of such models is that they allow the sorts of algorithms $A$ which arise most often in mathematical practice to be implemented in a manner that makes it possible to construct $M \in \mathfrak{M}$ such that $A$ and $M$ not only compute the same function (possibly up to an efficiently computable encoding of inputs and outputs), but also such that $time_M(n) \in O(time_A(n))$—i.e. although $M$ may have a larger *exact* running time than $A$ (e.g. by virtue of needing to explicitly compute the value of a function which is assumed as a primitive in the informal specification of $A$), it will have the same *asymptotic* running time.[25]

The foregoing observations provide further justification for the use of asymptotic notation to report the running time of individual algorithms. However, they also further highlight the significance of the implementation problem—i.e. in order for $M \in \mathfrak{M}$ to count as an implementation of $A$, not only must it mimic its operation in a step-by-step sense, but it must also do so in a manner which preserves its asymptotic running time complexity. The fact that we are able to construct such implementations using the models $\mathfrak{R}$ or $\mathfrak{T}_k$, for the sorts of procedures which have traditionally been valued as effective methods in mathematical practice, is suggestive of the fact that there may be a much closer connection between the specific properties of these models and our detailed intuitions about effectivity than is often acknowledged.[26]

With these details in place, it may finally be observed that algorithmic analysis is concerned not only with the sorts of attributions of running time complexity as reported by (3b,c), but also with proving so-called *lower-bound* results which report that a given algorithm is optimal in terms of time or space complexity. Recall, for instance, that INSERTIONSORT is known as a *comparison sorting algorithm* by virtue of the fact that it sorts the array $B$ using comparisons performed with $\leq$ but without making use, e.g., of the size or multiplicity of the elements which comprise this array. For such methods it is possible to prove the following:

(6) There is no comparison sorting algorithm with running time complexity asymptotically less than $\Omega(n \log_2(n))$.[27]

In terms of its logical structure (6) should be compared with other limitative propositions about computability, like (1) and (4). Suppose, however, our ultimate goal is to show—in conformity with the thesis (A)—that the quantifier over algorithms in (6) can be replaced with a quantifier over some other well-defined class of mathematical objects in a manner which preserves the meaning of the proposition expressed. In this case, we are faced with a refinement of the problems which we face with respect to (1) and (4). For recall that finding such a paraphrase for (1) requires only that we have at our disposal a Turing complete model of computation—i.e. one that per $\text{CT}_a$ contains a machine computing every function which is computable by an algorithm, whereas in the case of (4) we must also at least ensure that this model is in the first machine class. But now note that providing an adequate paraphrase of (6) requires that we consider a model of computation $\mathfrak{M}_2$ which satisfies at least the following properties: (i) $\mathfrak{M}_2$ is in the first machine class; (ii) the implementation problem for sorting algorithms can be solved for $\mathfrak{M}_2$ in a manner which preserves asymptotic running time complexity; (iii) it is still meaningful to talk about what it means for a machine $M \in \mathfrak{M}_2$ to be an intuitively faithful implementation of a procedures such as INSERTIONSORT and MERGESORT as comparison sorting algorithms.

## 2.4 In Search of a Foundational Framework

Given that complexity theory and algorithmic analysis represent our most refined methods for reasoning on the types of effective procedures which arise in mathematical practice, the sorts of considerations surveyed in §2.3 are typical of the primary linguistic and technical data to which a foundational theory of algorithms should be responsive. On this basis, it seems reasonable to extract the following general observations about the status we accord algorithms in our mathematical and computational practices:

(I) Algorithms are mathematical procedures which may be described either informally using standard mathematical prose or through the use of a pseudocode specification. Such procedures may be *executed* or *carried out* for a given *input*, resulting in a sequence of intermediate states which leads to the calculation of an *output*.

(II) Algorithms can be *implemented* by members of models of computation $\mathfrak{M}$. To implement an algorithm $A$ is to specify a machine $M \in \mathfrak{M}$ which not only computes the same function as $A$ (possibly up to an efficient encoding of inputs and outputs), but also operates in the same step-by-step manner.

(III) Algorithms possess their *asymptotic running time complexity* intrinsically. This imposes additional constraints on the implementation relation—e.g. (i) $\mathfrak{M}$ must be a model in the first machine class; (ii) if $M \in \mathfrak{M}$ implements $A$, then $M$ and $A$ must have the same asymptotic running time complexity.

The question which we must now confront is how a proponent of algorithmic realism might develop a general theory which simultaneously accords algorithms the status of mathematics objects, and also provides a satisfactory account of constraints like (I)–(III). In order to address this question it will be useful to consider three views about the nature of algorithms, two of which were briefly discussed in §2.1:

**Direct algorithmic realism**—Algorithms comprise a class $\mathcal{A}$ of intrinsically intensional objects which are distinct from those traditionally recognized in classical mathematics.

**Strong Church's Thesis**—Algorithms may be identified with (or *reduced to*) the members of a class of mathematical objects $\mathcal{M}$.

**Algorithms as abstracts**—Algorithms correspond to equivalence classes defined over some class of mathematical objects $\mathcal{M}$ factored by an appropriate equivalence relation.

In §2.1, the second (and also implicitly the first) of these options were dismissed summarily. It will now behoove us to backtrack slightly and examine them again in light of the considerations adduced in §2.2 and §2.3. This will allow for a better appreciation of why the algorithms-as-abstracts view appears to be the most plausible of the enumerated options.

As noted previously, direct algorithmic realism might seem like the most straightforward means of making sense of the realistic tone of much of our informal discourse about algorithms (as typified by the passage cited at the beginning of §2.3). For it seems that this discourse possesses many of the characteristics which philosophers often associate with ontological commitment to a category of objects—e.g. use of singular terms such as LUCAS, INSERTIONSORT, etc. to make (apparent) singular reference, the use of quantification over algorithms to express limitative results, etc. And thus when taken in conjunction with the apparent incongruity flagged in §2.2 between the abstractness of mathematical objects and the use of temporal language to describe algorithms and their executions, one might reasonably conclude that there is at least *prima facie* justification for considering algorithms to comprise a class of freestanding "procedural" entities which are unlike those traditionally countenanced as mathematical objects.

Another potential motivation for exploring this view is the observation that algorithms have traditionally been considered to be *intensional* entities. Part of the basis for such a classification derives from the observation that algorithms evidently cannot be assimilated to functions understood in the extensional sense. For whatever we take algorithms to be, it must be acknowledged that there can exist distinct algorithms for computing the same extensional function—e.g. although INSERTIONSORT and MERGESORT both compute the function $sort(B)$, they cannot both be equal to this function by virtue of the fact that we regard them as having distinct computational properties (e.g. running time complexity). For reasons in this vicinity, it might seem that there is some hope of developing direct algorithmic realism along the same lines as philosophers have occasionally attempted to develop theories for other classes of intensionally individuated objects—e.g. Fregean senses, propositions, constructive proofs, etc.

While it seems that there is nothing which blocks proceeding in this manner, the prospects for developing and defending such a theory of algorithms along these lines begins to seem less plausible when we start to look at the methodologies of complexity theory and algorithmic analysis in detail. For on the one hand, while there do exist formal theories of notions like senses (e.g. [12], [76], [40]), propositions (e.g. [4], [16], [83]), and constructive proofs (e.g. [50], [31]), many of these make use of modal apparatus which appears out of place in a mathematical context, or are otherwise parasitic on technical notions which are already employed in logic or theoretical computer science (e.g. the typed lambda calculus or operational semantics for programming languages of the sort to be discussed below). Such observations serve to deflate the hope that extant theories of intensional entities can be used to explain the sense in which there is something *distinctive* about algorithms which which provides a principled explanation of why they are distinct from previously defined classes of mathematical objects.[28] And on the other hand, none of these philosophically motivated accounts offers a explanation of the complex relationships which obtain between algorithms, antecedently defined mathematical problems, and specific kinds of models of computation described in §2.3.

Turning now to strong Church's thesis [SCT], another *prima facie* plausible view which one might attempt to develop is that it is possible to construct a mathematical object which could be taken to correspond to an algorithm, starting from its pseudocode representation. An obvious point of comparison for such a view is the so-called *sententialist* theory of propositions, according to which the proposition expressed by a natural language sentence just is that sentence itself. By analogy, a straightforward version of SCT might propose that we could similarly take an algorithm as nothing other than the text comprising its pseudocode specification.

Note, however, that sententialism is generally put forth as a form of *eliminativism* about propositions (a sentiment which is presumably counter to the spirit of algorithmic realism). And it is also typically held to suffer from a variety of problems—e.g. if unamended it leaves no room for two sentences in different natural languages to express the same proposition—versions of which can readily be seen to beset the analogous view about algorithms. A more plausible view is thus that algorithms should not be identified with linguistic descriptions such as pseudocode specifications themselves, but rather with some other sort of object which is derived by interpreting such specifications according to a form of compositional semantics. Such a view has obvious affinities to the conventional view that the proposition expressed by a sentence of a natural language is determined by a compositional semantic theory—e.g. of the sort originally proposed by Montague [53] and now widely studied in formal linguistics.

Note, however, that in developing such a proposal we face the initial problem that pseudocode cannot be understood to be a part of any extant natural language. For instance, it employs constructions—e.g. those used to describe iteration, recursion, and flow control—which appear to have no natural language counterparts. But at the same time, pseudocode is also not typically taken to be a fully fledged formal language with a precisely defined syntax and semantics. It would thus appear that there is little hope that current semantic theories for natural language can be directly applied to pseudocode specifications without substantial modifications and other precifications, which would most likely have to be supplied by the practice of computer science itself.

A related view is that algorithms are the sorts of entities which are expressed, not by pseudocode specifications themselves, but rather by their regimentations expressed using a formal programming language. It is often said that such languages were originally developed in part to provide a medium for expressing algorithms in a manner which is similar to pseudocode, but which has a sufficiently precise syntax to allow direct translation into the primitive instruction sets which can be carried out by conventional digital computers. But although it can be reasonably maintained that this is true of familiar "high level" languages like `Algol`, `C`, or `LISP`, it should also be kept in mind that unlike pseudocode specifications—which are at least intended to be continuous with informal mathematical language—programs constructed over such languages are originally *uninterpreted*.

It is now routinely acknowledged that formal semantics for programming languages are needed to enable rigorous correctness proofs similar to that considered in §2.2—i.e. that when a program is carried out in accordance with the intended interpretation of the constructs in which it is specified, it computes the values of an independently defined mathematical function. Recognition of this fact (which began in the 1960s) gave rise to the subject now known as *programming language semantics*. For present purposes, we may take a formal semantics for a programming language L to be a function $[\![\cdot]\!]^{\mathtt{L}}$ which maps $\mathtt{Prog}_{\mathtt{L}}$—the class of programs over L—onto some domain $\mathcal{D}$ of mathematical objects which in some (yet to be specified sense) are taken to be the interpretations of L-programs.

Suppose we now also assume that given an informally specified algorithm $A$, it is unproblematic to construct a program $\Pi \in \mathtt{Prog}_{\mathtt{L}}$ which we take to be an adequate expression of $A$. We might then attempt to understand expressions of the form

(7) $A$ is the algorithm expressed by $\Pi$.

as a sort of canonical means of referring to algorithms. And on this basis, we might also propose that the reference of terms of the form "the algorithm expressed by $\Pi$" is given by applying the function $[\![\cdot]\!]^{\mathtt{L}}$ to the program $\Pi$.

It is, however, by no means a trivial question what sort of objects should be taken to comprise the class $\mathcal{D}$ which forms the range of $[\![\cdot]\!]^{\mathtt{L}}$, or how the value of $[\![\Pi]\!]^{\mathtt{L}}$ ought to be determined according to the structure of $\Pi$. Abstracting away from many details, it is possible to identify two broad approaches to these problems—known respectively as *denotational semantics* and *operational semantics*. In the first case, $\mathcal{D}$ is taken to be an appropriately defined function space $X^Y$ such that the value of $[\![\Pi]\!]^{\mathtt{L}}_{den}$ will be a (possibly partial) function of type $X \to Y$, which is intended to correspond to the function induced by executing $\Pi$ for all inputs in $X$. From this it follows that a denotation semantics will associate any two programs $\Pi_1$ and $\Pi_2$ which compute the same function with the same extensional object. But this is presumably an unsatisfactory result in the current context given that e.g. we can easily construct programs which naturally express intuitively distinct algorithms (e.g. INSERTIONSORT and MERGESORT), but which compute the same function.

The specification of an operational semantics for L can be understood to address this issue by assigning to a program $\Pi$, not a function but rather an abstract mathematical representation of its mode of operation, as determined compositionally according to the interpretation assigned by basic constructs made available by L. Such a representation can take a number of forms.[29] What is significant for our current purposes, however, is that in the case of operational semantics, the members of $\mathcal{D}$ are recognizable as instances of *models of computation* in the sense introduced in §2.3. Such a model is a class of mathematical structures $\mathfrak{M}$ together with a definition of a (possibly partial) function, $App : \mathfrak{M} \times X \to Y$, which determines the output of applying $M \in \mathfrak{M}$ to input $x \in X$. In familiar cases like $\mathfrak{T}_k$ and $\mathfrak{R}$ the members of $\mathfrak{M}$ can be reasonably described as *machines*—i.e. they may be described as classes of *computational states*, over which various operations are defined and which induce a *transition* between states—while the $App(M, x)$ can be seen as providing a mathematical analysis of what it means to *carry out* or *execute* the computation which is induced by applying $M$ to $x$.

In order to make these notions somewhat more precise it will be useful to introduce Moschovakis's [56] notion of an iterator as was described informally in §2.1:

**Definition 2.4.1** *An iterator from a set X to a set Y* $\phi : X \to Y$ *is a quintuple* $\phi = \langle in, St, \sigma, T, out \rangle$ *such that*

- *St is an arbitrary non-empty set, the* computational states *of* $\phi$;
- $in : X \to St$ *is the* input function *of* $\phi$;
- $\sigma : St \to St$ *is the* transition function *of* $\phi$;
- $T \subseteq St$ *is the set of* terminal states *of* $\phi$ *and* $s \in T$ *implies* $\sigma(s) \in T$;
- $out : St \to Y$ *is the* output function *of* $\phi$.

A *computation* induced by the iterator $\phi$ on an input $x \in X$ is a sequence of states $s_0(x), s_1(x), s_2(x), \ldots$ such that $s_0(x) = in(x)$ and $s_{n+1}(x) = \sigma(s_n(x))$ if $s_n(x) \notin T$ and is undefined otherwise. The *length* of such a computation is given by $len_\phi(x) = n + 1$, where $n$ is the least such that $s_n(x) \in T$. The result of applying $\phi$ to $x$ induces a (possibly partial) function, $App(\phi, x) =_{df} out(s_{len_\phi}(x))$.

It is a straightforward but useful exercise to see how familiar models of computation like $\mathfrak{T}_k$ and $\mathfrak{R}$ can be modified so that their instances satisfy the definition of an iterator. It is evident, however, that this definition cannot itself be taken to serve as an adequate analysis of either the general notion of an *effective* model of computation (as is needed for the development of computability theory), or for the definition of the first machine (as is needed for the development of complexity theory and algorithmic analysis).[30] Nonetheless, Definition 2.4.1 is still useful in the sense that it allows us to speak uniformly about notions like computational states and transition functions across different models of computation.

Let us now return to assess the plausibility of the version of SCT considered previously. Relative to this proposal the reference of an expression of the form "the algorithm expressed by $\Pi$" is asserted to correspond to the value of $[\![\Pi]\!]^{\mathbb{L}}_{op}$, where $[\![\cdot]\!]^{\mathbb{L}}_{op}$ is the denotation function of some form of operational semantics for a programming language $\mathbb{L}$ for which $\Pi \in \texttt{Prog}_{\mathbb{L}}$. Based on the foregoing discussion, we can now see that (at least in prototypical cases), we will have $[\![\Pi]\!]^{\mathbb{L}}_{op} \in \mathfrak{M}$, where $\mathfrak{M}$ is some fixed model of computation. The question which we must now ask is whether this view gives rise to a version of algorithmic realism which satisfies the constraints described at the beginning of this section.

To see that it does not, it is sufficient to observe that, although the proposal in question succeeds in providing an account of how an informally specified algorithm $A$ can be associated with a mathematical object—i.e. by first regimenting a pseudocode specification of $A$ as a program $\Pi$ over a language $\mathbb{L}$, and then identifying $A$ with $[\![\Pi]\!]^{\mathbb{L}}_{op}$—it still leaves the choice of this object highly unconstrained. For note that even if we fix the schema (7) as a canonical means of making reference to algorithms, choices of both the programming language L and the precise form of the denotational semantics $[\![\cdot]\!]^{\mathrm{L}}_{op}$ are still left undetermined. But as will be evident to anyone who has ever attempted to express an algorithm *A* of even moderate complexity in a formal programming language, different languages lend themselves in different ways to regimenting the basic operation and flow control devices employed in *A*. There will thus typically be many programs over the same language which we will accept as equally "apt" expressions of *A*.

When compounded with the further variation which may be introduced by different forms of operational semantics we might adopt for the various choices of L, it would seem that the current proposal provides little insight into why we are justified (e.g.) in attributing a particular running time complexity (either exact or asymptotic) to the algorithm *A* itself. As such, it seems that the current proposal is unable to account for the constraints imposed on how we interpret our informal discourse about algorithms in a manner which is compatible with constraints imposed (per requirement III) above) by complexity theory and algorithmic analysis.

The difficulty just described evidently represents a variant of the familiar *what-numbers-could-not-be problem* described in §2.1—i.e. when we attempt to identify algorithms directly with either their linguistic specifications or the mathematical objects which we might take to be the semantic interpretations of these specifications, we are faced with an abundance of seemingly arbitrary choices about which objects should be taken as their "canonical" representations. A well-known reply to this problem comes in the form of the structuralist proposal (cf., e.g., [72]) which holds that "algebraic" objects like groups or graphs should not be indentified with individual sets, but rather with *structures* (conceived roughly as in model theory), which themselves are identified only up to isomorphism. An analogous refinement of the foregoing proposal is as follows: (i) rather than seeking to identify algorithms with particular programs or their interpretations, we seek to define an appropriate equivalence relation $\approx$ over the class $\mathfrak{P}$ of programs drawn from the different programming languages $\mathrm{L}_1, \mathrm{L}_2, \ldots$, which we take to be adequate media for expressing the sorts of algorithms we encounter in practice; (ii) we then identify an algorithm with the $\approx$-equivalence class of some program we take to express it.

This proposal leads us away from SCT toward the second form of the algorithms-as-abstracts view, discussed at the end of §2.1—i.e. the view that algorithms should be understood as the "logical objects" obtained via the abstraction principle (PP)—i.e.

(PP) the algorithm expressed by $\Pi_1$ = the algorithm expressed by $\Pi_2$ if and only if $\Pi_1 \approx \Pi_2$

Recall, however, that it was suggested there that this view also admits to a machine-based variant according to which the equivalence relation in question should be defined, not over programs, but over an appropriate class $\mathfrak{M}$ of machine models. This gives rise to the rival abstraction principle,

(MP) the algorithm implemented by $M_1$ = the algorithm implemented by $M_2$ if and only if $M_1 \simeq M_2$

where $\simeq$ is an appropriate equivalence relation defined over the class $\mathfrak{M}$.

Given that the algorithms-as-abstracts view appears to have significant advantages over direct algorithmic realism or SCT, it seems that the best prospects for an algorithmic realist lie with developing the form of this view based on (PP) or (MP) so as to provide a foundational account of algorithms which is in conformity with (I–III). But before entering into the details of how this might be accomplished in §2.5, it will also be useful to observe that there are a number of reasons to think that the the machine-based variant based on (MP) will ultimately fare better than the program-based variant based on (PP).[31]

In order to see why this is so, it is useful to briefly return to the comparison between the algorithms-as-abstracts view and the neo-logicist view of number theory mentioned in §2.1. Recall in particular that part of the standard defense of neo-logicism is that Hume's principle—i.e.

(HP) the number of $F$ = the number of $G$ if and only if $F \equiv G$

where $\equiv$ denotes the second-order definable relation of *equinumerosity*—represents an *analytic* feature of our concept of natural number. Although this claim has often been challenged (e.g. [8]), neo-logicists at least start out with a precise mathematical definition of the relation by which they propose to factor the class of finite sets into equivalence classes. And additionally, HP itself at least seems to have the following in its favor: (1) it is *extensionally adequate* in the sense that it explains both the truth of "the number of days in a fortnight = the number of moons of Neptune" and the falsity of "the number of planets = the number of US supreme court justices"; (2) even if it is not an *analytic* feature of our basic understanding of number, HP can plausibly be maintained to follow from such an understanding plus suitable definitions (cf., e.g., [38]).[32]

I will ultimately argue that no foreseeable version of the algorithms-as-abstracts views fares particularly well with respect to either of these criteria. But in order to see that the machine-based variant may at least fare better, note that a proponent of either form of the view faces the initial challenge of providing a precise definition of either $\mathfrak{M}$ and $\simeq$ or of $\mathfrak{P}$ and $\approx$. As we have seen, although there are many different models of computation that have been introduced, their members can typically be described in a transparent manner as iterators. This provides the proponent of the machine-based variant with a natural suggestion as to the appropriate choice of $\mathfrak{M}$—e.g. for maximal generality, they can just take $\mathfrak{M}$ to coincide with the (proper) class of sets satisfying a canonical translation of Definition 2.4.1 into the language of set theory. And as we will see in §2.5, over such a class it is at least possible to provide a general definition of equivalence—i.e. that of *mutual simulation*—which provides a *prima facie* plausible means of responding to concerns of type (1) (i.e. extensional adequacy) and type (2) (i.e. fidelity to an antecedently recognized notion of "procedural equivalence").

But matters would appear to stand somewhat differently with respect to the program-based variant of the algorithms-as-abstracts view. To get an impression of why this is so, note first that as with the notion of machine model, the number of programming languages which have been introduced in computer science numbers at least into the hundreds. However, the characteristics which qualify a formalism as a programming language seem to be less well defined than in the case of machine models. This is witnessed, for instance, by the existence of different *programming paradigms*—e.g. declarative, functional, object-oriented, etc.—each of which can be understood to be based on a different conception of what is involved with providing a linguistic description of a mathematical procedure (cf., e.g., [70]). Given that these languages employ primitive constructs drawn from a wide range of developments in logic and mathematics—e.g. the typed and untyped-lambda calculus (LISP and Haskell), first- and higher-order logic (PROLOG, HILOG), graph rewriting (GP), linear algebra (FORTRAN)—it seems unlikely that we can find an overarching mathematical definition analogous to that of an iterator relative to which all of the programming languages studied in computer science may be naturally circumscribed.

A proponent of the program-based variant might still attempt to nominate some specific language L relative to which it might be argued that all algorithms may be naturally expressed.[33] Putting aside the question of what could justify the choice of L over other languages which might be put forth for this purpose, one might then attempt to explicitly define the relation, $\approx$ over $\texttt{Prog}_{\texttt{L}}$, by attempting to identify pairs of programs (or potentially even individual programming language constructs) which satisfy some antecedent intuitions about when two programs (or constructions) express the same algorithm, or as we might put it, are "procedurally equivalent."

Such an approach bears an evident affinity to the notion of *synonymy isomorphism*, explored by Carnap [11] and Church [13] in the case of sentence meanings. Various approaches to defining procedural equivalence have also been studied in the literature on programming language semantics. But such proposals often have an *ad hoc* flavor in the sense that they concentrate on only particular species of "procedure preserving" transformations, such as renaming of bound variables, substitution of evaluable expressions, or transformation of one form of flow control construction into another.[34] They thus provide little confidence that a notion adequate to meeting the needs of algorithmic realism—e.g. with respect to extensional adequacy or the justification of attributing asymptotic running times directly to algorithms—can be defined in a manner which does not ultimately rely on the same intuitions which underlie the machine-based variant of the algorithms-as-abstracts view, which will be considered in §2.5.[35]

## 2.5 Procedural Equivalence

The view which will be discussed in this section is a further refinement of the proposal which I have just argued represents the most promising means of defending algorithmic realism—i.e. the machine-based variant of the algorithms-as-abstracts view. In order to develop such a proposal in greater detail, recall that one needs to put forth both a class of machine models $\mathfrak{M}$ and an equivalence relation $\simeq$ over $\mathfrak{M}$ such that if one then identifies algorithms with $\simeq$-equivalence classes over $\mathfrak{M}$ determined by their implementations, we may then provide a plausible account of requirements such as (I–III) from §2.4.

Such proposals have historically attracted the most attention among proponents of algorithmic realism. For unlike the situation we face with respect to the program-based variant of the algorithms-as-abstracts view, it seems possible to nominate definitions of $\mathfrak{M}$ and $\simeq$ which are both technically and conceptually plausible. With respect to the former, for instance, we can see that both complexity theory and algorithmic analysis provide a principled basis for requiring an implementation of an algorithm to be a member of a model of computation from the first machine class. Thus although the general definition of an iterator, introduced, previously is it itself too broad to serve as a definition of $\mathfrak{M}$, we might plausibly try to develop the machine based view by considering the class $\mathfrak{M}^1$ comprised of the iterator representations of the union of all models in the first machine class.[36]

Since machines in $\mathfrak{M}^1$ can be naturally represented as iterators, they share a common mathematical structure which is not possessed by programs formulated over different languages. Not only might we hope to take advantage of this common structure so as to provide a definition of $\simeq$ which somehow engages with the specification of machines by virtue of their properties as iterators, but there might even be some hope that this can be accomplished in a manner which can be understood to analyze a salient *pretheoretical* notion of computational equivalence. This might in turn be thought to improve the chances that the principle (MP) can be regarded as akin to a logical or conceptual truth which grounds our understanding of algorithms in something like the way the neo-logistics claim that Hume's principle grounds our understanding of the notion of natural number. It has in fact been repeatedly claimed that $\simeq$ can be taken to be a form of simulation equivalence of the sort which has been extensively studied in a variety of contexts in theoretical computer science. I will introduce and motivate this notion in §2.5.1, before arguing in §2.5.2 that it is also unlikely that a version of (MP) based on simulation equivalence will be able to meet the needs of algorithmic realism.

### 2.5.1 *Simulation Equivalence*

The origins of the notion of simulation which will be at issue in this section can be traced back to the results demonstrating that the models of computation originally defined in the 1930s all determine the class of partial recursive functions. Such results are *extensional* in the sense that they pertain to the class of functions which are determined by the machines comprising a model, rather than to how they are computed. However, they are typically demonstrated in a paradigmatically *intensional* manner, as described in the following passage:

> The proofs for the results ... have the following common structure. In every instance, the fact that one formally characterized class of partial functions is contained in another is demonstrated by supplying and justifying a uniform procedure according to which, given any [machine $M_1$] from the first characterization, we can find a [machine $M_2$] from the second characterization for the same partial function. ... [Results of this type show] that there is a sense in which each standard characterization appears to include all possible algorithms ... For, given a formal characterization ... there is a uniform effective way to "translate" any set of instructions (i.e. algorithm) of that characterization into a set of instructions of one of the standard formal characterizations.
>
> [66], p. 19.

Consideration of the equivalence proofs in question also makes clear that the notion of a "translation" between models is what would later come to be known as a *simulation*—i.e. that of a uniform transition-preserving mapping between the states comprising the computations induced by two machines $M_1$ and $M_2$ from different models.[37] The proposal that it is possible to *define* the notion of algorithm in terms of an appropriate refinement of this notion appears to have originated in the early development of the field now known as *process algebra* (cf. [67]), wherein the notion of simulation was first rigorously defined.

In one of the founding papers in this subject, Milner writes as follows:

> One aim ... is to make precise a sense in which two programs may be said to be realizations of the same algorithm. We can say loosely that for this to be true it is sufficient though perhaps not necessary that the programs do the same "important" computations in the same sequence, even though they differ in other ways: for example 1) we may disregard other computations perhaps different in the two programs, which are "unimportant" in the sense that they are only concerned with controlling the "important" ones, (2) the data may flow differently through the variables or registers, (3) the data may be differently represented in the two programs. ... [W]e give a relation of simulation between programs which may fairly be said to match [this description] ... Mutual simulation is an equivalence relation, and it is the equivalence classes under this relation which may be regarded as algorithms – at least this is an approximation to a definition of algorithm.
>
> [51] p. 2.

What Milner refers to here as "programs" are in fact structures very similar to iterators. If we let $\phi_1 = \langle in_1, St_1, \Sigma_1, T_1, out_1\rangle$ and $\phi_2 = \langle in_2, St_2, \Sigma_2, T_2, out_2\rangle$ be iterators of type $X \to Y$, Milner's proposal may thus be understood as the claim that such structures should be taken to implement the same algorithm just in case there exists a relation which correlates the steps in the computations induced by the input $x$ $\vec{s}(x) = s_0(x), s_1(x), s_2(x), \ldots$ of $\phi_1$ and $\vec{t}(x) = t_0(x), t_1(x), t_2(x), \ldots$ and of $\phi_2$ which satisfies certain additional properties. I will refer to these properties respectively as the *transitional* and *representational* conditions on a definition of simulation.

The transitional requirement is intended to analyze the intuition that iterators $\phi_1$ and $\phi_2$ corresponding to machines which implement the same algorithm $A$ ought to each perform the same sequences of operations which must be performed to carry out $A$ in the same order. In the most straightforward situation, this could be formalized by requiring that there exists a *simulation relation*, $R \subseteq St_1 \times St_2$, such that every transition between states $s, s' \in S_1$ mediated by $\sigma_1$ in a computation of $\phi_1$ is matched by a similar transition between $R$-related states mediated by $\sigma_2$ in the corresponding computation of $\phi_2$.[38] Writing $s \overset{i}{\mapsto} s'$ for $\sigma_i(s) = s'$ $(i \in \{1, 2\})$, such a condition can be schematized as follows:

$$(8)\quad \forall s \in St_1\, \forall s' \in S_2\, \forall t \in St_2[(s \overset{1}{\mapsto} s' \wedge R(s,t)) \rightarrow \exists t' \in S_2(t \overset{2}{\mapsto} t' \wedge R(s',t'))]$$

It is a condition of this type which is typically demonstrated to hold between machines from different models in the course of demonstrating both the sort of extensional equivalence and complexity overhead results mentioned previously.

But it is also not hard to see that as stated (8) is almost certainly too restrictive to apply to many cases in which we wish to regard $\phi_1$ and $\phi_2$ as implementations of the same algorithm. Milner flags this problem by noting that a simulation should only be required to relate sequences of states corresponding to "important" subcomputations but (presumably) not the "unimportant" sequences of states which comprise them. We can, for instance, imagine that the former correspond to implementations of the steps of the algorithm $A$ which we take $\phi_1$ and $\phi_2$ to mutually implement, while the latter correspond to the sequences of "finer grained" steps by which the iterators carry out these molar steps.

If we write $s \overset{i\,*}{\mapsto} s'$ to denote that the state $s'$ is derivable by some finite number of iterations of $\sigma_i$ from $s$, then one way of revising (8) to take this observation into account is as follows:

$$(9)\quad \forall s \in St_1\ \forall s' \in St_2\ \forall t \in St_2[(s \overset{1\,*}{\mapsto} s' \wedge R(s,t)) \rightarrow \exists t' \in St_2(t \overset{2\,*}{\mapsto} t' \wedge R(s',t'))]$$

As van Emde Boas [79] observes, however, in moving from (8) to (9) there is a risk that we will make the transitional condition too weak by allowing, e.g., a simulation which links only the initial and final states in each computation of $\phi_1$ and $\phi_2$. It is hence generally acknowledged that an adequate formalization of the transitional condition must find a compromise between these constraints which is compatible with the extensional adequacy condition that $\simeq$ holds between $\phi_1$ and $\phi_2$, just in case they derive from machines $M_1$ and $M_2$ which we are prepared to accept as implementations of the same algorithm.

The representational requirement on the analysis of simulation is intended to formalize the fact that if we regard the iterators $\phi_1$ and $\phi_2$ as implementations of the same algorithm, then a simulation between them ought to relate states which represent the same mathematical structures (or "data") on which we understand them to operate. Suppose, for instance, that we are willing to regard $\phi_1$ and $\phi_2$ as implementations of Euclid's algorithm, but that neither is based on a machine which is able to operate directly on natural numbers. (For instance, $\phi_1$ might be a Turing machine and $\phi_2$ might be a graph rewriting machine similar to that introduced by Schönhage [68].) In this case, both $\phi_1$ and $\phi_2$ will need to use some form of alternative encoding to represent both the numerical inputs $n, m$ to Euclid's algorithm, as well as the sequence of intermediate values (i.e. $r_0 = n \mod m, r_1 = m \mod r_0, r_2 = r_1 \mod r_0, \ldots$) which are computed during the course of its operation. (We might, for instance, imagine that $\phi_1$ uses binary strings written on its tape to store these values, whereas $\phi_2$ might uses a sequence of nodes and pointers.) In such a case the representational requirement imposes the condition that a simulation $R$ between $\phi_1$ and $\phi_2$ must relate states which represent the same numerical values for each computation of Euclid's algorithm as implemented by these iterators.

### 2.5.2 *The Exigencies of Simulation*

While the foregoing conditions characterize the conceptual role which the notion of simulation is expected to play, they fall short of providing a precise definition of an equivalence relation $\simeq$ defined on $\mathfrak{M}^1$. A variety of different technical definitions of simulation have been proposed in computer science. Many of these are designed specifically to treat non-deterministic or concurrent computation. But since all of the models we have been considering are deterministic and non-concurrent, it will be useful to take as our paradigm Moschovakis's [56] definition of *iterator isomorphism.*

**Definition 2.5.1** *An* isomorphism *between two iterators* $\phi_1$ *and* $\phi_2$ *is a bijection* $\rho : St_1 \to St_2$ *such that* $\rho(in_1(x)) = in_2(x)$ *for all* $x \in X$, $\rho[T_1] = T_2$, $\rho(\sigma_1(s)) = \sigma_2(\rho(s))$ *for all* $s \in S_1$ *and* $out_1(s) = out_2(\rho(s))$, *for every* $s \in T_1$ *which is* input-accessible *(i.e. such that for some* $x \in X$ *and some* $n \in \mathbb{N}$, $s = \sigma_1^n(in_1(x))$*). We say* $\phi_1$ *and* $\phi_2$ *are* isomorphic *just in case there exists a mapping with these properties.*

Requiring simulation relations to be bijections has the advantage of ensuring that simulation is an equivalence relation.[39] Clearly this definition can be modified in a number of ways—e.g. by relaxing the requirement that $\rho(x)$ be a function, or requiring only that it satisfies a requirement analogous to (8). But as I will now argue, there appear to be several reasons why any definition in this vicinity is still likely to fail to satisfy one or more of the requirements which would be necessary to vindicate the version of the algorithms-as-abstracts view under consideration.

#### 2.5.2.1 Formalizing the Transitional Condition

In order to satisfy Milner's transitional requirement we have observed that we must find a definition of simulation which strikes a balance between (8) and (9). This raises the concern that no definition in the spirit of iterator isomorphism can be found which induces the correct identity conditions for the algorithms we already employ (let alone those which we have yet to discover or explicitly define). For, on the one hand it is easy to construct pairs of such structures which intuitively implement the same algorithm but for which this condition does not hold. And on the other hand, note that once we have decided to uniformly represent machines as iterators, it follows that the running time complexity, $time_M(n)$, of a machine $M$ will be given by $\max\{len_\phi(x) : |x| = n\}$. But also note that if $\phi_1$ and $\phi_2$ are isomorphic in the sense of Definition 2.5.1, then it will follow that $len_{\phi_1}(x) = len_{\phi_2}(x)$ for all $x \in X$. Thus, if $\phi_1$ and $\phi_2$ are isomorphic in this sense, it will follow that they have the same exact running time complexity.[40] And this would appear to imply—counter to the argument given in §2.3—that *exact* as opposed to *asymptotic* running time is an intrinsic property of individual algorithms.

These observations can be taken to illustrate another way in which the classical problem of "grain"—which is often pressed against theories of intensional entities like Fregean senses and propositions—also arises when we attempt to provide a definition of $\simeq$ which matches our intuitions about algorithmic identity and non-identity.[41] But such concerns can also be understood to point toward a yet more general worry for any view which attempts to analyze the notion of algorithm in terms of a principle akin to (MP), wherein $\simeq$ is taken to be a form of simulation equivalence similar to iterator isomorphism.

The underlying difficulty may be illustrated by first recalling that for a given model $\mathfrak{M}$ in the first machine class and algorithm $A$, there will often be multiple equally "apt" or "faithful" ways of implementing $A$ as instances of $\mathfrak{M}$. To better appreciate the scope of this concern, note that many of the algorithms we employ in mathematics operate on structured entities like polynomials, matrices, or graphs. The operation of such procedures may also involve auxiliary data structures like trees, stacks, or heaps. Practical experience again bears out the fact that efficient implementations of such structures can be founded using the RAM model $\mathfrak{R}$.[42] But predictably, these representations are almost never unique—e.g. a matrix can be represented as a sequence of appropriately indexed registers or (if sufficiently sparse) as a hash table; a graph may be represented either as a list of adjacent vertices or as an adjacency matrix; a tree may be represented either as a prefix-closed list of sequences or as a graph, etc. And in addition, operations on these structures, which we often treat as unmediated operations when specifying algorithms informally—e.g. looking up an entry in a matrix, adding or deleting a node from a graph—will generally be associated with operations which require multiple steps when implemented using a RAM machine.

Consider, for instance, a specific algorithm $A$ for operating on graphs—e.g. Kruskal's minimum spanning tree algorithm.[43] A basic operation employed in this algorithm is that of adjoining an edge $(u, v)$ to a set of edges $E'$ of the graph $G' = \langle V, E' \rangle$, which will ultimately form a minimal spanning tree for the input graph $G = \langle V, E \rangle$. But now, suppose $\phi_1$ and $\phi_2$ are iterators representing RAM machines which implement $A$ in different ways—say $\phi_1$ represents G and $G'$ as adjacency matrices and $\phi_2$ represents them as lists. And finally, consider the sequence of steps $\vec{s}(G) = s_0(G), \ldots, s_{c_1(n)}(G)$ of $\vec{t}(G) = t_0(G), \ldots, t_{c_2(n)}(G)$ of $\phi_1$ and $\phi_2$ which will be required for $A$ to adjoin the edge $(u, v)$ to a particular spanning tree $U$ which it is constructing.

Although it is reasonable to assume that the lengths of $\vec{s}(G)$ and $\vec{t}(G)$ will vary as linear functions $c_1(n)$ and $c_2(n)$ for $n = |V|$, there is no *a priori* reason to suspect that these functions will be identical. Nonetheless, Milner's transitional condition on the definition of simulation requires that any simulation between $\phi_1$ and $\phi_2$ ought to correlate, not the individual steps of $\vec{s}(G)$ and $\vec{t}(G)$ (which will be impossible if $c_1(n)$ and $c_2(n)$ ever differ in value), but rather these sequences themselves. This suggests that in order to satisfy the spirit of the transitional requirement, we must find an appropriate way of partitioning the sequences of states corresponding to the computations of $\phi_1$ and $\phi_2$ into subsequences, to take into account how these iterators (or more accurately, the machines from they are derived) implement the basic steps in terms of which $A$ is defined.

In specific cases, it will often be clear how this can be accomplished by modifying Definition 2.5.1 so that a simulation $\rho(x)$ is now understood as correlating, not individual states, but rather sequences of states determined by the relevant partitioning.[44] But such a modification will obviously depend on the particular partitioning of states required to ensure that the definition holds between the pair of iterators $\phi_1$ and $\phi_2$ in question. It thus seems that the relevant modification to the definition of iterator can only be formulated *after* we have decided that $\phi_1$ and $\phi_2$ should both be understood as implementing the same algorithm.

Recall, however, that according to the algorithms-as-abstracts view it is only through a principle such as (MP) that we understand what algorithms are in the first place. But if this is so, then the definition of $\simeq$ must presumably be fixed in a manner which is independent of the details of specific algorithms and the class of machines which we are willing to regard as their implementations. The proposed modification to the definition of iterator isomorphism using partition functions thus highlights that, in moving away from a purely structural account of Milner's transitional requirement as exemplified by (8), there appears to be no natural stopping point short of (9) which is fixed independently of our prior understanding of *specific* algorithms and implementations. But it would seem that if a proponent of the algorithms-as-abstract view has any hope of maintaining that (MP) reflects a conceptual or analytic feature of our informal notion of algorithm, they cannot allow that the definition of $\simeq$ is fixed on a case-by-case basis in this manner.

#### 2.5.2.2 Formalizing the Representational Requirement

In order to formalize Milner's representational requirement, we must find some relation between the computational states of iterators which formalizes the fact that they encode the same mathematical "data" on which a single algorithm $A$ operates. As a first step in this direction, suppose that $\phi = \langle in, St, \Sigma, T, out \rangle$ and that $s \in St$ is one of the states of $\phi$. Then we will typically be able to think of $s$ as a mathematical structure, akin to a Turing or RAM machine configuration. Such a structure can in turn be viewed in the conventional manner of first-order model theory—i.e. as a model of a first-order language $\mathcal{L}_s$ of the form $\mathcal{A}_s = \langle A, R_1, \ldots, R_n, f_1, \ldots, f_n, \{c_i : i \in I\} \rangle$ consisting of a non-empty domain $A$, a set of relations $R_i$, functions $f_j$, together with a designated collection of constants $\{c_i : i \in I\} \subseteq A$.[45]

In this setting, formalizing the representational requirement amounts to saying when the structures $\mathcal{A}_s$ and $\mathcal{A}_t$ represent the same configuration of data (and presumably also control parameters such as the values of loop variables) on which $A$ operates. Since these structures are conventional first-order models, it is natural to first ask if any of the familiar notions of structural equivalence from model theory—e.g. isomorphism, elementary equivalence, back-and-forth equivalence—can be of use in this regard. Note, however, that if $\phi_1$ and $\phi_2$ are based on machines from different models of computation (e.g. one is derived from a Turing machine, and one from a RAM machine), then $\mathcal{A}_s$ and $\mathcal{A}_t$ will generally interpret distinct languages $\mathcal{L}_s$ and $\mathcal{L}_t$. But in this case, these structures cannot stand in any of the aforementioned relations.

In the case where $\phi_1$ and $\phi_2$ are based on machines from a single model of computation $\mathfrak{M}$, the situation is likely to be more complex. To continue with the prior example, for instance, suppose that $\phi_1$ and $\phi_2$ are both based on RAM machines which we wish to interpret as implementing an algorithm which operates on a graph $G = \langle V, E \rangle$. A concrete illustration of the problem we face in formalizing the representational condition is to nominate some relation, $U \subseteq St_1 \times St_2$, which holds just in case they contain sequences of registers encoding graphs which are isomorphic. Note, however, that by proceeding in this way, we will almost certainly incur the same problem highlighted previously with respect to formalizing the representational condition. For if $\phi_1$ and $\phi_2$ represent $G$ differently in their computational states—e.g. one uses a matrix representation, and the other a list representation—then we will be forced to define $U$ in a manner that will not generalize to arbitrary algorithms operating on arbitrary data structures.

We might attempt to circumvent both of the foregoing problems by trying to identify a more general definition which analyzes what it means for one structure $\mathcal{A}_t$ to be representable in another $\mathcal{A}_s$, even if these structures are for different signatures $\mathcal{L}_1$ and $\mathcal{L}_2$. One possibility is the notion of *definable interpretability* from model theory, which holds between structures $\mathcal{A}_s$ and $\mathcal{A}_t$, just in case it is possible to define an $\mathcal{L}_1$-structure $\mathcal{N}$ whose domain is an $\mathcal{L}_1$-definable subset $X$ of the domain of $\mathcal{M}_1$ consisting of $\mathcal{L}_1$-definable elements, subsets, and functions of $X$ so that $\mathcal{N}$ is isomorphic to $\mathcal{A}_t$.[46] We might then attempt to analyze the representational condition by requiring that computational states correlated by a simulation relation are mutually interpretable in one another in this sense.

But if we attempt to apply this definition to the sorts of cases likely to be encountered in practice, we quickly run into what appears to be a complication arising by virtue of the way first-order model theory interacts with the mathematical definitions of common models of computation. For, note that in order to describe the relation $U$ which a state $s$ of iterator $\phi_1$, described earlier, bears to a state $t$ of $\phi_2$ by virtue of representing the same graph G, it is not sufficient to work in a language which describes $s$ and $t$ simply as vectors of natural numbers representing the contents of the registers of $\phi_1$ and $\phi_2$. Rather, we must work over a language containing additional arithmetical vocabulary sufficient for describing how these iterators encode and operate on finite graphs as sequences of natural numbers stored in their registers.

The most natural way to do this is to view $s$ and $t$ as structures of the form $\mathcal{A}_s = \langle \mathbb{N}, R_1, 0, s, +, \times \rangle$ and $\mathcal{A}_t = \langle \mathbb{N}, R_2, 0, s, +, \times \rangle$, where $R_1, R_2$ are binary relations such that $R_i(n, m)$ holds just in the case when $m$ is stored in the $n$th register of $\phi_i$ as represented by $s$ (if $i = 1$) or $t$ (if $i = 2$). It is now easy to see that $\mathcal{A}_s$ and $\mathcal{A}_t$ are mutually interpretable in one another, since we can find formulas in either language which allow us to represent either structure in terms of the other, using standard techniques for coding finite sequences to construct an $\mathcal{L}_1$-formula $\psi_{R_2}(x, y)$ which defines $R_2$ up to isomorphism as a subset of $\mathbb{N}^2$ and an $\mathcal{L}_2$-formula $\psi_{R_1}(x, y)$ which defines $R_1$ upto isomorphism also as a subset of $\mathbb{N}^2$. But since $R_1$ and $R_2$ will only encode a finite number of register-value pairs which will be employed in the computations of $\phi_1$ and $\phi_2$ on a fixed graph, we can see that $\psi_{R_1}(x, y)$ and $\psi_{R_2}(x, y)$ can be taken to be purely *arithmetical* formulas—i.e. to simply show that $\mathcal{A}_t$ is *definable* over $\mathcal{A}_s$ does not require that we make use of the relation $R_2$ itself (and conversely for $\mathcal{A}_s$ and $R_2$).

In fact, since $\mathcal{A}_s$ and $\mathcal{A}_t$ are structures over a language which extends that of first-order arithmetic, it is also easy to see (again using standard coding techniques) that there will be a vast array of other structures of the form $\langle \mathbb{N}, R, 0, s, +, \times \rangle$ with which they will be mutually interpretable, but for which $R$ will not bear any discernible relation of structural similarity to the graph $G$, which we are assuming the states $s$ and $s'$ encode. This would seem to illustrate another fundamental limitation on our ability to use traditional notions of sameness of structure taken from model theory to analyze Milner's representational condition.

#### 2.5.2.3 Implementing Recursion

The foregoing observations raise the general concern that despite the intuitive appeal of Milner's original analysis of simulation in terms of transitional and representational conditions, his characterization does not succeed in implicitly defining a relation (or even a family of relations) which holds between iterators by virtue of what we might broadly describe as their general structural properties, which also induces an extensionally adequate criterion of identity for algorithms. But as I will now attempt to illustrate, there are also instances in which these conditions can themselves be seen as pulling in opposite directions on the form that a definition of simulation ought to take.

Most of the procedures previously mentioned would traditionally be classified as *iterative* (or *sequential*) algorithms, in the sense that the computations induced by their operation can be understood as sequences of states induced by the repeated application of a transition function in the sense formalized by the notion of an iterator. There is, however, another class of procedures studied in algorithmic analysis known as *recursive* (or *divide and conquer*) algorithms, which operate by successively decomposing their input into two or more *subproblems* (i.e. structural components) on which they then repeatedly call themselves until a base case is reached.

A commonly cited example is the sorting algorithm known as MergeSort. This procedure can be described informally as follows: (1) divide the input array $B$ into two subarrays $B_1$ and $B_2$ comprising the first and second halves of $B$ (rounding indices as needed); (2) if either of these arrays is of length greater than 1, recursively call MergeSort (i.e. the procedure here described) on $B_1$ and $B_2$; (3) combine the sorted arrays $B_1'$ and $B_2'$ which result from this into a single sorted array by calling the auxiliary procedure Merge.[47]

Suppose we write MergeSort$(B, i, j)$ to express a call to MergeSort on the sublist $B[i], \ldots, B[j]$ and Merge$(i, j, k)$ to express a call to Megre on the sublists $B[i], \ldots, B[j]$ and $B[j+1], \ldots, B[k]$. Then MergeSort can be expressed in pseudocode as follows:

```
MergeSort(B, p, r)
if p < r
   then q ← ⌊(p + r)/2⌋
        MergeSort(B, p, q)
        MergeSort(B, q + 1, r)
        Merge(B, p, q, r)
```

It will be useful to illustrate the operation of MergeSort by considering an example. To this end consider the initial array $B = [8, 7, 6, 5, 4, 3, 2, 1]$, as depicted at the bottom of Figure 2.1. The bottom four levels depict the decomposition of $B$ into halves, as effected by the recursive calls to MergeSort. This process terminates after calls of the form MergeSort$(B, i, i + 1)$ (as the next calls would be of the form MergeSort$(B, i, i)$ and MergeSort$(B, i + 1, i + 1)$ for which the condition $p < r$ will fail). At this point, control passes back to the point of the recursive call and the operations Merge$(B, i, i, i + 1)$ are performed. This results in the fifth level from the bottom. After these values have been computed, control passes to the prior call to MergeSort, by which the resulting values and the subarrays of length two are merged by performing operations of the form Merge$(B, i, i + 1, i + 3)$, yielding the second row from the top. Finally, control passes to the initial calls of the form MergeSort$(B, 1, 4)$ and MergeSort$(B, 5, 8)$, which leads to the computation of Merge$(B, 1, 4, 8)$, yielding the sorted array at the top.

As is evident from the foregoing description, MergeSort is most naturally described as operating in a manner which assumes that certain of its steps can be carried out in parallel—e.g the calls to the Merge operation for the inputs $[7, 8]$, $[5, 6]$ and $[3, 4]$, $[1, 2]$ corresponding to the third row from the top are understood to be carried out simultaneously. A

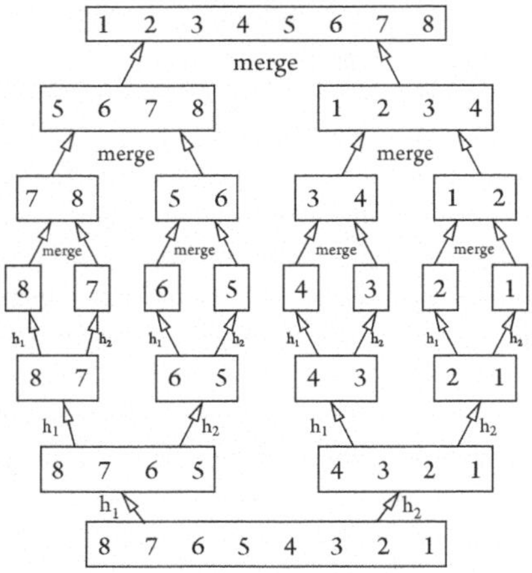


**Figure 2.1** The operation of MERGESORT on the array $[8, 7, 6, 5, 4, 3, 2, 1]$. $h_1$ denotes the operation of taking the first half of an array, $h_2$ the second half, and *merge* the merging operation described in the text.

question which often arises in computational practice, however, is whether such a procedure can be implemented using a machine drawn from a model in the first machine class. For as we have observed, such machines do not allow for unbounded parallelism of the sort which would be required to construct a direct implementation of the procedure just described.

In order to demonstrate that such implementations are available, it is typical to employ a sequential model of computation equipped with an auxiliary data structure known as a *stack*. Such a device can be used to keep track of both the subarrays which are being operated on and the location of a particular splitting or merging operation in the structure depicted in Figure 2.1. A method for transforming recursive specifications of procedures into machines in a class such as $\mathfrak{R}$ (to which a stack may either be added or implemented in terms of its other primitives) is known as an *implementation of recursion*. The details of how this is accomplished in practice are often quite complex (cf., e.g., [1]) and need not concern us here. What is of current concern, however, is that there will often not be a *unique* way of implementing a recursive algorithm specified using a sequential model such as $\mathfrak{R}$. For instance, there are distinct stack-based implementations of MERGESORT, $R_{left}$ and $R_{right}$, such that the former always performs the left-most possible merge operation first (relative to the sort of computation tree depicted in Figure 2.1) and the latter always performs the right-most possible merge operation first (as well as a variety of other intermediate possibilities).

These complications notwithstanding, the relation borne by both $R_{left}$ and $R_{right}$ to MERGESORT would conventionally be regarded as paradigmatic of the implementation relation in computational practice. There is thus a strong *prima facie* reason to think that any adequate definition of $\simeq$ ought to hold between $R_{left}$ and $R_{right}$. As should now be evident, however, no relation satisfying these properties can satisfy both the transitional and representational requirements of a definition of simulation simultaneously. For suppose that we let $\vec{l}_1, \ldots, \vec{l}_7$ and $\vec{r}_1, \ldots, \vec{r}_7$ respectively represent the sequences of transitions of $R_{left}$ and $R_{right}$ responsible for carrying out the merges depicted in Figure 2.1. According to the transitional condition, a simulation relation of type $R^* \subseteq St^*_{left} \times St^*_{right}$ ought to relate sequences of states which occur in the same order—i.e. $R^*(\vec{l}_1, \vec{r}_1), \ldots, R^*(\vec{l}_7, \vec{r}_7)$. But on the other hand, according to the representational condition, $R^*$ ought to relate sequences corresponding to the operations which MERGESORT performs on the same data—e.g. $R^*(\vec{l}_1, \vec{r}_5), R^*(\vec{l}_2, \vec{r}_4), R^*(\vec{l}_3, \vec{r}_7), \ldots$ The transitional and representational requirements thus cannot be simultaneously satisfied by any single-valued relation on $St^*_1 \times St^*_2$. As such, it would seem that there are instances in which the transitional and representational conditions on the definition of simulation are in genuine conflict with respect to certain of our pretheoretical intuitions about computational equivalence.

## 2.6 Taking Stock

### 2.6.1 *Moschovakis, Gurevich, and the Level-Relativity of Algorithms*

As I mentioned in §2.1, the task of providing a foundational account of the nature of algorithms of the sort we have been considering here has been largely overlooked by both mathematicians and computer scientists. Two notable exceptions are the theories of algorithms which been developed by Yiannis Moschovakis and Yuri Gurevich, each of whom can be understood as offering an extended defense of algorithmic realism. There are, however, several respects in which their views differ not only from one another, but also from the general framework I have argued in §2.4, and §2.5 represents the most promising means of developing this view. It will thus be useful to begin by investigating their proposals in more detail.

In a series of papers spanning more than 20 years (including [36], [37], [35]), Gurevich has developed the proposal that individual algorithms should be identified with instances of a class of formalisms known as abstract state machines [ASMs]—a proposal he refers to as the *ASM thesis*:

> The ASM thesis is that every sequential algorithm, on any level of abstraction, can be viewed as a sequential abstract state machine. [37], p. 1.

The definition of an ASM is very similar to that of an iterator—e.g. in [35] (p. 7) such a model $M$ is defined to consist of a class of *states* $\mathcal{S}(M)$ (a subset, $\mathcal{I}(M) \subseteq \mathcal{S}(M)$, of which are referred to as *initial*), and a transition function, $\tau_M : \mathcal{S}(M) \rightarrow \mathcal{S}(M)$. He also provides a definition similar to that given above of the computation induced by $M$ on input $x_0 \in \mathcal{I}(M)$ as the sequence derived by iterating $\tau(x)$ on $x_0$.

Moschovakis's proposal is in many respects similar, but is developed in terms of a formalism known as a *recursor*. Unlike any of the models we have considered thus far, this model takes recursion (rather than iteration) as its basic computational paradigm. A recursor is formally defined ([56], p. 85) to be a triple $\alpha = \langle \mathcal{D}, \tau, value \rangle$, such that $\mathcal{D} = \langle D, < \rangle$ is an inductive partially ordered set, $\tau : X \times D \rightarrow D$, and the result $\alpha(x)$ of applying $\alpha$ to $x \in X$ is defined to be value$(d)$, where $d$ is the $<$-least element of $D$ satisfying the fixed-point equation $d = \tau(x, d)$.[48] Two recursors, $\alpha_1 = \langle D_1, \tau_1, \text{value}_1 \rangle$ and $\alpha_2 = \langle D_2, \tau_2, \text{value}_2 \rangle$, are defined to be isomorphic just in case there exists an order-preserving bijection, $\pi : D_1 \rightarrow D_2$, for which $\pi(\tau_1(x, d)) = \tau_2(x, \pi(d))$ and $\text{value}_1(x, d) = \text{value}_2(x, \pi(d))$ ([56], pp. 86–87). On the basis of these definitions, Moschovakis proposes a view which might aptly be called the *recursor thesis*—i.e.

> *Proposal: Algorithms are recursors.* The mathematical structure of every algorithm on a poset $X$ to a set $W$ is modeled faithfully by some recursor $\alpha : X \rightarrow W$; and two recursors model the same algorithm if they are isomorphic. [56], p. 86.

Relative to the terminology adopted in §2.4, such remarks make clear that Gurevich is a proponent of Strong Church's Thesis,[49] while Moschovakis can be understood to be a proponent of the machine-based variant of the algorithms-as-abstracts view, where $\mathfrak{M}$ is taken to be the class of recursors and $\simeq$ is the relation of recursor isomorphism. Moschovakis and Gurevich can therefore both be reasonably classified as algorithmic realists. But it at first appears that they disagree in respect of how they would answer the following questions: (1) should algorithms be identified directly with the instances of a model of computation or rather with equivalence classes thereof? (2) should the mathematical class to which algorithms are reduced (either by direct identification or by factoring by a notion of procedural equivalence) be comprised of machines which take iteration or recursion as a basic computational paradigm?

Further reflection on their views would seem to suggest that their disagreement over (1) is not as substantial as it might initially appear. For instance, Moschovakis himself issues the following caveat about the definition of recursor isomorphism:

> [R]ecursor isomorphism is a very fine equivalence relationship which is not preserved by many useful algorithm transformations (optimizations, refinements, etc.), and we must often introduce 'coarser' equivalences ... to express important facts of informal algorithm analysis. [56], p. 87.

Some experimentation with formalizing algorithms as recursors bears out this point—i.e. in the course of constructing recursor representations of informally specified algorithms, we are often confronted with arbitrary choices about how to represent the data on which an algorithm operates, together with its computational state as a partially ordered set. And since different choices will lead to different recursors among which the defined isomorphism relation need not hold, it would seem that factoring representations of algorithms by recursor isomorphism does not lead to a satisfactory resolution to the problem of "grain" discussed in §2.5. It is, moreover, easy to see that isomorphic recursors will have the same running time complexity relative to the analysis Moschovakis provides. Both his view and Gurevich's hence predict that exact (as opposed to asymptotic) running time complexities are directly attributable to individual algorithms.

Moschovakis and Gurevich's proposals do, however, make different predictions about the identity conditions of certain recursive algorithms. Recall for instance, that we saw in §2.5.2.3 that in order to implement MERGESORT as a RAM machine we needed to make a seemingly arbitrary decision about how the merge operations it performs in parallel are put into sequential order, leading to distinct implementations, $R_{left}$ and $R_{right}$, which cannot stand in a simulation relation which simultaneously satisfies Milner's transitional and representational conditions. On the other hand, it is possible to canonically define a recursor $\alpha_{merge}$ whose operation can be understood to formalize the sort of parallel computation tree for MERGESORT depicted in Figure 2.1. On this basis Moscovakis ([56], p. 90) suggests that $\alpha_{merge}$ is a more apt mathematical representation of MERGESORT than iterators such as $R_{left}$ and $R_{right}$, deriving from sequential models of computation like $\mathfrak{R}$.

This example illustrates that the fundamental distinction between the proposals of Gurevich and Moschovakis is not so much that one prefers a "reductionist" ontology for algorithms, while the other prefers an "abstractionist" one, but rather whether the sort of objects with which they seek to identify algorithms derive from a model which is based on iteration or recursion. Taking into account the argument I made against taking simulation to be an adequate analysis of computational equivalence in §2.5.2, the foregoing observations may reasonably be taken to suggest that Moschovakis's proposal fares better than Gurevich's in accounting for the identity conditions of recursive algorithms like MERGESORT.[50]

There is, however, another sense in which the proposals of Moschovakis and Gurevich are alike, in the sense that they depart in the same way from the strategy which I argued at the beginning of §2.5 constitutes the most promising way of vindicating algorithmic realism. In particular, rather than attempting to define an algorithm in terms of its implementations from the class of machines $\mathfrak{M}^1$ from the first machine class, they take individual algorithms to be defined only relative to the primitive operations in terms of which they are specified. (Such a specification of primitives approximates a "level of abstraction" in the sense of Gurevich's formulation of the ASM thesis.) This highlights that both the ASM and recursor models have more in common with the general definition of an iterator than they with the specific models $\mathfrak{T}_k$ or $\mathfrak{R}$. For, whereas arbitrary mathematical operations may be used to define the states and transition functions of the former, the latter make available only very restrictive classes of functions and primitive data structures.

One potential advantage of the level-relative approach is borne out by the elegant reconstruction which Moscovakis ([56], pp. 83–84) provides of the sort of informal running time analysis of MERGESORT, as would typically be given in algorithmic analysis. As his analysis is developed directly by reasoning about the properties of the primitive mathematical notions in terms of which this algorithm is specified, it hence avoids the necessity of reasoning about the details of how machines like $R_{left}$ and $R_{right}$, discussed in §2.5.2.3, operate on registers or stacks. But in the case of many sequential algorithms which operate (e.g.) on graphs, matrices, or polynomials, the same will also be true of the ASM model by virtue of the considerations described in §2.5.2.1 and 2.5.2.2—i.e. since an ASM can be specified at an arbitrary "level of abstraction" it is also possible to simplify running time analyses by working with ASMs instead of (e.g.) RAM machines.

These considerations aside, however, the question also arises as to the utility of applying such analyses directly to "level-relative" models such as ASMs or recursors. For although it may be that an ASM $M$ or recursor $\alpha$ which we employ to model an informally specified algorithm $A$ will introduce *fewer* arbitrary properties into its mathematical representation than, say, an implementation $R$ of $A$ as a RAM machine, neither Moschovakis nor Gurevich provides an argument that $M$ or $\alpha$ must be *unique*. As such, their proposals must still presumably face variants of the what-numbers-could-not-be problem introduced in §2.1. But much more seriously than this, the generality of the ASM and recursor theses is purchased precisely at the price of severing the link between our practice of assigning running time complexities to informally specified algorithms and the underlying complexity costs of models like $\mathfrak{R}$ and $\mathfrak{T}_k$.

As I argued in §2.3, it is our ability to implement algorithms with respect to these models which ultimately explains how it is that complexity theory and algorithmic analysis provide a means by which we can meaningfully measure and compare the utility of different computational procedures in practice. As such, it seems reasonable to conclude that the level-relative view of algorithms advocated by Moschovakis and Gurevich does not provide an account which makes good on all of the foundational aspirations of algorithmic realism previously described.

### 2.6.2 *Algorithms, Identity, and Mathematical Practice*

I began this chapter by highlighting the import of foundational questions about algorithms to the task of accounting for the role of computational methods in mathematical proof. The last two sections have then been devoted to arguing that there is no apparent means of substantiating the algorithmic realist's claim that algorithms are freestanding objects which is simultaneously compatible with the methodologies of complexity theory and algorithmic analysis. In order to better orient ourselves with respect to how the diminished prospects of algorithmic realism bear on how one might ultimately go about answering questions of sorts (Q1) and (Q2) it will be useful to first recall that much of the evidence that was initially marshaled in favor of the plausibility of this view derives from our willingness to *speak* of algorithms as if they were abstract objects on a par with those studied in mathematics—e.g. natural numbers, graphs, groups, etc. This is evident not only from our practice of introducing algorithmic names to refer to individual procedures, but also in our willingness to quantify over certain classes of algorithms, as exemplified by statements like (1), (4), and (6).

Note, however, that another linguistic phenomenon which is often taken to be diagnostic of whether a class of expressions $t_1, t_2, \ldots$ is properly regarded as denoting objects, is our willingness to use them in identity statements of the forms $t_i =$ *the* $\varphi$, $t_i = t_j$, or $t_i \neq t_j$. We have seen some evidence that algorithmic names can be used in the first of these schema in the course of considering statements such as

(10) a) INSERTIONSORT = the algorithm expressed by program $\Pi$

b) MERGESORT = the algorithm implemented by machine $M$

Proponents of the algorithms-as-abstracts view will presumably take such statements as paradigmatic of the way in which algorithmic names are introduced into our mathematical language.

Based on their grammatical form, it might seem reasonable to take these statements to be a piece together with the following:

(11) a) $3 =$ the number of roots of $x^3 - 6x^2 + 11x - 6$
b) $D_4 =$ the group represented by $\langle x, y \mid x^4 = 1, y^2 = 1, yxy = x^{-1} \rangle$

But in the case of *bona fide* mathematical items like natural numbers and groups, it is also typical to find within our discourse statements instances of another form of identify statement—i.e. *the* $\varphi_1 =$ *the* $\varphi_2$—as exemplified by

(12) a) the number of primes less than $6 =$ the number of roots of $x^3 - 6x^2 + 11x - 6$
b) the group represented by $\langle x, y \mid x^4 = 1, y^4 = 1, xyxy = 1 \rangle =$
the group represented by $\langle x, y \mid x^4 = 1, y^2 = 1, yxy = x^{-1} \rangle$

It is the analogues of such statements involving the expressions "the algorithm implemented by $M$" and "the algorithm expressed by $\Pi$" which the two variants of the algorithms-as-abstracts views suggest ought to be regarded as fundamental to our general understanding of what it is to be an algorithm. In particular, it is precisely this kind of statement which figures on the left-hand side of the principles (MP) or (PP), which proponents of this view regard as an implicit definition of algorithmic identity. It is hence a significant observation that our everyday computational-cum-mathematical discourse seems to provide few examples of such statements about which we appear to have firm convictions. As the case described in §2.5.2.2 attests, we do occasionally attempt to adjudicate the status of statements like "the algorithm implemented $R_{left}$ = the algorithm implemented by $R_{right}$." But it would seem that our judgement about the truth or falsity of such statements is conventional at least to the extent that we do not expect that their truth values are fixed relative to other pre-established mathematical facts. On the other hand, it seems that not only do we regard statements like (12a,b) as potentially informative (e.g. in the sense that they admit to non-trivial mathematical proof) but also that their truth values are fixed in advance by our prior understanding of the meanings of the expressions in terms of which they are formulated.

Reflection on this point goes a fair distance towards explaining why we do not need to be realists about algorithms in order to explain the manifest utility of subjects like complexity theory and algorithmic analysis. For, recall the example of the algorithm LUCAS, discussed in §2.1. As we saw, it was by applying this algorithm that Lucas was able to accomplish the otherwise infeasible task of demonstrating the primality of $m_{127}$. Nonetheless, there is no evident sense in which the algorithm LUCAS itself figures in the proposition expressed by "$m_{127}$ is prime." Rather, the algorithm functions as an auxiliary device which aids in our demonstration of the theorem.

Note, however, that it was by similar routes that algebraic entities like groups also began their life in mathematics—i.e. as auxiliary devices figuring in the proof of purely number theoretic propositions.[51] Although the contemporary definition of a group had not been formulated at the time when algebraic methods started to be used, such structures are now universally recognized as bona fide mathematical objects. There thus appears to be little reason to regard an algebraic proof of a result in number theory to be in any sense "extra-mathematical." As we have seen, algorithmic realism seeks to secure the same status for results derived by computational methods by assimilating algorithms into the framework of classical mathematics.

If algorithmic realism is true, it thus seems reasonable to think that we should regard the status of algorithms in mathematics as something like that of groups. As we have seen, the general notion of an algorithm has a considerably older pedigree than that of a group. And although it is somewhat newer in origin than group theory, we also possess a theory which takes algorithms as its ostensible subject matter—i.e. algorithmic analysis. It is notable, however, that no consensus has arisen *within* this field as to the appropriate criterion of identity for algorithms, or even as to the form such a condition might take. In comparing algorithmic analysis and group theory, it is thus reasonable to wonder why the development of the latter has prompted the formulation of a family of notions of structural equivalence (e.g. homomorphism, isomorphism, embedding, etc.) for groups, while no family of similar notions has developed in the context of algorithmic analysis to serve the role of canonically relating one algorithm (or representation thereof) to another.

The answer to this question ultimately seems straightforward: the primary significance we place in the discovery of an algorithm $A$ is likely to be in applying it to solve instances of an antecedently identified mathematical problem—e.g. computing the values of a function $f(x)$. For this reason, our mathematical interest in $A$ is also likely to be exhausted by proving its correctness with respect to $f(x)$ and possibly of comparing its efficiency with that of other means of computing the values of this function.

If we are to perform the latter task in a manner which is responsive to the analysis of feasibility provided by complexity theory and algorithmic analysis, it seems that we have no choice but to work with a representation $M$ of $A$ drawn from a model $\mathfrak{M}$ in the first machine class. When working with such models, we have also seen that the implementation problem is often non-trivial—i.e. based on the informal specification of $A$; it may not initially be obvious how to employ the basic operations and data structures made available by $\mathfrak{M}$ to construct $M$ so as to mimic $A$'s operation in a step-by-step manner.

Given that we are able to construct such a machine, however, it is also, very likely that we will also be able to construct other formally distinct $M' \in \mathfrak{M}$, which implement $A$ in a different manner (e.g. by using a different representation of the data structures on which $M$ operates, as discussed in §2.5). But once we have constructed our initial representation $M$ (thereby making it possible to give rigorous proofs both of $A$'s correctness and its running time complexity), our interest will typically shift from the properties of $M$ to that of employing it to compute the values of $f(x)$. Our potential willingness to acknowledge that $M$ and $M'$ are both implementations of $A$ notwithstanding, mathematical considerations alone thus do not dictate that we should return to consider what (if anything) about their structural properties might account for this fact.

Thus although there will be many instances in which our practices may dispose us to regard two programs or machines as representing the same algorithm, the question of whether such judgements can be made precise does not seem to be of any independent mathematical significance. There thus seems to be no abiding reason why ontological questions of type (Q1) need to be answered before we can provide detailed answers to epistemological questions of type (Q2). At least to date, the story of algorithmic realism might thus be taken to represent a cautionary tale for the study of ontological commitment. For it is indeed difficult to deny that our discourse is highly suggestive of the fact that parts of theoretical computer science are engaged in the study of a class of procedural entities, closely related to but yet somehow distinct from, those studied in classical mathematics. But at the same time, a more detailed appraisal of the methodologies of the relevant subjects suggests that there may be no way to take this language at face value without contravening some of the assumptions about the nature of algorithms in which these fields themselves seem to be grounded.

This conclusion may at first sound negative with respect to the suggestion formulated in §2.1 that the general notion of algorithm deserves a place alongside concepts such as number, set, and function, as substrates of contemporary mathematics. However, it should now also be evident that many of the considerations which have been touched on in this chapter point to the import of questions refining (Q2). Some of these are as follows:

(Q3) What does it mean for a formal model to provide a representation of the constraints which we face in concretely embodied mathematical? Given that two functions are both computable (i.e. recursive), what does it mean to say that one is more intrinsically difficult to compute than the other? How can the utility of different methods for solving the same mathematical problem be meaningfully measured and compared? Is it possible to provide rigorous mathematical definitions of intrinsic computational difficulty and infeasibility?

It would seem that such questions must be subjected to closer philosophical scrutiny before it can reasonably be said that we have a satisfactory foundational account of the growing role which computational techniques play in mathematical practice. Beyond what has been said in §2.3, however, a more thorough consideration of these issues must await another occasion.

## Acknowledgement

This chapter derives from my Rutgers dissertation [17]. Thanks are owed to the (many) people who supported me during the (long) course of its preparation and afterwards. A proper subset of these is as follows: Robert Matthews, Paul Benacerraf, Sergei Artemov, Rohit Parikh, Yoson Yanofsky, Joel Hamkins, Roman Kossak, Haim Gaifman, Kit Fine, Michael Detlefsen, Susan Schweitzer, Hidenori Kurokawa, Roman Kuznets, Eric Pacuit, Evangelia Antonakos, Karen Kletter, Sean Walsh, Andrew Arana, and Chris Porter.

## Notes

1. Buchberger's algorithm can be understood as generalizing both Euclid's algorithm and Gaussian elimination. It finds a variety of applications in both pure and applied mathematics—Cf., e.g., [15].
2. For instance, consider Sturm's theorem—"The number of distinct real roots of a square-free polynomial $p(x)$ located in the half-open interval $(a, b]$ is given by $\sigma(a) - \sigma(b)$ where $\sigma(x)$ is the number of sign changes in the Sturm sequences $p_0(x), p_1, (x), \ldots, p_m(x)$ for $p(x)$." Although the classical proof of this result consists largely in specifying an algorithm for computing the

Sturm's sequence for $p(x)$ (and thereby the value of $\sigma(a) - \sigma(b)$) and proving it correct, the statement of the theorem itself makes no mention of this.

3. Of course one might also think that a view deserving the name "realism" about algorithms ought to be involved with providing a *non-reductive* account of our discourse about them—i.e. one which identifies algorithms not with mathematical objects as traditionally conceived, but rather with a distinctive class of entities which might (for instance) possess genuine spatio-temporal properties. Although such a view—which I will discuss further in §2.4 under the name *direct algorithmic realism*—does not appear to be widely held in mainstream computer science, it bears an obvious affinity to intuitionistic or constructive mathematics. For instance, the understanding of mathematical objects as idealized mental constructions within intuitionism may itself be taken to form a kind of converse to (A)—i.e. rather than seeking to identify procedural entities with mathematical ones, it seeks to identify mathematical entities with procedural ones. Unlike intuitionism, however, algorithmic realism is not intended to be a foundational thesis about mathematics as a whole, but rather a thesis about the relationship of procedural discourse—especially as it figures in theoretical computer science—to mathematics.
4. Donald Knuth—who is generally credited with founding (as well as naming) the field now known as *algorithmic analysis*—has been one of the most outspoken proponents of algorithmic realism. One of his most explicit formulations of this view is as follows: "Algorithms are concepts which have existence apart from any programming language ... I believe algorithms were present long before Turing et al. formulated them, just as the concept of the number 'two' was in existence long before the writers of first grade textbooks and other mathematical logicians gave it a certain precise definition." [45], p. 654.
5. Inchoate versions of strong Church's thesis can already be found in Post [64] and in Kolmorogov and Uspensky [46]. More explicit formulations are given by Knuth [41], Gandy [28], Sieg and Byrnes [74], and Dershowitz and Gurevich [18].
6. Inchoate forms of the algorithms-as-abstracts view can be found in Rogers [66] and Kreisel [48], whereas more refined versions have been developed by Milner [51], Moschovakis (for example) [56], and Yanofsky [82].
7. See, e.g., [38] for the definition and substantial discussion of these problems.
8. The version of the proof which was originally published included an exhaustive case analysis, conducted by computer, to check whether a collection of almost 2000 graphs possessed a property known as *reducibility*. Although this number has been reduced in subsequent proofs, all known demonstrations of the four color theorem rely on verifying a number of decidable graph theoretic statements, greater than the number which can be feasibly tested (or even surveyed) by a human mathematician. On this basis, Tymoczko [78] famously argued that the Appel and Haken demonstration is not a mathematical proof in the traditional sense, as its correctness relies on the empirical (and hence *a posteriori*) assumption that the computing device which is employed to carry out the relevant calculations has operated correctly.
9. The task of primality testing was described by Gauss [30] (p. 329) as "the most important and useful in arithmetic." Anticipating later developments in complexity theory, he went on to say that "nevertheless we must confess that all methods that have been proposed thus far are either restricted to very special cases or are so laborious and difficult that even for numbers that do not exceed the limits of tables constructed by estimable men, they try the patience of even the practiced calculator." The significance of primality testing—and in particular of the Lucas–Lehmer test, described later on—to the status of the debate about "computer proofs" was first observed by Detlfesen and Luker [19].
10. Note, this is often true regardless of whether the algorithm is carried out by hand or with a computer. For even using a computer, our ability to practically apply an algorithm to solve particular instances of a problem will still be determined by its running time complexity. As such,

Lucas's algorithm remains a significant improvement over the naive primality algorithm for testing Mersenne numbers, even when carried out mechanically rather than with paper and pencil calculation.

11. Observations about the tension between the abstract subject matter of mathematics and the use of dynamic language in the formulation of mathematical proofs are at least as old as Plato's critique of the geometers in the *Republic* VII.1 [62]. The resolution typically favored by proponents of classical mathematics is to interpret talk of constructions (e.g. of a line bisecting an angle, the minimal closure of a set) as a heuristic way of expressing atemporal existence assertions which may in some cases lack constructive proof—c.f., e.g., [72]. But as the methods which will be at issue in this chapter are all effective, the issue at stake will not be whether dynamic discourse involving algorithms has genuine constructive content. Rather, it is whether classical mathematics possesses the resources to interpret the operational discourse itself in a manner which is faithful to our understanding of the mode of operation of the individual algorithms in question and, more broadly, of our means of classifying them according to their computational properties.
12. The use of the term "iterator" to refer to such structures is due to Moschovakis [56]. However, similar models have been independently proposed by many other theorists—e.g. Kolmogorov and Uspensky [47], Knuth [41], and Gandy [28].
13. For instance, the computation induced by $\phi_{\text{LUCAS}}$ on the input 5 is the sequence of states $\langle 5, 4\rangle, \langle 4, 17\rangle, \langle 3, 8\rangle, \langle 2, 0\rangle$, and $App(\phi_{\text{LUCAS}}, 5) = 1$.
14. Standard references for these subjects are as follows: complexity theory [29], [60], [20]; algorithmic analysis [41], [42], [43], [71], [14]; semantics of programming languages [75], [58], [52].
15. For a history of such methods and their role in the development of algorithmic analysis, see [41].
16. As is evident from this example, pseudocode employs a combination of natural language and formal programming constructions. Many of the latter express instructions which are awkward to describe in natural language—e.g. that several operations should be iterated until a specified condition is met, or that a subprocedure should be called recursively. The conventions for interpreting pseudocode thus involve many conventions about how scope and variable binding are to be interpreted, which are akin to those employed in formal logic. See [14] pp. 19–20 for a more detailed discussion of such conventions.
17. More precisely, this is the so-called *worst case running time* of *A*. Although it is possible to also study the best and average case running times (relative to some probability distribution on inputs) of *A*, the worst case metric is often taken to be the most useful in practice—see, e.g. [71] for discussion.
18. For present purposes I will assume that it is clear what it means for a particular mathematical formalism—e.g. the lambda calculus or the class of Turing machines—to serve as such a model. Matters of family resemblance aside, however, an important foundational question which is shared by computability theory, complexity theory, and algorithmic analysis, is that of making the notion of a model of computation precise. I will return to this issue in §2.4 (see also [28], [73], [79]).
19. For instance, the former restriction excludes the variant of the Turing machine model with only a single non-blank symbol in its tape alphabet, while the latter excludes the so-called parallel RAM [PRAM] model, or the so-called MBRAM variant of the traditional sequential model, which allows for unit time multiplication. Although these models are Turing complete—and hence sufficient for the development of computability theory—they are members of the so-called *second machine class* for which the complexity classes **P** and **PSpace**, defined later, coincide. See [79] for details.

20. For instance, in distinguishing the first machine class from the second, van Emde Boas states "The first machine class represents the class of reasonable sequential models. However, for the machine models in the second machine class it is not clear at all that they can be considered to be reasonable. It seems that the marvelous speed-ups obtained by the parallel models of the second machine class require severe violations of basic laws of nature. Stated differently: if physical constraints are taken into account, all gains of parallelism seem to be lost." [79], p. 14.
21. See the later discussion on the use of the order of growth notation $O(f(n))$. Also note that many of the problems mentioned above pertain to the classification of mathematical objects (e.g. logical formulas or graphs) which are not natural numbers. As such, suitable encodings of these items will often need to be found in order that they may be presented as inputs to an appropriate $M \in \mathfrak{M}$. In the case where $\mathfrak{M}$ is taken to be the Turing machine model, $\mathfrak{T}_k$, it is typical to assume problem instances are encoded as binary strings. In this case a computational problem $X$ will be a subset of $\{0, 1\}^*$.
22. See, e.g., chapter 8 of [60] for the relevant definitions and examples.
23. Note, however, that unlike the classical limitative results of computability theory, the statement of (4) is premised on the non-coincidence of the complexity classes **P** and **NP**. Such a qualification can be removed by considering a computational problem which is known to be complete for a complexity class such as **NExp** (i.e. non-deterministic exponential time; see chapter 20 of [60] for examples) which can be proven to properly contain **P**.
24. The use of asymptotic notation in computer science was originally introduced and popularized by Knuth—see., e.g., [44].
25. For further discussion of the status of these assumptions as well as the choice of specific models for the purpose of implementing algorithms arising in mathematical practice, see, e.g., [14] pp. 23–28 and [41], §1.3.
26. It is, for instance, by no means a trivial observation, that the sort of primitive algebraic operations in terms of which procedures like Gaussian elimination or the Gröbner basis algorithm are stated can be implemented using the limited set of operations and data structures made available by the standard RAM model, in a manner which preserves informal estimates of their running time complexity. In particular, such implementations often exist only by virtue of sophisticated techniques (such as hashing) which allows rapid look up of complex data objects in a manner which is independent of their size.
27. Here $\Omega(f(n))$ denotes the *asymptotic lower bound* for $f(x)$—i.e. the class of functions which $c \cdot f(n)$ eventually dominates for some constant $c$.
28. These considerations notwithstanding, it has been repeatedly suggested that the relationship between functions (understood extensionally) and algorithms can be taken to mirror that of the relationship between the reference of an expression and its Fregean sense—e.g [80], [21], [55], [40]. However, this represents at best a partial reconstruction of Frege's own understanding of the sense of a functional expression. For on the one hand, while an algorithm for computing a function $f(x)$ prototypically provides an *effective* means of determining the denotation of $f(a)$, a mathematical function can be introduced by a non-constructive definition (e.g. $f(x) = 1$ if the Riemann hypothesis is true and 0 otherwise) which still might be taken to give its sense. And on the other hand, while we have seen that we typically speak of the application of an algorithm to an argument as something which happens *in time*, Frege explicitly rejects the view that the composition of functions and arguments in the domain of sense should be understood as a temporal process (cf. [27]).
29. For instance, in the *structural operational semantics* of Plotkin [63], the class $\mathcal{D}$ is comprised of labeled trees whose structure represents the compositional structure of $\Pi$ and whose nodes

are labeled with representations of state transitions of the form $\langle \alpha, s \rangle \rightarrow s'$, with the intended interpretation "if statement $\alpha$ is executed in state $s$, then the resulting computation will terminate in state $s'$." On the other hand, in the so-called *abstract machine semantics* of Nielson and Nielson [58], $\mathcal{D}$ is comprised of instances of a simple RAM model.

30. For on the one hand, it is evident that since Definition 2.4.1 itself places no constraints on the mapping $\tau$, there will exist iterators $\phi$ which compute (e.g.) all functions of type $\mathbb{N} \rightarrow \mathbb{N}$ (inclusive of non-recursive ones). And on the other hand, even if various so-called *locality* and *boundedness* constraints are placed on this definition to rule out such examples (e.g. in the manner of [28] or [73]), it will still be possible to canonically represent various models from the second machine class as iterators.
31. I will discuss the machine-based variant in greater depth in §2.5. For general discussion of the program-based variant, see, e.g., [65], [39], [22], [77]. For a well-worked-out technical proposal along these lines, see Yanofsky [82], who writes "For us, an algorithm is the sum total of all the programs that express it. In other words, we look at all computer programs and partition them into different subsets. Two programs in the same subset will be two implementations of the same algorithm. These two programs are "essentially" the same." [82], p. 3.
32. Another dimension along which the neo-logicist project may be usefully compared with the algorithms-as-abstracts view pertains to the fact that we possess a prior axiomatic theory of the natural numbers whose interpretability in a putatively analytic extension of second-order logic can be used as a criterion of success. In particular, the adjunction of HP to pure second-order logic yields a system in which it is possible to derive the second-order Peano axioms $\mathsf{PA}^2$ once zero and successor are analyzed in the manner which Frege suggests (i.e. "Frege's Theroem"—cf., e.g., [10]). Note on the other hand that we do not at present possess anything like an axiomatic theory of algorithms (over what language would such a theory be formulated? what would its axioms be?) relative to which the success of principles like PP and MP in accounting for our procedural discourse can be judged.
33. Understood in the current context, such a claim is considerably stronger than Church's thesis. For in order to show that $\mathsf{L}$ is "expressively complete" in the relevant sense, it must be shown that not only does it allow for the formulation of programs which compute all recursive functions, but also that these formulations are adequately representative of the relevant class of all algorithms. And it would appear that it would be hard to make a systematic case for this (even in the case of widely used languages like C) without having at our disposal a detailed account of which sorts of computational properties apply in the first instance to algorithms as opposed (e.g.) to their implementations.
34. The technical goal of such accounts is typically not that of providing a conceptual analysis of "procedural equivalence," but the much more modest one of determining (e.g.) when two program fragments can be substituted for one another in a manner which does not affect the execution of a program into which they are substituted (see, e.g., [61]).
35. To take just one example (derived from [82]), consider the following two program fragments: (1) for $i = 1$ to $n + 1$ $\{x = f(x)\}$, and (2) for $i = 1$ to $n$ $\{x = f(x)\}$; $x = f(x)$. One might claim that (1) and (2) express the same "procedural meaning" in the sense that they each describe a method for iterating the application of function $f(x)$ to itself $n + 1$ times (where in the latter case, the iteration by which this is achieved is simply "unraveled" by one step). In speaking in this manner, however, we seem to rely on *extra-linguistic* intuitions about the effect of executing the program in question—e.g. based on how data is stored in the registers of an abstract machine relative to which it is carried out. And as I will suggest in §2.5, such intuitions seem to be most naturally accounted for in terms of the notion of simulation equivalence, which is defined on machines rather than programs.

36. In [79] the following precise definition of $\mathfrak{M}^1$ is given: a model $\mathfrak{M}$ is in the first machine class just in case there exists a polynomial time, linear space overhead simulation of $\mathfrak{M}$ relative to the single-tape Turing machine model $\mathfrak{T}_1$. (It may be seen that this definition is sufficiently broad to encompass $\mathfrak{T}_k$ (for $k > 1$) and the standard RAM model $\mathfrak{R}$, as well as many more refined models based on computational architectures which may be concretely embodied as practical microprocessor designs.) On the other hand, the second machine class $\mathfrak{M}^2$ is defined to contain those models $\mathfrak{M}$ such that $\mathbf{P}^{\mathfrak{M}} = \mathbf{PSpace}^{\mathfrak{M}}$—i.e. for which the classes of polynomial time and polynomial space coincide when these notions are analyzed relative to $\mathfrak{M}$. Since it is currently unknown even if $\mathbf{P}^{\mathfrak{T}_1} \neq \mathbf{PSpace}^{\mathfrak{T}_1}$, it is in principle possible that $\mathfrak{M}^1$ and $\mathfrak{M}^2$ coincide. Although, of course, this is taken to be exceedingly unlikely, any indeterminacy in the definition of $\mathfrak{M}^1$ thus lies with the status of open separation questions in complexity theory and not with the imprecision of notions like "reasonableness" by which this class might be initially characterized.
37. See Kreisel [48], pp. 177–178 for similar comments about the significance of such "translation-based" proofs in regard to strong Church's thesis.
38. Since the models we are considering here are all deterministic, the relations $R$ in question will all be single-valued—a point which will be taken into account when we later adopt the definition of iterator isomorphism. I have introduced the notion of simulation in terms of relations so as to conform to the tradition of process algebra wherein non-deterministic and concurrent models are also typically considered.
39. Milner originally achieved this by requiring that there exist a pair of relations $R$ and $R'$ such that $R$ satisfies (8) between $\phi_1$ and $\phi_2$, and conversely, $R'$ satisfies (8) between $\phi_2$ and $\phi_1$. These conditions respectively correspond to the back and forth clauses of the familiar definition of *bisimulation*, which is studied widely in process algebra and modal logic—cf., e.g., [6]. However, we are only interested in deterministic algorithms whose computations can be represented as linear sequences of states rather than trees. In such cases, it is more straightforward to adopt a definition like 2.5.1, in which simulations are functions rather than relations.
40. In other words, iterator isomorphism in the sense of the definition 2.5.1 is a *congruence* with respect to the property of the exact running time complexity, $time_M(f(n))$.
41. For a number of concrete examples of this type, see [17] and [7].
42. Techniques for constructing RAM implementations of such structures are considered systematically in the context of Knuth's meticulous implementations of informally specified algorithms as so-called MIX machines (a variant of the RAM model). See in particular Chapter II of [42].
43. See, e.g., [14] pp. 631–633.
44. For instance, in the case under consideration, there would presumably exist monotone functions, $\tau_1, \tau_2 : X \times \mathbb{N} \to \mathbb{N}$, which uniformly partition the computations $\vec{s}(x)$ and $\vec{t}(x)$ of $\phi_1$ and $\phi_2$ into subsequences $\vec{s}(x) = s_{\tau_1(x,0)}(x), \ldots, s_{\tau_1(x,1)}(x), s_{\tau_1(x,1)+1}(x), \ldots, s_{\tau_1(x,2)}(x), \ldots, s_{\tau_1(x,k-1)+1}(x), \ldots, s_{\tau_1(x,k)}(x)$ and $\vec{t}(x) = t_{\tau_2(x,0)}(x), \ldots, t_{\tau_2(x,1)}(x), t_{\tau_2(x,1)+1}(x), \ldots, t_{\tau_2(x,2)}(x), \ldots, t_{\tau_2(x,k-1)+1}(x), \ldots, t_{\tau_2(x,k)}(x)$, such that the subsequences $s_{\tau_1(x,i)}(x), \ldots, s_{\tau_1(x,i)}(x)$ and $t_{\tau_2(x,i)}(x), \ldots, t_{\tau_2(x,i)}(x)$ correspond to the "unimportant" computations which implement the same step in the operation of $A$. On this basis, we could imagine redefining an isomorphism between $\phi_1$ and $\phi_2$ to be a function $\rho^* : St_1^* \to St_2^*$ mapping all and only sequences of the form $s_{\tau_1(x,i)}(x), \ldots, s_{\tau_1(x,i)}(x)$ to $t_{\tau_2(x,i)}(x), \ldots, t_{\tau_2(x,i)}(x)$ for $i \leq time_A(x)$
45. For instance, in the case of the Turing machine model, we could represent states as structures of the form $\langle \mathbb{N}, f, c \rangle$, where $f : \mathbb{N} \to \{0, \ldots, k-1\}$ and $f(n)$ represents the contents of the $n$th tape cell, and $c$ records the location of the machine's head.
46. See [49] for definitions and examples of this relation.

47. MERGE takes two sublists $B_1$ and $B_2$ as inputs and operates by repeatedly comparing the first elements of these lists relative to $<$. It selects the smaller of these elements, removes it from the relevant list, and adds it to the end of a new list $C$ which is returned out as output. If $B1$ and $B2$ are assumed to be sorted, then it is easy to see that $C$ will be sorted as well after this procedure is applied.
48. Less formally, a recursor can be understood as the interpretation of a set of mutually recursive equation definitions $f_1(\vec{x}), \ldots, f_n(\vec{x})$ (or more generally as a *recursion scheme* in the sense of Greibach [34]) relative to a fixed-point semantics of the sort originally proposed by Scott [69] in the form of *domain theory*.
49. Or at least this appears to be true of Gurevich's views up to his joint 2008 paper with Dershowitz [18]. In his joint subsequent 2009 paper with Blass and Dershowitz [7], it is argued that the program-based variant of the algorithms-as-abstracts view is unsustainable by virtue of our inability to define an appropriate equivalence relation $\approx$ on programs for reasons resembling those discussed in §2.4. For instance it is repeatedly suggested in this paper that the relation between the programs of "expressing the same algorithm" is subjective or interest relative. But since no account of why the same problem does not also beset the machine-based variant of this view is given in this paper; it is unclear at present whether Gurevich should be understood to be continuing to support the ASM thesis in its original form.
50. There is, however, some question of how general this advantage can be taken to be. For on the one hand, there are standard techniques which allow for the uniform translation between members of iterative and recursive models of computation (cf., e.g., the *flowchart* and *recursion scheme* models developed in [34]). And on the other, an open but widely believed conjecture in complexity theory (i.e. $\mathbf{NC} \neq \mathbf{P}$, cf. [33]) predicts that there is a wide class of naturally occurring computational problems (i.e. those which are complete for the class $\mathbf{P}$) which can only be solved by sequential algorithms. If this is the case, then problems like sorting (for which parallel processing of logarithmic depth can be shown to provide faster algorithms) may represent only an exceptional case of the sort of problems we confront in mathematical practice.
51. A prototypical example is Gauss's [30] proof of Fermat's little theorem. We would now describe this proof as relying on the fact that the set of integers $\{1, \ldots, p\}$ forms a group under multiplication which is isomorphic to the cyclic group $\mathbb{Z}_p$. However, no mention of this group appears in the statement of the theorem (i.e. "if $p$ is prime, then for any integer $a$, $a^p - a$ is divisible by $p$"), nor had the modern notion of group been defined at the time of Gauss's proof.

# References


[1] H. Abelson and G.J. Sussman. *Structure and Interpretation of Computer Programs*. MIT Press, Cambridge, MA, 1985.

[2] K. Appel and W. Haken. Every planar map is four colorable. Part i: Discharging. *Illinois Journal of Mathematics*, **21**(3):429–490, 1977.

[3] K. Appel, W. Haken, and J. Koch. Every planar map is four colorable. Part II: Reducibility. *Illinois Journal of Mathematics*, **21**(3):491–567, 1977.

[4] G. Bealer. A solution to Frege's puzzle. In *Philosophical Perspectives, Volume 7: Language and Logic*, pp. 17–60. Blackwell, Oxford, 1993.

[5] P. Benacerraf. What numbers could not be. *Philosophical Review*, **74**(1):47–73, 1965.

[6] P. Blackburn, M. de Rijke, and Y. Venema. *Modal Logic*. Cambridge University Press, Cambridge, 2001.

[7] A. Blass, N. Dershowitz, and Y. Gurevich. When are two algorithms the same? *Bulletin of Symbolic Logic*, **15**(2):145–168, 2009.

[8] G. Boolos. Is Hume's Principle Analytic? In *Language, Thought, and Logic: Essays in Honour of Michael Dummett*, Ed. Richard G. Heck Jr. pp. 245–261. Oxford University Press, Oxford, 1997.
[9] B. Buchberger. A theoretical basis for the reduction of polynomials to canonical forms. *ACM SIGSAM Bulletin*, **10**(3):19–29, 1976.
[10] J.P. Burgess. *Fixing Frege*. Princeton Monographs in Philosophy. Princeton University Press, Princeton, 2005.
[11] R. Carnap. *Meaning and Necessity*. The University of Chicago Press, Chicago, 1947.
[12] A. Church. A formulation of the logic of sense and denotation. *Structure, Method, and Meaning: Essays in Honor of Henry M. Sheffer*, Liberal Arts Press, New York, pp. 3–24, 1951.
[13] A. Church. Intensional isomorphism and identity of belief. *Philosophical Studies*, **5**(5):65–73, 1954.
[14] T. Cormen, C. Leiserson, and R. Rivest. *Introduction to Algorithms*. MIT Press, second edition, 2005.
[15] D. Cox, J. Little, and D. O'Shea. *Ideals, Varieties, and Algorithms*. Springer-Verlag, New York, 1992.
[16] M.J. Cresswell. *Structured Meanings: The Semantics of Propositional Attitudes*. MIT Press, 1985.
[17] W. Dean. *What algorithms could not be*. PhD thesis, Rutgers University, 2007.
[18] N. Dershowitz and Y. Gurevich. A natural axiomatization of computability and proof of Church's Thesis. *Bulletin of Symbolic Logic*, **14**(3):299–350, 2008.
[19] M. Detlefsen and M. Luker. The four-color theorem and mathematical proof. *Journal of Philosophy*, 77, 1980.
[20] D. Du and K. Ko. *Theory of Computational Complexity*. John Wiley & Sons, 2000.
[21] M. Dummett. Frege's distinction between sense and reference. In *Truth and Other Enigmas*, pp. 116–144. Harvard University Press, 1978.
[22] L. Floridi. Philosophy and computing: an introduction. *Ethics and Information Technology*, **2**(2):137–138, 2000.
[23] C. Foster. *Algorithms, Abstraction and Implementation: Levels of Detail in Cognitive Science*. Academic Press, New York, 1992.
[24] G. Frege. *Die Grundlagen der Arithmetik*. Koebner, Breslau, 1884.
[25] G. Frege. *Grundgesetze der Arithmetik: begriffsschriftlich abgeleitet*. Pohle, Jena, 1893, 1903. Two volumes. Reprinted in [26].
[26] G. Frege. *Grundgesetze der Arithmetik*. Olms, Hildesheim, 1962.
[27] G. Frege. Compound thoughts. *Mind*, **72**(285):1–17, 1963.
[28] R. Gandy. Church's Thesis and principles for mechanisms. In H.K.J. Barwise and K. Kunen, eds, *The Kleene Symposium*, volume 101, pp. 123–148. North-Holland, Amsterdam, 1980.
[29] M. Garey and D. Johnson. *Computers and Intractability. A Guide to the Theory of NP-Completeness*. W.H. Freeman and Company, New York, 1979.
[30] C.F. Gauss. *Disquisitiones Arithmeticae*. Springer, New York, 1986. Translated by Arthur A. Clarke.
[31] J. Girard, Y. Lafont, and P. Taylor. *Proofs and Types*. Cambridge University Press, 1989.
[32] K. Gödel. On undecidable propositions of formal mathematical systems (with postcript dated 1964). In S. Feferman, ed, *Kurt Gödel Collected Works. Vol. I. Publications 1929-1936*, pp. 346–371. Oxford University Press, Oxford, 1986.
[33] R. Greenlaw, H. Hoover, and W. Ruzzo. *Limits to Parallel Computation:* **P**-*Completeness Theory*. Oxford University Press, USA, Berlin, 1995.

[34] S.A. Greibach. *Theory of Program Structures: Schemes, Semantics, Verification*, volume 36 of *Lecture Notes in Computer Science*. Springer, 1975.
[35] Gurevich. Sequential abstract-state machines capture sequential algorithms. *ACM Transactions on Computational Logic*, **1**(1):77–111, 2000.
[36] Y. Gurevich. Evolving algebras 1993: Lipari Guide. Specification and ValidationMethods. *In Specification and Validation Methods*, Ed. E. Börger, Oxford University Press, 1995, 231–243.
[37] Y. Gurevich. The sequential ASM thesis. *Bulletin of the EATCS*, **67**:93–124, 1999.
[38] B. Hale and C. Wright. *The Reason's Proper Study*. Oxford University Press, Oxford, 2001.
[39] D. Harel. *Algorithmics: the spirit of computing*. Addison-Wesley, Reading, MA, 2006.
[40] J.F. Horty. *Frege on Definitions: A Case Study of Semantic Content*. Oxford University Press, Oxford, 2007.
[41] D. Knuth. *The Art of Computer Programming, Volume I: Fundamental Algorithms*. Addison Wesley, New York, 1973.
[42] D. Knuth. *The Art of Computer Programming, Volume II: Seminumerical Algorithms*. Addison Wesley, New York, 1973.
[43] D. Knuth. *The Art of Computer Programming, Volume III: Searching and Sorting*. Addison Wesley, New York, 1973.
[44] D. Knuth. Big omicron and big omega and big theta. *ACM Sigact News*, **8**(2):18–24, 1976.
[45] D.E. Knuth. Algorithm and program; information and data. *Communications of the ACM*, **9**(9):654, 1966.
[46] A. Kolmogorov and V. Uspensky. On the notion of algorithm. *Uspekhi Matematicheskikh Nauk*, **8**(4/56):175–176, 1953.
[47] A. Kolmogorov and V. Uspensky. To the definition of algorithms. *Uspekhi Matematicheskikh Nauk*, **13**(4):3–28, 1958.
[48] G. Kreisel. Some reasons for generalizing recursion theory. In R. Gandy and C. Yates, eds, *Logic Colloquium 69*. North-Holland, Amsterdam, 1971.
[49] D. Marker. *Model Theory*, volume 217 of *Graduate Texts in Mathematics*. Springer-Verlag, New York, 2002.
[50] P. Martin-Löf. *Intuitionistic Type Theory*. Bibliopolis, Naples, 1984.
[51] R. Milner. An algebraic definition of simulation between programs. Technical Report CS-205, Department of Computer Science Stanford University, 1971.
[52] J. Mitchell. *Foundations for Programming Languages*, volume 1. MIT Press Cambridge, MA, 2002.
[53] R. Montague. The proper treatment of quantification in ordinary english. In J.M.J. Hintikka and P. Suppes, eds, *Approaches to Natural Language: Proceedings of the 1970 Stanford Workshop on Grammar and Semantics*, pp. 221–242. Reidel, 1970.
[54] Y.N. Moschovakis. Abstract recursion as a foundation for the theory of algorithms. *Computation and Proof Theory*, pp. 289–364, 1984.
[55] Y.N. Moschovakis. Sense and denotation as algorithm and value. In *Logic Colloquium '90*, pp. 210–249. North-Holland, Amsterdam, 1990.
[56] Y.N. Moschovakis. On the founding of a theory of algorithms. In H. Dales and G. Oliveri, eds, *Truth in Mathematics*, pp. 71–104. Clarendon Press, Oxford, 1998.
[57] Y.N. Moschovakis. What is an algorithm? In B. Engquist and W. Schmid, eds, *Mathematics Unlimited—2001 and Beyond*, pp. 919–936. Springer-Verlag, Berlin, 2001.
[58] F. Nielson and H.R. Neilson. *Semantics with Applications – A Formal Introduction*. John Wiley & Sons, New York, NY, 1992.
[59] P. Odifreddi. Classical Recursion Theory. Vol. I, volume 125 of *Studies in Logic and the Foundations of Mathematics*. North-Holland, Amsterdam, 1989.
[60] C. Papadimitriou. *Computational Complexity*. Addison-Wesley, New York, 1994.

[61] A.M. Pitts. Operationally-based theories of program equivalence. *Semantics and Logics of Computation*, **14**:241, 1997.
[62] Plato. *Republic*. Hackett Publishing Company, Indianapolis, 2004. Translated by C.D.C. Reeve.
[63] G. Plotkin. A structural approach to operational semantics. Report DAIMI FN-19. Computer Science Department, Aarhus University, 1981.
[64] E. Post. Finite combinatory processes-formulation 1. *Journal of Symbolic Logic*, **1**(3):103–105, 1936.
[65] Z. Pylyshyn. *Computation and Cognition: Toward a Foundation for Cognitive Science*. MIT Press, Cambridge, MA, 1984.
[66] H. Rogers. *Theory of Recursive Functions and Effective Computability*. MIT Press, Cambridge, MA, second edition, 1987.
[67] D. Sangiorgi and J. Rutten. Origins of bisimulation and coinduction. In *Advanced Topics in Bisimulation and Coinduction*. Cambridge University Press, Cambridge, 2011.
[68] A. Schönhage. Storage modification machines. *SIAM Journal of Computing*, **9**(3):490–508, 1980.
[69] D. Scott. Outline of a mathematical theory of computation. Technical report, Oxford University Computing Laboratory, 1970.
[70] R.W. Sebesta. *Concepts of Programming Languages*, volume 4. Addison-Wesley, New York, 2002.
[71] R. Sedgewick. *Algorithms*. Addison-Wesley, New York, 1983.
[72] S. Shapiro. *Philosophy of Mathematics: Structure and Ontology*. Oxford University Press, Oxford, 1997.
[73] W. Sieg. On computability. In A. Irvine, ed, *Philosophy of Mathematics*, volume 4 of *Handbook of the Philosophy of Science*, pp. 549–630. North-Holland, Amsterdam, 2009.
[74] W. Sieg and J. Byrnes. *K*-graph machines: generalizing Turing's machines and arguments. *Gödel'96: Logical Foundations of Mathematics, Computer Science, and Physics*, pp. 98–119, 2001.
[75] J.E. Stoy. *Denotational Semantics: The Scott-Strachey Approach to Programming Language Theory*. MIT Press, Cambridge, MA, 1981.
[76] P. Tichý. *The Foundations of Frege's Logic*. Walter de Gruyter, New York, 1988.
[77] R. Turner and A. Eden. The philosophy of computer science. In E.N. Zalta, ed, *The Stanford Encyclopedia of Philosophy*. Summer 2009 edition, 2009.
[78] T. Tymoczko. The four-color problem and its philosophical significance. *Journal of Philosophy*, **76**(57–83), 1979.
[79] P. van Emde Boas. Machine models and simulations. In J. Van Leeuwen, ed, *Handbook of Theoretical Computer Science (vol. A): Algorithms and Complexity*. MIT Press, Cambridge, MA, 1990.
[80] J. Van Heijenoort. Frege on sense identity. *Journal of Philosophical Logic*, **6**(1):103–108, 1977.
[81] C. Wright. *Frege's Conception of Numbers as Objects*, Vol. 2 of *Scots Philosophical Monographs*. Aberdeen University Press, Aberdeen, 1983.
[82] N.S. Yanofsky. Towards a definition of an algorithm. *Journal of Logic and Computation*, **21**(2):253–286, 2011.
[83] E.N. Zalta. *Intensional Logic and the Metaphysics of Intentionality*, volume 92. MIT Press, Cambridge, MA, 1988.

# 3

# The Second Incompleteness Theorem: Reflections and Ruminations

ALBERT VISSER

## 3.1 Introduction

It is abundantly clear that the usual statements of Gödel's second incompleteness theorem will not do. *If a theory is sufficiently strong, then it does not prove its own consistency.* The formulation has (at least) two problems.

(1) The usual formulation is not very clear concerning the precise scope of the result. What does "sufficiently strong" mean? To this question there could very well be several answers depending on which form we give the theorem. We formulate some versions of the theorem in Section 3.2. We zoom in on a different kind of answer in Section 3.3.1.

(2) The second problem is that we have to explain what a consistency statement of a theory is. As a first step we could produce a specific implementation of a consistency statement of a specific theory. For example, we could use the (arithmetization of) syntax developed in [Bus87] to produce a consistency statement for, say, elementary arithmetic (EA).[1] At this point our problem splits into three.

   (i) First, what does make this sentence into a *bona fide* consistency statement and why is e.g. a Feferman style consistency statement for Peano arithmetic (or: PA) not right? And what shall we say about a statement of cutfree consistency for EA? This problem is the problem of *intensionality in metamathematics.*

   (ii) Second, if we use a consistency statement either in the context of a weak theory or of an unsound theory, can we still reasonably say that it expresses what it pretends to express, since—in the weak case—we cannot verify basic properties of the proof predicate, or—in the unsound case—the statement is necessarily about a totality that contains non-standard proofs?

*Gödel's Disjunction.* First Edition. Leon Horsten & Philip Welch (eds).

(iii) Third, can we state generally what the class of admissible consistency statements is?

In this chapter, we will zoom in on two questions (ii) and (iii). The driving force behind both questions (i) and (ii) is, of course, that we do not want to read the second theorem as saying that, for any consistent computably enumerable (CE) theory there is a certain class of strings such that the theory does not derive any element of that class. We want the foundational insight that the theory does not prove *its own consistency*. We will propose an answer to problem (ii) that does justice to this intuition.

Problem (iii) looks more like a mathematical problem that is independent of specifically philosophical worries. Harvey Friedman proposed some approaches to this problem. See his:

http://www.cs.nyu.edu/pipermail/fom/2010-January/014290.html.

In response to Harvey's ideas, I wrote the papers [Vis11] and [Vis12]. In Section 3.5, we will sketch the ideas of those papers.

One conclusion that we can draw from the discussion in this chapter is that it is good to carefully distinguish the various questions concerning the second incompleteness theorem. There is no good reason that these questions should be in sync. For example, not all generalizations help with understanding the meaning question. Conversely, the meaning question need not be a guide to interesting generalization.

### 3.1.1 *Status of the Technical Results in this Chapter*

The main aim of this chapter is to offer a philosophical discussion of the second incompleteness theorem and to survey a number of approaches to it. The logico-technical results of Section 3.3 are original for this chapter. However, they overlap in part with results in my paper [Vis14a], which was written later, but appeared as a preprint earlier. The present chapter does a bit more than [Vis14a], since it develops *uniform* L-predicates rather than just L-predicates. The other paper does a bit more, since it contains an extension of the characterization result, Theorem 3.3.7 of this chapter, to a more general case. Section 3.5 does not contain new results, but rather gives an overview of ideas and results from [Vis11] and [Vis12].

## 3.2 Versions of the Second Incompleteness Theorem

In this section, we will briefly consider some modern formulations of the second incompleteness theorem. Theories in this chapter will always be theories of one-sorted predicate logic in a finite signature with a p-time decidable axiomatization. See Section A.1 of the Appendix. We assume that a $\Delta_1^{\mathrm{b}}$-formula $\alpha_U$ defining the set of axioms of $U$ is part of the specification of $U$, so theories are more than just sets of theorems.

We write $N : U \rhd V$ or $N : V \lhd U$ for: $N$ is an interpretation of $V$ in $U$. We write $U \rhd V$ for: there is an interpretation $N$ such that $N : U \rhd V$. We use $V \lhd U$ as an equivalent of $U \rhd V$. We pronounce $U \rhd V$ as: $U$ interprets $V$, and $V \lhd U$ as: $V$ is interpretable in $U$. We use $U \equiv V$ for: $U \rhd V$ and $U \lhd V$. We pronounce $U \equiv V$ as: $U$ is mutually interpretable with $V$. Interpretability is explained in Section A.2 of the Appendix.

### 3.2.1 A Basic Version of G2

Here is my first version of G2. Let $\mathsf{S}^1_2$ be the weak arithmetic given in [Bus86]. See also [HP93]. The provably total recursive functions of $\mathsf{S}^1_2$ are precisely the p-time computable functions.

**Theorem 3.2.1** *For any consistent theory U, we have:* $U \not\rhd (\mathsf{S}^1_2 + \mathsf{con}(U))$.

This version rests on ideas of Solovay, Friedman, and Pudlák. For some details, see e.g. [Vis11]. We note that $\mathsf{con}(U)$ is really $\mathsf{con}(\alpha_U)$. Instead of $\mathsf{S}^1_2$ we could as well have taken Robinson's Arithmetic $\mathsf{Q}$ (see [TMR53]), or $\mathsf{PA}^-$, the theory of discretely ordered commutative semirings with a least element (see [Kay91, Jeř12]), or $\mathrm{I}\Delta_0 + \Omega_1$ (see [WP87]). (A good textbook to find information about all these theories is [HP93].) Moreover, instead of an arithmetical theory we could have employed a string theory like Grzegorczyk's theory $\mathsf{TC}$ (see [Grz05, GZ08, Šve07, Vis09b, vPR⁺10]) or a theory of (possibly finite) sets like adjunctive set theory $\mathsf{AS}$ (see [MPS90, Vis13]). All these theories are 'the same' in the sense that they are mutually interpretable. In some cases we do even have a stronger claim: e.g. we have a fixed translation that witnesses that for every $\Pi_1$-sentence $P$, we have $(\mathsf{Q} + P) \equiv (\mathsf{S}^1_2 + P)$.

Let's highlight two things about this version. First, it demands *nothing* of the theory $U$, except our conditions on theories, noted previously.[2] Second, the main relation involved is not provability (a relation between a theory and a sentence) but interpretability (a relation between theories). We return to this last observation in Section 3.3.4.

### 3.2.2 G2 as a Statement of Interpretability Power

The Gödel–Hilbert–Bernays–Wang–Henkin–Feferman theorem, appropriately generalized using ideas of Solovay, Pudlák, Friedman, and Wilkie and Paris, tells us that $(\mathsf{S}^1_2 + \mathsf{con}(U)) \rhd U$. See e.g. [Vis11] for some details. Using this fact we get an appealing form of the second incompleteness theorem. We have writing $\top$ for a suitable designated inconsistent theory:

**Theorem 3.2.2** *For any consistent theory U, we have:*

$$\textit{if } U \ntrianglelefteq \top, \textit{ then } U \ntrianglelefteq (\mathsf{S}^1_2 + \mathsf{con}(U)) \ntrianglelefteq \top.$$

So, the second incompleteness theorem tells us that the degrees of interpretability do not have co-atoms. This version highlights the 'interpretability power' of consistency statements. Friedman's idea of finding coordinate-free versions of G2 is roughly to replace consistency by interpretability power. We treat my two variants of Friedman's idea in Section 3.5. See also [Vis11, Vis12].

### 3.2.3 *Feferman's Theorem*

In his classical paper [Fef60], Sol Feferman gave a beautiful version of the second incompleteness theorem. Here is my general formulation of the Feferman version. We write $A^N$ for the result of applying the translation associated with the interpretation $N$ to $A$. See also Appendix A.2.

**Theorem 3.2.3** *We have:* $N : U \rhd \mathsf{S}^1_2 \quad \Rightarrow \quad U \rhd (U + \mathsf{incon}^N(U))$.

A proof of Feferman's theorem will be given in Section 3.3.2 (Theorem 3.3.6(b)), where we also provide something like a converse of the theorem (Theorem 3.3.7).

We note that Theorem 3.2.3 immediately implies Theorem 3.2.2. Suppose we have $N : U \rhd (\mathsf{S}^1_2 + \mathsf{con}(U))$. Then, by Theorem 3.2.3, we find:

$$U \rhd (U + \mathsf{con}^N(U) + \mathsf{incon}^N(U)) \rhd \mathbb{T}.$$

Hence $U$ is inconsistent.

### 3.2.4 *G2 as an Admissible Rule*

Given a theory $U$ of signature $\Sigma$ and an arbitrary second signature $\Theta$. An inference $A/B$ is admissible for $U, \Theta$ iff, for all translations $\tau : \Theta \to \Sigma$, we have $U \vdash A^\tau \Rightarrow U \vdash B^\tau$. We write $A \mathrel{|\!\!\sim}^{\Theta}_{U} B$ for: $A/B$ is admissible for $U, \Theta$. Let $\mathcal{A}$ be the signature of arithmetic. Theorem 3.2.2 yields, for each theory $U$, the following admissible rule:

**Theorem 3.2.4** *We have:* $\mathsf{S}^1_2 + \mathsf{con}(U) \mathrel{|\!\!\sim}^{\mathcal{A}}_{U} \bot$.

We note that in some cases the formula, $(\bigwedge \mathsf{S}^1_2 \wedge \mathsf{con}(U)) \to \bot$, is in $\Lambda^{\mathcal{A}}_U$, the logic of all $\mathcal{A}$-schemes valid in $U$; for example, in the case where $U$ is decidable.[3] However, for many $U$, the G2-rule will be non-trivial in the sense that it is not in $\Lambda^{\mathcal{A}}_U$.

In Sections 3.3 and 3.5, we will formulate versions of G2 in a somewhat different style. Well, ... it will be part of the discussion whether these 'versions' are really versions.

## 3.3 Meaning as Conceptual Role

What does it mean that a sentence $A$ in the language of $U$ is a *consistency statement* for $U$? In this section we discuss one possible attempt to answer the question: the idea that what makes something into a consistency statement for $U$ is the fact that it is a statement of the form 'falsum is not provable in $U$', where something is a provability predicate iff $U$ verifies a number of global properties of the predicate under consideration. Thus, this strategy has two ingredients: (i) something is a provability predicate if it has a certain conceptual role and (ii) the relevant conceptual role is theory-internal.

### 3.3.1 L-Predicates

For concreteness, we will consider one specific variant of the strategy to treat provability as internal conceptual role. Consider a theory $U$ and an interpretation $N$ of the Tarski–Mostowski–Robinson theory $\mathsf{R}$ in $U$. A $U$-predicate $\triangle$ is *an L-predicate* for $U, N$ if it satisfies the Löb conditions, given below. We write $\triangle A$ for: $\triangle(\underline{\ulcorner A \urcorner})$, where $\underline{\ulcorner A \urcorner}$ is the numeral of the Gödel number of $A$, where we interpret the numbers via $N$. The Gödel numbering is supposed to be fixed and standard. We will write $\triangledown$ for $\neg\triangle\neg$. Here are the conditions:

L1. $\vdash A \;\Rightarrow\; \vdash \triangle A$
L2. $\triangle A, \triangle(A \to B) \vdash \triangle B$
L3. $\triangle A \vdash \triangle\triangle A$

Here we read the quantifiers over sentences *externally*. We can also formulate uniform versions of the notion of L-predicate with internal quantifiers. In this chapter, we give just one possibility, which seems fairly natural. Consider a theory $U$ and an interpretation $N$ of $\mathsf{S}^1_2$ in $U$. A $U$-predicate $\triangle$ is *a uniform L-predicate* for $U, N$ if it satisfies the uniform Löb conditions: Here are the conditions:

$\mathsf{L_u}1.$ $\vdash A \;\Rightarrow\; \vdash \triangle A$
$\mathsf{L_u}2.$ $\vdash \forall A, B \in \mathsf{sent}^N_U\, ((\triangle A \wedge \triangle(A \to B)) \to \triangle B)$
$\mathsf{L_u}3.$ $\vdash \forall A \in \mathsf{sent}^N_U\, (\triangle A \to \triangle\triangle A)$
$\mathsf{L_u}4.$ $\vdash \forall A \in \mathsf{sent}^N_U\, (\Box^N_{\mathsf{pred}} A \to \triangle A)$

As usual we interpret the numbers via $N$. The Gödel numbering is supposed to be fixed and standard. The predicate $\Box_{\mathsf{pred}}$ stands for a standard arithmetization of provability in predicate logic w.r.t. the signature of $U$. In the context of uniform L-predicates, we will always employ efficient numerals and a Smullyan-style arithmetization. Efficient numerals are numerals in the style of the binary notation system. The reason is that we need to use an arithmetization of the transformation $A \mapsto \underline{\ulcorner A \urcorner}$. Using efficient numerals and Smullyan-style arithmetization, this transformation is p-time.

Here is a list of examples of L-predicates and uniform L-predicates.

A. Let $\Box_{\mathsf{R}}$ be a usual arithmetization of provability in $\mathsf{R}$. Then, $\bigwedge \mathsf{S}^1_2 \to \Box_{\mathsf{R}} A$ is an L-predicate for $\mathsf{R}, \mathsf{ID}_{\mathsf{R}}$.[4]

B. Suppose $U$ is a theory and $N$ is an interpretation of $\mathsf{R}$ in $U$. Then $\top$, or, if you wish $x = x$ is an L-predicate for $U, N$. If $N$ is an interpretation of $\mathsf{S}^1_2$, then $\top$ or $x = x$ is a uniform L-predicate for $U, N$.

C. Suppose $V$ is an extension in the same language of $U$, and suppose $N$ is an interpretation of $\mathsf{S}^1_2$ in $U$. Let $\Box_V$ be the ordinary provability predicate for $V$. Then, $\Box^N_V$ is a uniform L-predicate for $U, N$.

D. Suppose $N$ is an interpretation of $\mathsf{S}^1_2$ in $U$ and suppose $J$ is a $U$-definable, $U$-verifiable cut of $N$ which is closed under the function $\omega_1$. (Here $\omega_1$ is the function

$x \mapsto 2^{(^2\log(x))^2}$.) Then $\square^J_U$ is an L-predicate for $U, N$. We note that $\square^J_U$ will not generally be uniform. A variant of this example is to consider $\square^{[J]}_U$, where we allow the witnessing proofs to be in $N$; only their length is constrained to be in $J$.

E. In EA with the identity interpretation we can consider both ordinary provability and cutfree provability in EA. These two are not provably extensionally equal. We need the totality of superexponentiation to show that they are extensionally equal. Ordinary provability is a uniform L-predicate and cutfree provability is an L-predicate. Instead of cutfree provability we can also consider Herbrand provability. See [Vis90] and [Kal91] for some basic facts about the two predicates.

F. In EA, with the identity interpretation, we can consider both ordinary provability in EA and provability in EA with an oracle for $\Sigma_1$-truth. These two are not provably extensionally equal. We need the $\Sigma_1$-collection to show that they are extensionally equal. Both predicates are uniform L-predicates.

G. Consider PA with the identity interpretation. The predicate 'provable in PA with an oracle for $\Pi_{n+1}$-truth' is a uniform L-predicate. The logic of the hierarchy of such predicates is Japaridze's Logic GLP. See [Jap85]. See also [Boo93]. This logic was used by Lev Beklemishev to extract proof theoretic ordinals from its closed fragment. See e.g. [Bek04, Bek05, BJV05, Bek06].

H. We take as theory ZF, as interpretation of arithmetic the interpretation of the natural numbers as finite von Neumann ordinals and as predicate *truth in all transitive models of* ZF. This example was studied by Solovay in [Sol76]. See also [Boo93]. A closely related example is to consider truth in all $V_\kappa$ where $\kappa$ is inaccessible. See [Boo93].

I. Let $U$ be a one-sorted, first-order version of second-order arithmetic, let $N$ be relativization to the predicate representing the number-sort. Let $\triangle$ be the arithmetization of provability in $U$ with the $\omega$-rule. Then $\triangle$ is a uniform L-predicate for $U, N$. See [Boo93].

J. Recently Graham Leach-Krouse studied an internal version of $\Omega$-validity over ZFC with the von Neumann interpretation. He shows inter alia that this notion gives us a uniform L-predicate.

We formulate a version G2 as follows.

**Theorem 3.3.1** *For all consistent theories $U$ and all interpretations $N$ of* R *in $U$ and all L-predicates $\triangle$ for $U, N$, we have $U \nvdash \neg \triangle\bot$.*

The theorem follows from a version of Löb's theorem.

**Theorem 3.3.2 (Löb's theorem).** *We have:*

a. *Suppose that $N : \mathsf{R} \lhd U$ and $\triangle$ is an L-predicate for $U, N$. Then:*
   i. *For all $U$-sentences $A$ we have: $U \vdash \triangle A \to A \;\Rightarrow\; U \vdash A$*
   ii. *For all $U$-sentences $A$ we have: $U \vdash \triangle(\triangle A \to A) \to \triangle A$*

b. *Suppose that* $N : \mathsf{S}^1_2 \lhd U$ *and* $\triangle$ *is a uniform L-predicate for* $U, N$. *Then:*
   i *For all U-formulas* $A\vec{x}$ *we have:* $U \vdash \forall\vec{x}\,(\triangle A\vec{x} \to A\vec{x}) \;\Rightarrow\; U \vdash \forall\vec{x}\,A\vec{x}$
   ii. $U \vdash \forall A \in \mathsf{sent}^N_U\,(\triangle(\triangle A \to A) \to \triangle A)$

Before proving Löb's theorem we formulate some conventions and prove a useful lemma. Let $U_n$ be the theory axiomatized by the axioms of $U$ with Gödel number smaller than $n$. We write $\Box_V$ for ordinary provability in $V$.

**Theorem 3.3.3** *Suppose* $N : \mathsf{S}^1_2 \lhd U$ *and* $\triangle$ *is a uniform L-predicate for* $U, N$. *Then we have:*

a. *For each n, we have* $U \vdash \forall A \in \mathsf{sent}^N_U\,(\Box_n A \to \triangle A)$.
b. *For some* $n^*$, *we have:*
   $\mathsf{L}^*_{\mathsf{u}}1.\; U_{n^*} \vdash A \;\Rightarrow\; \vdash \triangle A$
   $\mathsf{L}^*_{\mathsf{u}}2.\; U_{n^*} \vdash \forall A, B \in \mathsf{sent}^N_U\,((\triangle A \wedge \triangle(A \to B)) \to \triangle B)$
   $\mathsf{L}^*_{\mathsf{u}}3.\; U_{n^*} \vdash \forall A \in \mathsf{sent}^N_U\,(\triangle A \to \triangle\triangle A)$
   $\mathsf{L}^*_{\mathsf{u}}4.\; U_{n^*} \vdash \forall A \in \mathsf{sent}^N_U\,(\Box^N_{U_{n^*}} A \to \triangle A)$
c. $U \vdash \forall\vec{x}\,A\vec{x} \;\Rightarrow\; U \vdash \forall\vec{x}\,\triangle A\vec{x}$.

***Proof*** We prove (a). Let $N : \mathsf{S}^1_2 \lhd U$ and let $\triangle$ be a uniform L-predicate for $U, N$. Let $B$ be the conjunction of the axioms of $U_n$. By $\mathsf{L}_{\mathsf{u}}1$, we have (\$) $U \vdash \triangle B$. Reason in $U$. Consider any $A$. Suppose $\Box_{U_n} A$. Then, $\Box_{\mathsf{pred}}(B \to A)$, It follows, by $\mathsf{L}_{\mathsf{u}}4$, that $\triangle(B \to A)$. By (\$) and $\mathsf{L}_{\mathsf{u}}2$, we find $\triangle A$.

Claim (b) is trivial.

We prove (c). Suppose $U \vdash \forall\vec{x}\,A\vec{x}$. Then, for some $n$, $U \vdash \Box_{U_n}\forall\vec{x}\,A\vec{x}$. It follows that $U \vdash \forall\vec{x}\,\Box_{U_n} A\vec{x}$. Hence, by (a), $U \vdash \forall\vec{x}\,\triangle A\vec{x}$. ■

We prove Löb's theorem.

***Proof*** We prove (b) and leave (a), which is somewhat simpler, to the reader.

First we prove b(ii). Suppose that $N : \mathsf{S}^1_2 \lhd U$ and $\triangle$ is a uniform L-predicate for $U, N$. Let $n$ be large enough to have the properties of Lemma 2.3.3(2). We reason in $U_n$. Consider any $C$ and $D$ in $\mathsf{sent}^N_U$. Suppose that we have $\triangle C$ and $\triangle(C \to (\triangle C \to D))$. It follows, by $\mathsf{L}_{\mathsf{u}}2$, that $\triangle(\triangle C \to D)$. Since, by $\mathsf{L}_{\mathsf{u}}3$, we also have $\triangle\triangle C$, we may conclude, by $\mathsf{L}_{\mathsf{u}}2$, that $\triangle D$. We leave $U_n$ again.

We have found:

$$U_n \vdash \forall C, D \in \mathsf{sent}^N_U\,(\triangle(C \to (\triangle C \to D)) \to (\triangle C \to \triangle D)).$$

It follows that

$$U_n \vdash \Box^N_{U_n}\forall C, D \in \mathsf{sent}^N_U\,(\triangle(C \to (\triangle C \to D)) \to (\triangle C \to \triangle D)).$$

Hence,

$$U_n \vdash \forall C, D \in \mathsf{sent}^N_U\,\Box^N_{U_n}(\triangle(C \to (\triangle C \to D)) \to (\triangle C \to \triangle D)).$$

It follows that

$$U_n \vdash \forall C, D \in \mathsf{sent}_U^N \, \triangle(\triangle(C \to (\triangle C \to D)) \to (\triangle C \to \triangle D)).$$

By $\mathsf{L_u}3$ and $\mathsf{L_u}2$, we may conclude that

$$(\dagger) \quad U_n \vdash \forall C, D \in \mathsf{sent}_U^N \, (\triangle(C \to (\triangle C \to D)) \to \triangle(\triangle C \to \triangle D)).$$

Clearly,

$$U_n \vdash \forall C, D \in \mathsf{sent}_U^N \, \Box_{U,n}((\triangle D \to D) \to ((\triangle C \to \triangle D) \to (\triangle C \to D))).$$

Hence,

$$U_n \vdash \forall C, D \in \mathsf{sent}_U^N \, \triangle((\triangle D \to D) \to ((\triangle C \to \triangle D) \to (\triangle C \to D))).$$

By $\mathsf{L_u}2$, we have

$$(\ddagger) \quad U_n \vdash \forall C, D \in \mathsf{sent}_U^N \, ((\triangle(\triangle D \to D) \wedge \triangle(\triangle C \to \triangle D)) \to \triangle(\triangle C \to D)).$$

Combining $(\dagger)$ and $(\ddagger)$, we get

$$U_n \vdash \forall C, D \in \mathsf{sent}_U^N \, ((\triangle(\triangle D \to D) \wedge \triangle(C \to (\triangle C \to D))) \to \triangle(\triangle C \to D)).$$

By $\mathsf{L_u}2$ it follows that

$$\begin{aligned} U_n \vdash \forall C, D \in \mathsf{sent}_U^N \\ ((\triangle(\triangle D \to D) \wedge \triangle(C \to (\triangle C \to D)) \wedge \triangle((\triangle C \to D) \to C)) \\ \to (\triangle(\triangle C \to D) \wedge \triangle C)). \end{aligned}$$

Applying $\mathsf{L_u}3$ and $\mathsf{L_u}2$ to the consequence of the previous equation, we find

$$\begin{aligned} (\$) \quad U_n \vdash \forall C, D \in \mathsf{sent}_U^N \\ ((\triangle(\triangle D \to D) \wedge \triangle(C \to (\triangle C \to D)) \wedge \triangle((\triangle C \to D) \to C)) \to \triangle D). \end{aligned}$$

We reason in $U_n$. Consider any $A$ in $\mathsf{sent}_U^N$. By the Gödel fixed point lemma, we can find a $B$ such that $\Box_{U_n}(B \to (\triangle B \to A))$ and $\Box_{U_n}((\triangle B \to A) \to B)$. Hence, $\triangle(B \to (\triangle B \to A))$ and $\triangle(B \to (\triangle B \to A))$. By $(\$)$, we may conclude that if $\triangle(\triangle A \to A)$, then $\triangle A$.

To prove b(i), suppose $(\alpha) \vdash \forall \vec{x} \, (\triangle A\vec{x} \to A\vec{x})$. By Lemma 2.3.3(3), we find that $\vdash \forall \vec{x} \, \triangle(\triangle A\vec{x} \to A\vec{x})$. Ergo, by b(ii), we find $(\beta) \vdash \forall \vec{x} \, \triangle A\vec{x}$. Combining $\alpha$ and $\beta$, we may conclude, $\vdash \forall \vec{x} \, A\vec{x}$. ∎

### 3.3.2 *On HBL-Predicates*

In this section, we address the question of what are the inconsistency statements over a given theory $U$. We do not have a general solution, but we will solve our problem for a special case. First we restrict ourselves to *essentially reflexive* pairs $U, N$, where $N : \mathsf{S}^1_2 \lhd U$.

A pair $U, N$ is *essentially reflexive* if $U$ proves the uniform reflection principle for every finite subtheory of $U$ w.r.t. provability in $N$. If $U, N$ is essentially reflexive, one can show that

$U$ satisfies full induction w.r.t. $N$. Conversely, consider a pair $W, K$, where $K : \mathsf{S}^1_2 \lhd W$. If $W$ satisfies full induction w.r.t. $K$ *and if $W$ is sequential*,[5] then $W, K$ is essentially reflexive.[6]

Examples of essentially reflexive pairs $U, N$ are $\mathsf{PA}$ with the identity interpretation and $\mathsf{ZF}$ with the von Neumann interpretations of the natural numbers as finite von Neumann ordinals.

Second, we restrict ourselves to either uniform L-predicates or to HBL-predicates. Here 'HBL' stands for: Hilbert–Bernays–Löb. An L-predicate for $U, N$ is HBL if it satisfies the following condition:

$\Sigma_1$-$\mathsf{C}$. $U \vdash S^N \to \Delta S^N$, where $S$ is a $\Sigma_1$-sentence.

We write $U_n$ for the theory axiomatized by the axioms of $U$ with Gödel number less then $n$. We write $\vdash A$ for $U \vdash A$, and $\Box$ for $\Box_U$, where $\Box$ is a standard arithmetization of provability, $\Box_x$ for provability in $U_x$, etcetera.

Here is a first result.

**Theorem 3.3.4** *Let $U, N$ be an essentially reflexive pair. Suppose $\Delta$ is a uniform L-predicate for $U, N$. Then $\Delta$ is a HBL-predicate for $U, N$.*

***Proof*** Suppose $\Delta$ is a uniform L-predicate for $U, N$. We choose $n$ to be so large that $U_n \vdash \mathsf{EA}^N$. Let $S$ be $\Sigma_1$. We have $\vdash S^N \to \Box^N_n S^N$, and hence $\vdash S^N \to \Delta S^N$. ■

Here is a further result.

**Theorem 3.3.5** *Suppose $U, N$ is an essentially reflexive pair. The following are equivalent.*

a. *$\Delta$ is a HBL-predicate for $U, N$.*
b. *$\Delta$ is an L-predicate for $U, N$ and, for all $U$-sentences $A$ and all $n$, $\vdash \Box^N_n A \to \Delta A$.*

***Proof*** (a) $\Rightarrow$ (b): Suppose $\Delta$ is a HBL-predicate for $U, N$. Let $A$ and $n$ be given. By essential reflexiveness, we have $\vdash \Box_n A \to A$. Hence, by $\mathsf{L}1$, we find (1) $\vdash \Delta(\Box_n A \to A)$. Also, by $\Sigma_1$-completeness, (2) $\vdash \Box_n A \to \Delta \Box_n A$. Applying $\mathsf{L}2$ to (1) and (2), we find the desired result $\vdash \Box_n A \to \Delta A$.

(b) $\Rightarrow$ (a). Suppose that $\Delta$ is an L-predicate for $U, N$ and that, for all $U$-sentences $A$ and all $n$, we have $\vdash \Box^N_n A \to \Delta A$. Consider any $\Sigma_1$-sentence $S$. We choose $n$ large enough so that $U_n \vdash \mathsf{EA}^N$. It follows that $\vdash S^N \to \Box^N_n S^N$ and, hence, $\vdash S^N \to \Delta S^N$. ■

We will employ the following definitions.

- A sentence $A$ in the language of $U$ is *interpretable over $U$* if $U$ interprets $U + A$.
- We write $A \rhd_U B$ for $U + A$ interprets $U + B$. So $A$ is interpretable over $U$ if $\top \rhd_U A$.

We will usually suppress the subscript $U$. We remind the reader that we write $\nabla C$ for $\neg\triangle\neg C$. The following result connects HBL-predicates with interpretability.

**Theorem 3.3.6** *Let $U, N$ be an essentially reflexive pair and let $\triangle$ be a HBL-predicate for $U, N$. We have:*

*a.* $\nabla A \rhd A$.

*b.* $A \rhd (A \wedge \triangle\neg A)$.

We note that Theorem 3.3.6(b) is a version of Feferman's theorem.

***Proof*** (a) By Theorem 3.3.5, we have $\vdash \nabla A \to \Diamond_n A$, for all $n$. The Orey–Hájek characterization tells us that $\{\Diamond_n A \mid n \in \omega\} \rhd A$. Hence $\nabla A \rhd A$.

We prove (b). We have $(A \wedge \triangle\neg A) \rhd (A \wedge \triangle\neg A)$ and, by (a), using Löb's principle for $\triangle$: $(A \wedge \nabla A) \rhd \nabla(A \wedge \triangle\neg A) \rhd (A \wedge \triangle\neg A)$. We employ a disjunctive interpretation to obtain

$$((A \wedge \triangle\neg A) \vee (A \wedge \nabla A)) \rhd (A \wedge \triangle\neg A).$$

Hence, $A \rhd (A \wedge \triangle\neg A)$. ∎

*Warning: Theorems 3.3.5 and 3.3.6 need not be $U$-verifiable since we need not have:* $\vdash \Box A \to \Box\triangle A$.

The next theorem gives the characterization of role inconsistency in terms of interpretability. Conversely, it characterizes interpretability over a theory in terms of HBL-predicates or of uniform L-predicates.

**Theorem 3.3.7** *Consider any essentially reflexive pair $U, N$. The following are equivalent.*

1. *$A$ is interpretable over $U$.*
2. *$A$ is $U$-provably equivalent to $\triangle\bot$, for some uniform L-predicate $\triangle$ for $U, N$.*
3. *$A$ is $U$-provably equivalent to $\triangle\bot$, for some HBL-predicate $\triangle$ for $U, N$.*

***Proof*** (1) $\Rightarrow$ (2): Suppose $\top \rhd A$. It follows, by the Orey–Hájek characterization, that, for all $n$, we have $\vdash \Diamond_n A$.

We define: $\triangle^{[A]}B :\leftrightarrow A \vee \exists x\,(\Box^N_x(A \to B) \wedge \Diamond^N_x A)$.[7] We write, locally, $\triangle$ for $\triangle^{[A]}$.

We verify $\mathsf{L_u}1$. Suppose $\vdash B$. Then, for some $n$, $\vdash \Box^N_n(A \to B) \wedge \Diamond^N_n A$. Hence $\vdash \triangle B$.

We verify $\mathsf{L_u}2$. We reason in $U$. Consider any $U$-sentences $B$ and $C$. Suppose $\triangle B$ and $\triangle(B \to C)$. In case $A$ we are immediately done. Suppose $\neg A$. We have (a) $\exists x\,(\Box^N_x(A \to B) \wedge \Diamond^N_x A)$ and (b) $\exists y\,(\Box^N_y(A \to (B \to C)) \wedge \Diamond^N_y A)$. Let $a$ witness (a) and let $b$ witness (b). Let $c := \mathsf{max}(a, b)$. Clearly, we have $\Box^N_c(A \to B) \wedge \Diamond^N_c A$ and $\Box^N_c(A \to (B \to C)) \wedge \Diamond^N_c A$. Hence, $\Box^N_c(A \to C) \wedge \Diamond^N_c A$, and, thus, $\triangle C$.

We verify $\mathsf{L_u}3$, to wit $\vdash \forall B \in \mathsf{sent}^N_U\,(\triangle B \to \triangle\triangle B)$. Thus it is sufficient to prove $\vdash \forall B \in \mathsf{sent}^N_U\,\triangle\triangle B$, in other words,

(†) $\vdash \forall B \in \mathsf{sent}_U^N\ (A \vee \exists x\,(\Box_x^N(A \rightarrow (A \vee \exists y\,(\Box_y^N(A \rightarrow B) \wedge \Diamond_y^N A))) \wedge \Diamond_x^N A))$.

Clearly (†) is equivalent to $\vdash A \vee \exists x\,(\Box_x^N \top \wedge \Diamond_x^N A)$. Since $\vdash \Box_0^N \top \wedge \Diamond_0^N A$, we are done.

We obtain $\mathsf{L_u}3$ from (†) by noting that $\vdash \Diamond_0^N A$. This observation entails that $\vdash \forall B \in \mathsf{sent}_U^N\ (\Box_0^N B \rightarrow \triangle B)$.

(2) ⇒ (3): This is immediate since, by Theorem 3.3.4, every uniform L-predicate is HBL.

(3) ⇒ (1): This is by Theorem 3.3.6(b). ■

We note that $\triangle^{[A]}$ need not be the only HBL-predicate for $U, N$ such that $A$ is its inconsistency statement.

From the proof of Theorem 3.3.7, we have the following further insight.

**Theorem 3.3.8** *Let $U, N$ be an essentially reflexive pair. We have $\triangle^{[A]}$ is a HBL-predicate (uniform L-predicate) iff $\top \rhd A$.*

Since, by a result of Lindström, we know that the set of $A$ that are interpretable over an essentially reflexive consistent theory $U$ is $\Pi_2$-complete, we have the following corollary.

**Corollary 3.3.9** *Suppose $U, N$ is an essentially reflexive pair and $U$ is consistent. Then, the set of HBL-predicates for $U, N$ is $\Pi_2$-complete. Similarly for the set of uniform L-predicates.*

The following corollary is a well-known result, but it is nice to see that Theorem 3.3.7, considered as a characterization of interpretability over a theory, is able to support a new proof.

**Corollary 3.3.10** *Suppose $U, N$ is an essentially reflexive pair. Suppose $\triangle_0$ and $\triangle_1$ are HBL-predicates for $U, N$ and $\vdash \triangle_0 A \rightarrow \triangle_1 \triangle_0 A$, for any $U$-sentence $A$. Then, we have $\top \rhd (\triangle_0 \bot \wedge \triangle_1 \bot)$. As a consequence, we have, for any $B$, if $\top \rhd B$, then $\top \rhd (B \wedge \Box^N \bot)$.*

***Proof*** Suppose that $\triangle_0$ and $\triangle_1$ are as stipulated in the theorem. We define

$$\triangle^\star A :\leftrightarrow \triangle_0 A \wedge \triangle_0 \triangle_1 A \wedge \triangle_1 A.$$

It is easy to see that $\triangle^\star$ is HBL. Moreover $\triangle^\star \bot$ is provably equivalent to $\triangle_0 \bot \wedge \triangle_1 \bot$.

To see that the promised consequence holds, we just need to note that for any HBL-predicate $\triangle$, the pair $\Box, \triangle$ satisfy the conditions on $\triangle_0, \triangle_1$. ■

Since, for essentially reflexive pairs $U, N$, there are Orey sentences $O$, such that $\top \rhd O$ and $\top \rhd \neg O$, we have the following corollary.

**Corollary 3.3.11** *Suppose $U, N$ is essentially reflexive. There are uniform L-predicates $\triangle_0$ and $\triangle_1$ such that $\vdash \triangle_0 \bot \leftrightarrow \neg \triangle_1 \bot$. As a consequence, there is a HBL-predicate $\triangle$ such that $\top \rhd \nabla \top$.*

The corollary shows a disanalogy between $\Diamond^N\top$ and $\triangledown\top$, for certain $\triangle$. We never have $\top \rhd \Diamond^N\top$. The reason that this basic insight fails for some statements of the form $\triangledown\top$ is due to the following. Suppose $K$ is an interpretation of $U$ in $U$. Then, we do not necessarily have that $\triangle^K$ is again uniform L or even HBL. All conditions are generally preserved, but for $\mathsf{L_u3}$ or even $\mathsf{L3}$. We do not get $\vdash \triangle^K A \to \triangle^K\triangle^K A$, but just $\vdash \triangle^K A \to \triangle^K\triangle A$. On the other hand we do have $\vdash \forall A\ (\Box^{NK} A \to \Box^{NK}\Box^{NK} A)$. Thus, adding a $\Diamond^N\top$ always increases the interpretability strength of $U$, but the same does not generally hold for the $\triangledown\top$.

**Remark 3.3.12** *In a sequel to this chapter [Vis14a] we expand the results of the present section to yield a full characterization of interpretability over a theory.*

### 3.3.3 *Feferman on L-Predicates*

In his classical paper [Fef60], Sol Feferman discusses HB-predicates. Clearly, his discussion could equally well apply to L-predicates or HBL-predicates. Feferman writes:

> One approach to the treatment of intensional problems in a general setting could be by means of three derivability conditions set down for a proof definition $\mathsf{Prf}_{\mathcal{T}}(x, y)$ by Hilbert and Bernays [. . . ]
>
> This approach suffers several limitations. First of all there is no simple method of seeing whether a given proof definition satisfies the derivability conditions. Further, the third derivability condition is peculiar to theories which contain a sufficient amount of arithmetic; it is thus not relevant to other problems of arithmetization, for example the treatment of Gödel's completeness theorem. On the other hand, the first two derivability conditions are rather too weak to be used by themselves with such other problems.; it simply happens that they are sufficient (when accompanied by the third condition) to be used for the proof of the proof of the second underivability theorem. Finally, one does not see clearly from the conditions what are the possibilities for manipulating those proof conditions which satisfy them.

Let us note that Feferman's discussion has nothing to do with the idea of analyzing meanings as roles. His objections are directed at the proposal to make HB-style predicates a central ingredient of a general framework to study intensional phenomena. As I understand it, for Feferman, such a framework is, for a substantial part, mathematically motivated.[8]

Feferman states his objections to the HB conditions and not to the conditions we consider. However, I think his discussion can be fruitfully transposed to the specific conditions we are interested in.

The (transposed version of the) first objection is that the verification that a predicate satisfies the Löb conditions is often a substantial amount of work. The point is well taken. However, the framework could consist of general theorems about all L-predicates, accompanied by detailed verifications that predicates from a salient list are indeed L-predicates (like our list in Section 3.3.1). The fact that such verifications are much work need not count against the framework. Sometimes it is hard to verify that a mathematical object is a category. This need not be held against category theory.

The second objection zooms in on $\Sigma_1$-completeness. We note that Hilbert and Bernays were assuming that their predicates are $\Sigma_1$. Since we are making no such assumption, for us there are two corresponding conditions, to wit $\Sigma_1$-C and L3. The principle $\Sigma_1$-C is indeed problematic since it usually needs the presence of the totality of exponentiation. We note, however, that the usual proof predicate also belongs to a much more restricted class, to wit $\exists\Sigma_1^{\mathsf{b}}$. The principle of $\exists\Sigma_1^{\mathsf{b}}$-completeness can be proved w.r.t. the usual provability predicates already in $\mathsf{S}_2^1$. So opting for the more restricted class increases the scope substantially.

We turn to L3. For the Feferman predicate over PA, we *do* have $\Sigma_1$-C, but L3 fails; this follows from the fact that PA proves its own Feferman consistency while the *other* Löb conditions are fulfilled. Still we do have the Gödel–Hilbert–Bernays–Wang–Henkin–Feferman theorem for the consistency statement based on the Feferman predicate.[9] Thus the second objection is indeed successful.

Note however that the second objection refutes a very strong claim for the Löb conditions: we demand that every salient predicate satisfies these conditions and that every important proof somehow employs the conditions. If we allow variants of the conditions, we see that they occur at least in some important proofs. For example, if we inspect the formalization of the proof of the Gödel–Hilbert–Bernays–Wang–Henkin–Feferman theorem, we will see that a *variant* L3 for ordinary provability is employed, to wit: $\vdash \Box^N A \to \exists x \in N\, \Box^N \Box^N_x A$.

Finally, the Feferman predicate does turn out to have an exciting modal analysis: see [Mon78], [Vis89], [Sha94]. Summarizing, I would say that Feferman has a good point with the objection, but that still an exciting and important role for the Löb conditions in a general framework is compatible with the objection.

Feferman's third objection is the most interesting, since it points to a programmatic idea: are there interesting constructions to build predicates fulfilling the Löb conditions? There is certainly not much work in this direction. I contribute to this question in a sequel to this chapter [Vis14a].

Concluding, we may say that Feferman's second objection has some force, but that it does aim at a rather strong claim on behalf of the use of the Löb conditions. Feferman's third objection is correct even for the current *status questionis*. It could be viewed, however, as a program for a line of study of the Löb conditions.

### 3.3.4 *Philosophical Discussion*

What has been achieved by our schematic recasting of G2 as Theorem 3.3.1? Well, first of all, it does not seem to be G2 at all. It only says *if* $\Delta$ *has certain properties, then* .... Here the answer clearly should be—as pointed out earlier in the discussion of Feferman's first objection—that the true theorem G2 is Theorem 3.3.1 plus an expanding list of examples of L-predicates.

How about the idea that the notion of L-predicate explicates the meaning of consistency? Consider any theory $U$ with an interpretation $N$ of R. We note that $x = x$ is an L-predicate. This makes $\underline{\ulcorner\bot\urcorner} \neq \underline{\ulcorner\bot\urcorner}$ a consistency statement. In general $U$ will have many non-equivalent consistency statements, as shown in Section 3.3.2. Can we really accept

$\ulcorner\bot\urcorner \neq \ulcorner\bot\urcorner$ as a consistency statement of $U$? What would be involved in such acceptance? We could perhaps say that *being a consistency statement* is not an absolute concept but a *role* w.r.t. a choice of an L-predicate. So we should say things like *A is a consistency statement in the context of* $\Delta$.

It seems to me that this answer cannot satisfy the *foundationally motivated* questioner who wants to lay down conditions for what it is to express consistency. We want to know whether a certain statement expresses consistency in *our* sense, not in a role sense that crucially depends on the theory at hand. There is nothing against calling role consistency statements *consistency statements*, from a mathematical point of view, given that such statements capture mathematically salient properties of samples from the original class. However, what is mathematically convenient does not *ipso facto* answer philosophical questions. Why should it?

## 3.4 Solution of the Meaning Problem

We turn to problem 2.ii: if we consider a theory $U$ that is either unsound (in the sense that it has no $\omega$-models) or too weak to verify basic facts about provability, does a statement $\mathsf{con}(U)$ still express consistency in the context of the theory?

My proposal is to deconstruct the question. I claim that the theory $U$ is not the *locus designatus* where the statement $\mathsf{con}(U)$ gets its meaning. This can be best understood by looking at one of our formulations of G2:

*for any consistent theory U, we have:* $U \not\rhd (\mathsf{S}^1_2 + \mathsf{con}(U))$.

Here we see that we can view the second incompleteness theorem, not as a statement concerning the relation of *a theory and a sentence*, but concerning *the relation between two theories*. It is a claim about the interpretability relation between an arbitrary theory $U$ and theory $\mathsf{S}^1_2 + \mathsf{con}(U)$. The theory $U$ could be any theory; it could be a theory for which we cannot imagine a decent interpretation. The theory $\mathsf{S}^1_2$, on the other hand, has a standard semantics: it is a theory about the natural numbers. In other words, the theory $\mathsf{S}^1_2$ *can be taken to be* a meaningful theory with an intended meaning. My proposal is to do precisely that. We view the second incompleteness theorem as the statement of non-interpretability between a merely formal theory $U$ and an interpreted theory $\mathsf{S}^1_2 + \mathsf{con}(U)$. Of course, the theorem-qua-mathematical-theorem is not concerned with the meaningfulness of $\mathsf{S}^1_2$. It is in the context of a philosophical reading of the second incompleteness theorem, that we view the theory $\mathsf{S}^1_2$ as meaningful, with as intended meaning the standard semantics in the natural numbers.

Isn't it strange to have a relationship between two things of such different character as a merely formal theory and a meaningful theory? No, the relationship is *carried* by a relationship between the merely formal theory and a merely formal theory. Here the second merely formal theory is associated to the meaningful theory as being its 'form'. In the same way, I am heavier than this book, even if, on a certain theory of object individuation, I and the book are quite different sorts of thing. I am heavier than the book because my body is heavier than the book. I simply inherit many properties from my body. I supervene on the body and inherit certain properties and relations from the body.

If we write $[\mathsf{S}^1_2 + A]$ for the meaningful theory with as outer form the merely formal theory $\mathsf{S}^1_2 + A$, and we write $\mathfrak{o}$ for the relation, *is the outer form of*, we can define $U \rhd^* [\mathsf{S}^1_2 + A]$ by: $U \rhd (\mathsf{S}^1_2 + A)$ and $(\mathsf{S}^1_2 + A)\ \mathfrak{o}\ [\mathsf{S}^1_2 + A]$. In this notation, our reading of the second incompleteness theorem would be: $U \not\rhd^* [\mathsf{S}^1_2 + \mathsf{con}(U)]$.

We note that if we replace $\mathsf{S}^1_2$ in the formulation by a meaningful syntactic theory, then the situation gets even better: we do not need to explain the role of Gödel coding any more.[10]

Our dissolution of the problem does not mean that there is no problem of *what it is to express consistency*, for example, in the case where we compare ordinary consistency with Feferman consistency. However, this question should be asked against the background of *the standard semantics*. This is a problem different from (ii). So, in a sense, the point of the dissolution is to put the meaning question in its proper place: the standard or intended semantics.

Our (dis)solution leaves entirely open how to think about the standard semantics. We could still give a Wittgensteinean language game explanation for this or what have you. The whole point—in game terms—is that the relevant game is neither given in terms of the formal game determined by $U$, nor the formal game determined by $\mathsf{S}^1_2 + \mathsf{con}(U)$.

I submit that the idea that a merely formal theory could give us the meaning—in the foundationally relevant sense of *meaning*—of, say, a predicate, was always a non-starter. The meaning of a predicate is given by its application conditions. There are simply no application conditions in a merely formal theory.

## 3.5 Abolishing Arbitrariness

In this section, we turn away from the problem of meaning and address the problem that there seems to be no treatment of G2 that does not involve arbitrary choices. No mathematical treatment of G2 correctly reflects our intuitive insight which is independent of the long sequence of design choices usually associated with the formulation of G2.

Sources of arbitrariness are: the chosen treatment of syntax, the proof system, the coding of the syntax and the proofs, and the representation of the set of axioms. For example, usually we develop the syntax inductively, but we could also describe the syntactical objects as labeled constrained graphs, or we could treat them as strings satisfying certain local conditions and certain global conditions on balance of brackets. Let's for example opt for an inductive treatment. We can realize this first inside the theory of strings, as was proposed both by Tarski and by Quine, but we could also opt for a theory of binary graphs or a theory of finite sets. Let's zoom in on strings. There are still a lot of choices: Polish notation, reverse Polish notation, the usual infix notation for e.g. conjunction and identity, etc. Are the indices of variables in unary or in binary of dyadic or . . . ? Now we have to translate the strings as numbers: here we can use Smullyan-style coding, but there is also Gödel's way and Kleene's . . .

In this section, we consider two strategies. In the first we represent consistency statements as the unique solutions of certain equations. In the second we generalize the notion of consistency statement to a more mathematical notion.

This section does not contain new results, but rather gives a quick overview of some of the results of [Vis11] (Section 3.5.1) and [Vis12] (Section 3.5.2). For details and careful proofs the reader is referred to these papers.

### 3.5.1 *Consistency Statements as Unique Solutions of Equations*

In this section we describe results of [Vis11]. Our treatment works for *finitely axiomatized sequential theories*. It can be extended to sequential CE theories, but we will not do that here. See the treatment in [Vis11].

Consider a sequential theory $U$. The theory $\mathsf{PC}(U)$ is obtained as follows. First we extend the signature with a new sort of classes. Second, we add predicative comprehension:

$$\vdash \exists X\, \forall x\, (x \in X \leftrightarrow Ax\vec{z}\vec{Z}).$$

Here the formula $A$ does not contain bound class variables. As is well known, over a sequential theory, we can finitely axiomatize this comprehension scheme.

We formulate a number of important insights.

**Theorem 3.5.1** *We have for $\Pi^0_1$-sentences $P$ and $P'$:*

$$(\mathsf{S}^1_2 + P) \rhd (\mathsf{S}^1_2 + P') \quad \Leftrightarrow \quad \mathsf{EA} \vdash P \to P'.$$

This result is due to Wilkie and Paris ([WP87]). It was independently known to Harvey Friedman.

**Theorem 3.5.2** *We have* $\mathsf{PC}(A) \equiv (\mathsf{S}^1_2 + \mathsf{con}(A))$.

The left to right direction of the proof uses the definability of a satisfaction predicate for the language of $A$ in the language of $\mathsf{PC}(A)$ with sufficiently good properties. From right to left, we use an adaptation of the Gödel–Hilbert–Bernays–Wang–Henkin–Feferman construction, which is a version of a proof of the model existence lemma. To be able to operate without having enough induction, one uses Solovay's method of shortening cuts.

The statement $\mathsf{con}(A)$ is of course not coordinate-free. The reader is to choose his or her favorite version. The elimination of the coordinates happens next. The statement $\mathsf{con}(A)$ is the ladder we throw away; it merely serves to witness the existence of a $\Pi_1$-sentence $P$ such that $\mathsf{PC}(A) \equiv (\mathsf{S}^1_2 + P)$.

We show that $\Pi_1$-sentences $P$ such that $\mathsf{PC}(A) \equiv (\mathsf{S}^1_2 + P)$ are unique modulo $\mathsf{EA}$-provable equivalence. Consider any $\Pi_1$-sentence $P$. Suppose that $\mathsf{PC}(A) \equiv (\mathsf{S}^1_2 + P)$. It follows, by Theorem 3.5.2, that $(\mathsf{S}^1_2 + \mathsf{con}(A)) \equiv (\mathsf{S}^1_2 + P)$. Hence, by Theorem 3.5.1, we may conclude: $\mathsf{EA} \vdash P \leftrightarrow \mathsf{con}(A)$.

Thus, we can characterize a consistency statement as the unique $\Pi_1$-sentence $P$ modulo $\mathsf{EA}$-provable equivalence, such that $\mathsf{PC}(A) \equiv (\mathsf{S}^1_2 + P)$.

Similarly, a cut-free consistency statement (or: restricted, tableaux, Herbrand consistency statement) is modulo $\mathsf{EA}$-provable equivalence, the unique $\Pi_1$-sentence $P$ such that $A \equiv (\mathsf{S}^1_2 + P)$.

**Open question 3.5.3** The multi-exponential transformations provided by our characterization theorems are somewhat unsatisfactory. Can we not improve these to p-time? Can we not replace EA by $S^1_2$? Well, already the transformation between on the one hand Smullyan coding and on the other hand Kleene coding or Gödel coding (using the $\beta$-function) is exponential. These last two historical codings are widely recognized as paradigmatic examples of Gödel coding. They are still used in recent textbooks.[11]

Yes, but what if we restrict ourselves to efficient coding schemes like Smullyan's? Is there any characterization of consistency statements with appropriate qualifications about the efficiency of the coding scheme that works modulo $S^1_2$-provable equivalence?

**Open question 3.5.4** The results of this section do not work if we turn to constructive theories. Is there an approach—perhaps via a generalization of the notion of interpretation—that preserves the essence of the results to the constructive case?

### 3.5.2 *Bounded Interpretations*

In this section, we summarize the results of [Vis12]. Let $\Gamma_k$ be the set of formulas of alternating quantifier depth $\leq k$.[12] An $n$-dimensional interpretation $K : U \rhd V$ is *bounded* when, for some $k$, and for all $V$-formulas $A(x_0, \ldots, x_{m-1})$, there is a $U$-formula $B(\vec{y}_0, \ldots, \vec{y}_{m-1})$ in $\Gamma_k$, such that:

$$U \vdash \forall \vec{y}_0 \in \delta_K, \ldots, \vec{y}_{m-1} \in \delta_K \; (B(\vec{y}_0, \ldots, \vec{y}_{m-1}) \leftrightarrow A^K(\vec{y}_0, \ldots, \vec{y}_{m-1})).$$

In short: $A^K$ is in $\Gamma^k$ modulo $U$-provable equivalence. We write $U \rhd_{\mathsf{bo}} V$ for: there is a bounded $K$ with $K : U \rhd V$. Let $U$ and $V$ be CE theories. We have:

i. If $U \rhd_{\mathsf{bo}} V \rhd W$, then $U \rhd_{\mathsf{bo}} W$.
ii. If $U \rhd V \rhd_{\mathsf{bo}} W$, then $U \rhd_{\mathsf{bo}} W$.

We can show:

**Theorem 3.5.5** *If $A$ is sequential then $A \not\rhd_{\mathsf{bo}} A$.*

The proof uses the existence of partial satisfaction predicates for formulas of $\Gamma_k$, in combination with a version of the first incompleteness theorem due to Pavel Pudlák.

The idea of our present approach to G2 is that (i) consistency statements are finitely axiomatized theories rather than sentences, and (ii) we do not define the notion of consistency statement per se, but what it is for a *sentence-qua-theory* to have *consistency power* over a given theory.[13] We define:

A finitely axiomatized sequential theory $A$ is a *pro-consistency statement* for $U$ iff $A \rhd_{\mathsf{bo}} U$.

We have:

**Theorem 3.5.6** *Suppose $U$ is a CE theory. Then $U$ does not interpret any pro-consistency statement for $U$.*

***Proof*** Suppose our theorem fails. Then, we would have a pro-consistency statement $A$ such that: $A \rhd_{\mathsf{bo}} U \rhd A$. And, hence $A \rhd_{\mathsf{bo}} A$. Quod non. ■

Thus, Theorem 3.5.6 is our new version of G2. Just as in the case of the L-predicates approach, it looks more like a lemma for G2 than G2 itself, since we need the extra information that the usual consistency statements are indeed pro-consistency statements in our sense. Thus, just as in the case of the L-predicates approach, the true version of G2 should be taken to be Theorem 3.5.6 plus a list of verified examples of pro-consistency statements which includes the ordinary consistency statements.

**Example 3.5.7** We provide an example of a pro-consistency statement that is not a consistency statement.

Let $U$ be a CE theory. We define the *super-logarithmically bounded consistency of $U$*, or *slb-consistency*, by:

$$\mathsf{slb\text{-}con}(U) :\leftrightarrow \forall x\ (\ \mathsf{supexp}(x){\downarrow} \rightarrow \mathsf{con}_x(U)\ ).$$

The restricted consistency in this case involves both a bound on the axioms used and a bound on the complexity of the formulas in the proofs. We have:

a. $\mathsf{slb\text{-}con}(U)$ is $\Pi^0_1$.

b. $\mathsf{S}^1_2 + \mathsf{slb\text{-}con}(U)$ is a pro-consistency, statement for $U$.

c. Let $A$ be a consistent sequential sentence. Then,

$$(\mathsf{S}^1_2 + \mathsf{con}(A)) \vdash (\mathsf{S}^1_2 + \mathsf{slb\text{-}con}(A)) \not\rhd (\mathsf{S}^1_2 + \mathsf{con}(A)).$$

So there are pro-consistency statements of $A$ that are essentially weaker than $\mathsf{con}(A)$ (over $\mathsf{S}^1_2$).

The present approach does not try to pin down consistency statements, but just aims to define relative consistency strength. In some senses, this is a weakness compared to the other approach that determines consistency statements modulo EA-provable equivalence. The strength of the present approach is that, conceivably, we could take some of the ideas of G2 to places where we do not have the usual coding devices.

One idea would be to look at theories of non-surjective pairing. The basic theory PAIR is:

pair1. $\exists x\, \forall y\ y \notin x$

pair2. $\forall x, y\, \exists z\, \forall u\, (u \in z \leftrightarrow (u = x \vee u = y))$

Can we repeat our development using theories that directly interpret PAIR?[14]

**Open question 3.5.8** It is easy to see that $\mathsf{PC}(\mathsf{PAIR}) \rhd \mathsf{S}^1_2$, but do we have $\mathsf{S}^1_2 \rhd \mathsf{PC}(\mathsf{PAIR})$?

We know that there are extensions $U$ of $\mathsf{PAIR}$ with quantifier elimination, so $U \rhd_{\mathsf{bo}} U$, but are (some of) these finitely axiomatizable? Do we have, for any finitely axiomatized $\mathsf{PAIR}$-theory $A$, $A \not\rhd_{\mathsf{bo}} A$?

**Open question 3.5.9** The results of this section do not work if we turn to constructive theories. Is there an approach—perhaps via a generalization of the notion of interpretation—that preserves the essence of the results to the constructive case?

# Notes

1. The theory $\mathsf{EA}$ is $\mathrm{I}\Delta_0 + \mathsf{exp}$, where $\mathsf{exp}$ is the axiom that exponentiation is total.
2. Of course, these conditions can be further relaxed to many-sorted, infinite signature (as long as it is recursively presented) and CE axiom set, etc.
3. Formally:
$\Lambda^{\Theta}_U := \{A \in \mathsf{sent}_\Theta \mid \top \mid\!\sim^{\Theta}_U A\} = \{A \in \mathsf{sent}_\Theta \mid \text{for all } \tau : \Theta \to \Sigma \text{ we have } U \vdash A^\tau\}$.
4. We assume that we have a formulation of $\mathsf{S}^1_2$ in the usual arithmetical signature.
5. For a definition of *sequential*, see Section A.3 of the Appendix.
6. The results of this section should also work for theories that are only essentially *sententially* reflexive, like Peano Corto. In these weak cases our definition of $\Box_n$ changes in that we also demand that in the proofs we only consider formulas of complexity below $n$. We do not explore this perspective here. See also [Vis14b].
7. Note that the second conjunct is the Feferman predicate for $U + A$.
8. One reason why it is somewhat difficult to respond to Feferman's arguments is that the scope, aims, and claims of the presumed framework are not very clear to me. Evidently the framework should treat the incompleteness theorems and the Gödel–Hilbert–Bernays–Wang–Henkin–Feferman theorem. But I am not sure whether e.g. methods for determining (upper and lower bounds) for the complexity of theories like Presburger arithmetic should be in. If we look at Feferman's development it is somewhat surprising that he makes no attempt to work systematically *modulo interpretability*. He mainly concentrates on extensions of $\mathsf{PA}$. Versions of G2 modulo interpretability are only developed in Section 8 of [Fef60], where they could have been given immediately. A framework that employs interpretability more directly was published by Montague one year later in [Mon61].
9. As we have seen in Corollary 3.3.11, for HBL-predicates the provability version of G2 always holds, but the interpretation version may fail.
10. But see [HV13] for some doubts concerning this point.
11. Our discussion suggests that we could undermine the intuitive appeal of our characterizations by looking at a coding scheme that uses superexponential resources. It would be interesting to develop some such scheme in detail. My prediction is that we will have difficulty accepting such a scheme as an 'honest' scheme for Gödel numbering.
12. See Section A.4 of the Appendix.
13. The basic form of the idea is due to Harvey Friedman. His way of realizing the idea is however quite different from ours.
14. See Section A.3 of the Appendix for the notion of *direct interpretation*.

15. See [Bus86] or [HP93] for an explanation of the relevant formula classes.
16. Vincent van Oostrom gave a variant of this formulation of Gerhardy's measure in conversation.

# References


[Bek04] L.D. Beklemishev. Provability algebras and proof-theoretic ordinals. *Annals of Pure and Applied Logic*, **128**:103–124, 2004.

[Bek05] L.D. Beklemishev. Reflection principles and provability algebras in formal arithmetic. *Russian Mathematical Surveys*, **60**(2):197–268, 2005.

[Bek06] L.D. Beklemishev. The worm principle. In Z. Chatzidakis, P. Koepke, and W. Pohlers, eds., *Logic Colloquium '02*, volume 27 of *Lecture Notes in Logic*, pp. 75–95. A.K. Peters and CRC press, Natick, MA, 2006.

[BJV05] L.D. Beklemishev, J.J. Joosten, and M. Vervoort. A finitary treatment of the closed fragment of Japaridze's provability logic. *Journal of Logic and Computation*, **15**(4):447–463, 2005.

[Boo93] G. Boolos. *The Logic of Provability*. Cambridge University Press, Cambridge, 1993.

[Bus86] S.R. Buss. *Bounded Arithmetic*. Bibliopolis, Napoli, 1986.

[Bus87] S.R. Buss. A conservation result concerning bounded theories and the collection axiom. *Proceedings of the American Mathematic Society*, **100**:709–716, 1987.

[Bus15] S. Buss. Cut Elimination In Situ. *Gentzen's Centenary*, Springer International Publishing, pp. 245–277, Kahle, Reinhard and Rathjen, Michael, http://dx.doi.org/10.1007/978-3-319-10103-3_10, 2015, doi: 10.1007/978-3-319-10103-3_10.

[Fef60] S. Feferman. Arithmetization of metamathematics in a general setting. *Fundamenta Mathematicae*, **49**:35–92, 1960.

[Ger03] P. Gerhardy. Refined complexity analysis of cut elimination. In Matthias Baaz and Johann Makowsky, eds., *Proceedings of the 17th International Workshop CSL 2003*, volume 2803 of *LNCS*, pp. 212–225. Springer-Verlag, Berlin, 2003.

[Ger05] P. Gerhardy. The role of quantifier alternations in cut elimination. *Notre Dame Journal of Formal Logic*, **46**(2):165–171, 2005.

[Grz05] Andrzej Grzegorczyk. Undecidability without arithmetization. *Studia Logica*, **79**(2):163–230, 2005.

[GZ08] Andrzej Grzegorczyk and Konrad Zdanowski. Undecidability and concatenation. In A. Ehrenfeucht, V.W. Marek, and M. Srebrny, eds., *Andrzej Mostowski and Foundational Studies*, pp. 72–91. IOS Press, Amsterdam, 2008.

[HP93] P. Hájek and P. Pudlák. *Metamathematics of First-Order Arithmetic*. Perspectives in Mathematical Logic. Springer, Berlin, 1993.

[HV13] V. Halbach and A. Visser. *Self-Reference in Arithmetic*. Logic Group Preprint Series 316, Faculty of Humanities, Philosophy, Utrecht University, Utrecht, http://www.phil.uu.nl/preprints/lgps/, 2013.

[Jap85] G. Japaridze. The polymodal logic of provability. In *Intensional Logics and Logical Structure of Theories: Material from the Fourth Soviet–Finnish Symposium on Logic, Telavi*, pp. 16–48, 1985.

[Jeř12] E. Jeřábek. Sequence encoding without induction. *Mathematical Logic Quarterly*, **58**(3):244–248, 2012.

[Kal91] M.B. Kalsbeek. *Towards the Interpretability Logic of* $I\Delta_0 + \mathsf{EXP}$. Logic Group Preprint Series 61, Faculty of Humanities, Philosophy, Utrecht University, Utrecht, http://www.phil.uu.nl/preprints/lgps/, 1991.

[Kay91] R. Kaye. *Models of Peano Arithmetic.* Oxford Logic Guides. Oxford University Press, Oxford, 1991.

[Mon61] R. Montague. Semantic closure and non-finite axiomatizability. In *Infinitistic Methods, Proceedings of the Symposium on Foundations of Mathematics, (Warsaw, 2-9 September 1959)*, pp. 45–69, Pergamon Oxford, 1961.

[Mon78] F. Montagna. On the algebraization of a Feferman's predicate (the algebraization of theories which express Theor; X). *Studia Logica,* **37**:221–236, 1978.

[MPS90] J. Mycielski, P. Pudlák, and A.S. Stern. A lattice of chapters of mathematics (interpretations between theorems), volume 426 of *Memoirs of the American Mathematical Society.* AMS, Providence, Rhode Island, 1990.

[Pud83] P. Pudlák. Some prime elements in the lattice of interpretability types. *Transactions of the American Mathematical Society,* **280**:255–275, 1983.

[Pud85] P. Pudlák. Cuts, consistency statements and interpretations. *Journal of Symbolic Logic,* **50**(2):423–441, 1985.

[Sha94] V.Yu. Shavrukov. A smart child of Peano's. *Notre Dame Journal of Formal Logic,* **35**:161–185, 1994.

[Sol76] R.M. Solovay. Provability interpretations of modal logic. *Israel Journal of Mathematics,* **25**:287–304, 1976.

[Šve07] V. Švejdar. An interpretation of Robinson's Arithmetic in its Grzegorczyk's weaker variant. *Fundamenta Informaticae,* **81**(1–3):347–354, 2007.

[TMR53] A. Tarski, A. Mostowski, and R.M. Robinson. *Undecidable Theories.* North–Holland, Amsterdam, 1953.

[Vis89] A. Visser. Peano's smart children: A provability logical study of systems with built-in consistency. *Notre Dame Journal of Formal Logic,* **30**(2):161–196, 1989.

[Vis90] A. Visser. Interpretability logic. In P.P. Petkov, ed., *Mathematical logic, Proceedings of the Heyting 1988 summer school in Varna, Bulgaria,* pp. 175–209. Plenum Press, Boston, 1990.

[Vis93] A. Visser. The unprovability of small inconsistency. *Archive for Mathematical Logic,* **32**(4):275–298, 1993.

[Vis09a] A. Visser. Cardinal arithmetic in the style of Baron von Münchhausen. *Review of Symbolic Logic,* **2**(3):570–589, 2009.

[Vis09b] A. Visser. Growing commas – a study of sequentiality and concatenation. *Notre Dame Journal of Formal Logic,* **50**(1):61–85, 2009.

[Vis11] A. Visser. Can we make the second incompleteness theorem coordinate free. *Journal of Logic and Computation,* **21**(4):543–560, 2011. First published online 12 August 2009, doi: 10.1093/logcom/exp048.

[Vis12] A. Visser. The second incompleteness theorem and bounded interpretations. *Studia Logica,* **100**(1-2):399–418, 2012.

[Vis13] A. Visser. What is sequentiality? In P. Cégielski, Ch. Cornaros, and C. Dimitracopoulos, eds., *New Studies in Weak Arithmetics,* volume 211 of *CSLI Lecture Notes,* pp. 229–269. CSLI Publications and Presses Universitaires du Pôle de Recherche et d'Enseingement Supérieur Paris-est, Stanford, 2013.

[Vis14a] A. Visser. *Transductions in arithmetic.* Annals of Pure and Applied Logic, Elsevier, **167**(3):211–234, 2016.

[Vis14b] A. Visser. Peano Corto and Peano Basso: A study of local induction in the context of weak theories. *Mathematical Logic Quarterly,* **60**(1–2):92–117, 2014.

[vPR+10] V. Čačić, P. Pudlák, G. Restall, A. Urquhart, and A. Visser. Decorated linear order types and the theory of concatenation. In F. Delon, U. Kohlenbach, P. Maddy, and F. Stephan, eds., *Logic Colloquium 2007*, volume 35 of *Lecture Notes in Logic*, pp. 1–13. ASL and Cambridge University Press, New York, 2010.

[WP87] A. Wilkie and J.B. Paris. On the scheme of induction for bounded arithmetic formulas. *Annals of Pure and Applied Logic*, 35:261–302, 1987.

# Appendix A Basic Facts and Definitions

In this appendix we explain some basic notions.

## *A.1 Theories*

Theories are, in this chapter, theories of first-order predicate logic, that have a finite signature and are axiomatized by an axiom set that is represented by a $\Delta_1^{\mathsf{b}}$-formula.[15] Theories will usually be one-sorted, but we will occasionally consider two-sorted theories.

The formula specifying the axiom set is part of the data for the theory. Thus, we treat theories *intensionally* and not as mere sets of theorems.

We say that a theory is *finitely axiomatized* if its axiomatization has the form $\bigvee_{i<n} x = \ulcorner A_i \urcorner$. Note that $\mathsf{S}^1_2$ may prove that a theory has an axiom-set of, say, less than two axioms, without being able to prove the equivalence of the formula defining the axiom set with any formula of the prescribed form.

Our official signatures are relational, however, via the term-unwinding algorithm, we can also accommodate signatures with functions.

## *A.2 Translations and Interpretations*

We present the notion of *m-dimensional interpretation without parameters*. There are two extensions of this notion: we can consider piecewise interpretations and we can add parameters. We will not treat these extensions in this chapter.

Consider two signatures $\Sigma$ and $\Theta$. An $m$-dimensional translation $\tau : \Sigma \to \Theta$ is a quadruple $\langle \Sigma, \delta, \mathcal{F}, \Theta \rangle$, where $\delta(v_0, \ldots, v_{m-1})$ is a $\Theta$-formula and where for any $n$-ary predicate $P$ of $\Sigma$, $\mathcal{F}(P)$ is a formula $A(\vec{v}_0, \ldots, \vec{v}_{n-1})$ in the language of signature $\Theta$, where $\vec{v}_i = v_{i0}, \ldots, v_{i(m-1)}$. Both in the case of $\delta$ and $A$ all free variables are among the variables shown. Moreover, if $i \neq j$ and $k \neq \ell$, then $v_{ik}$ is syntactically different from $v_{j\ell}$.

We demand that we have $\vdash \mathcal{F}(P)(\vec{v}_0, \ldots, \vec{v}_{n-1}) \to \bigwedge_{i<n} \delta(\vec{v}_i)$. Here $\vdash$ is provability in predicate logic. This demand is inessential, but it is convenient to have.

We define $B^\tau$ as follows:

- $(P(x_0, \ldots, x_{n-1}))^\tau := \mathcal{F}(P)(\vec{x}_0, \ldots, \vec{x}_{n-1})$.
- $(\cdot)^\tau$ commutes with the propositional connectives.
- $(\forall x\, A)^\tau := \forall \vec{x}\, (\delta(\vec{x}) \to A^\tau)$.
- $(\exists x\, A)^\tau := \exists \vec{x}\, (\delta(\vec{x}) \wedge A^\tau)$.

There are two worries about this definition. First, what variables $\vec{x}_i$ on the side of the translation $A^\tau$ correspond with $x_i$ in the original formula $A$? The second worry is that substitution of variables in $\delta$ and $\mathcal{F}(P)$ may cause variable clashes. These worries are never important in practice: we choose 'suitable' sequences $\vec{x}$ to correspond to variables $x$, and we avoid clashes by $\alpha$-conversions. However, if we want to give precise definitions of translations and, for example, of composition of translations, these problems come into play. These problems are clearly solvable, but they are beyond the scope of this chapter.

We allow identity to be translated to a formula that is not identity. There are several important operations on translations.

- $\mathsf{id}_\Sigma$ is the identity translation. We take $\delta_{\mathsf{id}_\Sigma}(v) := v = v$ and $\mathcal{F}(P) := P(\vec{v})$.
- We can compose translations. Suppose $\tau : \Sigma \to \Theta$ and $\nu : \Theta \to \Lambda$. Then $\nu \circ \tau$ or $\tau\nu$ is a translation from $\Sigma$ to $\Lambda$. We define:
  - $\delta_{\tau\nu}(\vec{v}_0, \ldots, \vec{v}_{m_\tau - 1}) := \bigwedge_{i<m_\tau} \delta_\nu(\vec{v}_i) \wedge (\delta_\tau(v_0, \ldots, v_{m_\tau-1}))^\nu$.
  - $P_{\tau\nu}(\vec{v}_{0,0}, \ldots, \vec{v}_{0,m_\tau-1}, \ldots \vec{v}_{n-1,0}, \ldots, \vec{v}_{n-1,m_\tau-1}) :=$
    $\bigwedge_{i<n, j<m_\tau} \delta_\nu(\vec{v}_{i,j}) \wedge (P(v_0, \ldots, v_{n-1})^\tau)^\nu$.
- Let $\tau, \nu : \Sigma \to \Theta$ and let $A$ be a sentence of signature $\Theta$. We define the disjunctive translation $\sigma := \tau\langle A\rangle\nu : \Sigma \to \Theta$ as follows. We take $m_\sigma := \mathsf{max}(m_\tau, m_\nu)$. We write $\vec{v} \restriction n$, for the restriction of $\vec{v}$ to the first $n$ variables, where $n \leq \mathsf{length}(\vec{v})$.
  - $\delta_\sigma(\vec{v}) := (A \wedge \delta_\tau(\vec{v} \restriction m_\tau)) \vee (\neg A \wedge \delta_\nu(\vec{v} \restriction m_\nu))$.
  - $P_\sigma(\vec{v}_0, \ldots, \vec{v}_{n-1}) := (A \wedge P_\tau(\vec{v}_0 \restriction m_\tau, \ldots, \vec{v}_{n-1} \restriction m_\tau)) \vee$
    $(\neg A \wedge P_\nu(\vec{v}_0 \restriction m_\nu, \ldots, \vec{v}_{n-1} \restriction m_\nu))$

Note that in the definition of $\tau\langle A\rangle\nu$ we used a padding mechanism. In the case, for example, $m_\tau < m_\nu$, the variables $v_{m_\tau}, \ldots, v_{m_\nu-1}$ are used 'vacuously' when we have $A$. If we had piecewise interpretations, where domains are built up from pieces with possibly different dimensions, we could avoid padding by building the domain of disjoint pieces with different dimensions.

A translation relates signatures; an interpretation relates theories. An interpretation $K : U \to V$ is a triple $\langle U, \tau, V\rangle$, where $U$ and $V$ are theories and $\tau : \Sigma_U \to \Sigma_V$. We demand: for all axioms $A$ of $U$, we have $V \vdash A^\tau$. Here are some further definitions.

- $\mathsf{ID}_U : U \to U$ is the interpretation $\langle U, \mathsf{id}_{\Sigma_U}, U\rangle$.
- Suppose $K : U \to V$ and $M : V \to W$. Then, $KM := M \circ K : U \to W$ is $\langle U, \tau_M \circ \tau_K, W\rangle$.
- Suppose $K : U \to (V + A)$ and $M : U \to (V + \neg A)$. Then $K\langle A\rangle M : U \to V$ is the interpretation $\langle U, \tau_K\langle A\rangle\tau_M, V\rangle$. In an appropriate category $K\langle A\rangle M$ is a special case of a product.

The notation $K : U \to V$ is inspired by the idea of interpretations as arrows in a category. There is also an intuition of interpretability as a generalization of provability. The traditional notations and notions associated with this intuition are:

- $K : U \lhd V$ stands for $K : U \to V$.

- $K : V \rhd U$ stands for $K : U \to V$.
- $U \lhd V$ stands for $\exists K\ K : U \lhd V$. We say: $U$ is *interpretable* in $V$.
- $V \rhd U$ stands for $\exists K\ K : V \rhd U$. We say: $V$ *interprets* $U$.
- $U \equiv V$ stands for $U \rhd V$ and $V \rhd U$. We say: $V$ and $U$ are *mutually interpretable.*

## A.3 Sequential Theories

A sequential theory provides an interpretation $N$ of a weak number theory, say $\mathsf{S}^1_2$, and sequences of all objects of the domain of the theories with projections in $N$. We can use these sequences to develop partial satisfaction predicates. Using these we can prove restricted consistency statements of $U$ in $U$.

The notion of sequential theory has a very simple definition, discovered by Pavel Pudlák. We first need the definition of a very weak set theory. The theory adjunctive set theory, or AS, is a one-sorted theory with a binary relation $\in$.

AS1 $\vdash \exists x\, \forall y\; y \notin x$,
AS2 $\vdash \forall x, y\, \exists z\, \forall u\, (u \in z \leftrightarrow (u \in x \vee u = y))$.

We note that we do not demand extensionality. For example, in AS we could have lots of 'empty sets'.

An interpretation is *direct* iff it is one-dimensional, unrelativized (that is, it has the trivial domain) and identity preserving (that is, it translates identity to identity).

A theory $U$ is sequential iff it directly interprets AS. By a substantial bootstrap, we can define, in a sequential theory $U$, an interpretation $N$ of a weak number theory, sequences of all objects, etc.

For details see, for example, [Pud83], [Pud85], [MPS90], [HP93], [Vis09a], and [Vis13].

We can generalize the notion of sequentiality a bit to *poly-sequentiality* by replacing *direct interpretation* in the definition by its obvious generalization to the $m$-dimensional case.

## A.4 Complexity Measures

In sequential theories we can define partial satisfaction predicates for formulas with complexity below $n$, for any $n$. The presence of these predicates has as a consequence that, for any sequential theory $U$ and for any $n$, we can find an interpretation $N$ of a weak arithmetic like Buss' $\mathsf{S}^1_2$ in $U$ such that $U \vdash \mathsf{con}^N_n(U)$. See, for example, [Vis93] for more details. We give the relevant definitions of complexity notions.

*Restricted provability* plays an important role in this chapter. An $n$-proof is a proof from axioms with Gödel number smaller than or equal to $n$ only involving formulas of complexity smaller than or equal to $n$. To work conveniently with this notion, a good complexity measure is needed. This should satisfy three conditions: (i) eliminating terms in favour of a relational formulation should raise the complexity only by a fixed standard number; (ii) translation of a formula via the translation corresponding to an interpretation $K$ should raise the complexity of the formula by a fixed standard number depending only on $K$; (iii) the tower of exponents involved in cut-elimination should be of height linear in the complexity of the formulas involved in the proof.

Such a good measure of complexity together with a verification of desideratum (iii)—a form of nesting degree of quantifier alternations—is supplied in the work of Philipp Gerhardy. See [Ger03] and [Ger05]. It is also provided by Samuel Buss in his preliminary draft [Bus15]. Buss also proves that (iii) is fulfilled.

Gerhardy's measure corresponds to the following formula classes:

- $\mathsf{AT}$ is the class of atomic formulas.
- $\mathsf{N}^\star_{-1} = \Sigma^\star_{-1} = \Pi^\star_{-1} := \emptyset$.
- $\mathsf{N}^\star_n ::= \mathsf{AT} \mid \neg\mathsf{N}^\star_n \mid (\mathsf{N}^\star_n \wedge \mathsf{N}^\star_n) \mid (\mathsf{N}^\star_n \vee \mathsf{N}^\star_n) \mid (\mathsf{N}^\star_n \to \mathsf{N}^\star_n) \mid \forall\,\Pi^\star_n \mid \exists\,\Sigma^\star_n$.
- $\Sigma^\star_n ::= \mathsf{AT} \mid \neg\Pi^\star_n \mid (\mathsf{N}^\star_{n-1} \wedge \mathsf{N}^\star_{n-1}) \mid (\Sigma^\star_n \vee \Sigma^\star_n) \mid (\Pi^\star_n \to \Sigma^\star_n) \mid \forall\,\Pi^\star_{n-1} \mid \exists\,\Sigma^\star_n$.
- $\Pi^\star_n ::= \mathsf{AT} \mid \neg\Sigma^\star_n \mid (\Pi^\star_n \wedge \Pi^\star_n) \mid (\mathsf{N}^\star_{n-1} \vee \mathsf{N}^\star_{n-1}) \mid (\mathsf{N}^\star_{n-1} \to \mathsf{N}^\star_{n-1}) \mid \forall\,\Pi^\star_n \mid \exists\,\Sigma^\star_{n-1}$.

We may define $\rho(A)$ as the minimal $n$ such that $A$ is in $\mathsf{N}^\star_n$.[16]

Samuel Buss gives the following formula classes.

- $\Sigma^*_0 = \Pi^*_0 =$ the class of quantifier-free formulas.
- $\Sigma^*_n ::= \Sigma^*_{n-1} \mid \Pi^*_{n-1} \mid \neg\,\Pi^*_n \mid (\Sigma^*_n \wedge \Sigma^*_n) \mid (\Sigma^*_n \vee \Sigma^*_n) \mid (\Pi^*_n \to \Sigma^*_n) \mid \exists\,\Sigma^*_n$.
- $\Pi^*_n ::= \Sigma^*_{n-1} \mid \Pi^*_{n-1} \mid \neg\,\Sigma^*_n \mid (\Pi^*_n \wedge \Pi^*_n) \mid (\Pi^*_n \vee \Pi^*_n) \mid (\Sigma^*_n \to \Pi^*_n) \mid \forall\,\Pi^*_n$.

We may define $\rho(A)$ as the smallest $n$ such that $A$ is in $\Sigma^*_n$. This is the same measure, as was employed in [Vis93]. For our purposes it does not matter whether we use Gerhardy's of Buss' definition.

We use $\mathsf{proof}_{U,n}$ for the proof predicate where only $U$-axioms with Gödel numbers $\leq n$ are allowed and where the formulas occurring in the proof are in the complexity class $\Gamma_n$ of all formulas of complexity $\leq n$. Similarly we use $U \vdash_n A$, $\mathsf{con}_n(U)$, $\Box_{U,m}A$, etc.

# 4

# Iterated Definability, Lawless Sequences, and Brouwer's Continuum

JOAN RAND MOSCHOVAKIS

## 4.1 Introduction

Gödel argued in [2] that the unprovability, in any consistent recursively enumerable extension of the arithmetic of addition and multiplication, of the consistency of that system strongly suggests that "the human mind ... infinitely surpasses the powers of any finite machine." Implicit in this argument is the assumption that arithmetic is in fact consistent, and that the human mind can recognize this fact.

Brouwer, for whom all mathematics was open ended, probably would have said that the consistency of intuitionistic arithmetic is a trivial consequence of its truth. Gödel's negative translation showed that "*the system of intuitionistic arithmetic and number theory is only apparently narrower than the classical one, and in truth contains it, albeit with a somewhat deviant interpretation.*"[1]

Formally, intuitionistic arithmetic is a proper subsystem of classical arithmetic. But intuitionistic analysis **FIM** includes a continuity principle incompatible with classical predicate logic. Kleene's function-realizability interpretation, a recursive implementation of the Brouwer–Heyting–Kolmogorov explanation of the intuitionistic connectives and quantifiers, proves **FIM** consistent relative to its classically correct subsystem **B**; so intuitionistic and classical analysis are equiconsistent.

The negative interpretation does not extend to analysis because the negative translation of the axiom of countable choice is not an intuitionistic principle. It is consistent with **FIM** by realizability (using a classical argument), and so classical analysis **R** has a "somewhat deviant" negative interpretation in a consistent extension of **FIM**.

Since classical arithmetic is consistent with **FIM**, one might hope for a different consistent extension of intuitionistic analysis containing classical analysis (not just its negative interpretation) as a part. In [4] Kleene defended his choice to formalize **FIM** using only variables over numbers and over arbitrary choice sequences, since the general recursive functions were the only constructive functions he needed and those could be coded

*Gödel's Disjunction*. First Edition. Leon Horsten & Philip Welch (eds).

by numbers. For a quasi-intuitionistic system including classical analysis a third sort of variable is indicated, as even the class of all hyperarithmetical sequences fails to satisfy the (classically and intutionistically correct) bar theorem, and if every choice sequence were to be a classical object the continuity principle would fail.

Brouwer's notion of "lawlike sequence" (choice sequence all of whose values are determined in advance) provides a solution. The lawlike sequences can play the role of the classical objects within the intuitionistic continuum, and this role can be formalized in a three-sorted axiomatic system **FIRM**, containing **R** and **FIM** as two-sorted subsystems, with extensional versions of Kreisel's "lawless sequence" axioms for a class of choice sequences defined in terms of the lawlike sequences. Consistency is established for a stronger system **FIRM**($\prec$) assuming an ordinal-definable, definably well-ordered collection of lawlike sequences is countable. For an intuitionist, the assumption that there are at most countably many lawlike sequences needs no justification.

## 4.2 Choice Sequences

### *4.2.1 Brouwer's Continuum*

For the intuitionistic theory of the continuum, Brouwer's concept of "choice sequence" replaces the classical notion of an arbitrary completed infinite sequence of natural numbers. Choice sequences are generated by more or less freely choosing one integer after another. At each stage, the chooser may or may not specify restrictions (consistent with those already made) on future choices.

Since the first $n$ values of a choice sequence $\alpha$ may be the only information available at the $n$th stage of its construction, every function defined on *all* choice sequences must be continuous, so for example if $A(\alpha)$ is $\forall x(\alpha(x) = 0)$, then $\neg\forall\alpha(A(\alpha) \vee \neg A(\alpha))$ holds.

Any choice sequence *all* of whose values are determined in advance according to some fixed law is called "lawlike" or a "sharp arrow." Brouwer did not specify which laws may determine lawlike sequences, but evidently every lawlike sequence is definable in some sense.

### *4.2.2 The Problem of Defining "Definability"*

Considering whether or not it is possible to define the concept of absolute mathematical definability, Gödel [3] observed that constructibility is not a "satisfactory formulation of definability" because "quantification is admitted only with respect to constructible sets and not with respect to sets in general." Instead he proposed "definability by expressions containing names of ordinal numbers and logical constants, including quantification referring to sets," adding

> ... "definability in terms of ordinals" even if it is not an adequate formulation for "comprehensibility by our mind", is at least an adequate formulation in an absolute sense for a closely related property ..., namely the property of "being formed according to a law" as opposed to "being formed by a random choice of the elements".

This idea can be adapted to define "lawlike sequence" by iterating definability, with quantification over numbers and lawlike sequences and (restricted) quantification over arbitrary choice sequences, to build a classical model of a three-sorted theory. The integers of the model are the standard integers, the arbitrary choice sequences of the model are all the classical one-place number-theoretic functions, and the lawlike sequences of the model form a definably well-ordered subclass of the choice sequences.

### 4.2.3 *"Lawlike" versus "Lawless" Sequences*

Kreisel [7] proposed axioms for numbers, "constructive" functions, and intensionally "lawless" sequences in which *"the simplest kind of restriction on restrictions is made, namely some finite initial segment of values is prescribed, and, beyond this, no restriction is to be made."* His constructive functions satisfy countable and dependent choice. Equality of lawless sequences is decidable. Distinct lawless sequences are independent and satisfy the axiom of open data and a strong form of continuous choice ("the only problematic axiom"). There are no variables over arbitrary choice sequences.

Troelstra [13] pointed out that Kreisel's problematic axiom entailed an "extension principle": every continuous number-valued functional defined on the lawless sequences has a continuous extension to all sequences. To justify his extension principle philosophically, he introduced a conceptual abstraction operator allowing *any* choice sequence to be viewed as lawless. He also filled in the details of Kreisel's proof that every sentence of his two-sorted language is formally equivalent to one without lawless sequence variables, so "lawless sequences can be regarded as a figure of speech." Troelstra used the term "lawlike" instead of "constructive," and noted that the lawlike sequence variables *may* be interpreted as ranging over "the classical universe of sequences" ([13] p. 4).

Kreisel's notion of lawless sequence collapsed because he identified independence with inequality. The collapse can be avoided by taking $\alpha, \beta$ to be independent lawless sequences if and only if their fair merge $[\alpha, \beta]$ is lawless. Then Kreisel's axioms can be forced to hold by defining "lawless" relative to a given well ordered collection of lawlike sequences whose intended interpretation is countable.

## 4.3 The Formal Systems RLS($\prec$) and FIRM($\prec$)

**RLS**($\prec$) is the system for which a classical model will be obtained by iterating definability, and **FIRM**($\prec$) is the quasi-intuitionistic extension of **RLS**($\prec$) whose consistency follows by realizability.

### 4.3.1 *The Three-Sorted Language $\mathcal{L}(\prec)$*

The language, extending the two-sorted language of [6] and [5], contains three sorts of variables with or without subscripts, also used as metavariables:

$i, j, k, \ldots, p, q, w, x, y, z$ over natural numbers,
$a, b, c, d, e, g, h$ over lawlike sequences,
$\alpha, \beta, \gamma, \ldots$ over arbitrary choice sequences;

finitely many constants $f_0 = 0$, $f_1 = {}'$ (successor), $f_2 = +$, $f_3 = \cdot$, $f_4 = \exp$, $f_5, \ldots, f_p$ for primitive recursive functions and functionals; two binary predicate constants $=, \prec$; Church's $\lambda$ denoting function abstraction; parentheses (,) denoting function application; and the logical symbols $\&, \vee, \rightarrow, \neg$ and quantifiers $\forall, \exists$ over each sort of variable.

*Terms* (of type 0) and *functors* (of type 1) are defined inductively. Number variables and 0 are terms. Sequence variables of both sorts, and unary function constants, are functors. If $f_i$ is a $k_i, m_i$-ary function constant, $u_1, \ldots, u_{k_i}$ are functors and $t_1, \ldots, t_{m_i}$ are terms, then $f_i(u_1, \ldots, u_{k_i}, t_1, \ldots, t_{m_i})$ is a term. If $u$ is a functor and $t$ is a term then $(u)(t)$ is a term. If $t$ is a term then $\lambda x(t)$ (also written $\lambda x.t$) is a functor.

*Prime formulas* are of the form $s = t$ where $s, t$ are terms, or $u \prec v$ where $u, v$ are functors. If $u, v$ are functors then $u = v$ abbreviates $\forall x u(x) = v(x)$. Composite formulas are formed as usual, with parentheses determining scopes.

Terms and functors with no occurrences of arbitrary choice sequence variables are *R-terms* and *R-functors* respectively. Formulas with no free occurrences of arbitrary choice sequence variables are *R-formulas.*

### 4.3.2 Axioms and Rules for Three-Sorted Intuitionistic Predicate Logic

The logical basis is intuitionistic three-sorted predicate logic, as in [6], but with additional rules and axiom schemas for the lawlike sequence variables:

$9^R$. $C \rightarrow A(b) \,/\, C \rightarrow \forall b A(b)$ if $b$ is not free in $C$.

$10^R$. $\forall b A(b) \rightarrow A(u)$ if $u$ is an $R$-functor free for $b$ in $A(b)$.

$11^R$. $A(u) \rightarrow \exists b A(b)$ if $u$ is an $R$-functor free for $b$ in $A(b)$.

$12^R$. $A(b) \rightarrow C \,/\, \exists b A(b) \rightarrow C$ if $b$ is not free in $C$.

$\forall a \exists! \beta (\forall x\, a(x) = \beta(x))$ is a theorem, so every lawlike sequence is a choice sequence.[2]

### 4.3.3 Axioms for Three-sorted Intuitionistic Number Theory

Equality axioms assert that $=$ is an equivalence relation and $x = y \rightarrow \alpha(x) = \alpha(y)$ (so $x = y \rightarrow a(x) = a(y)$). For terms $r(x), t$ with $t$ free for $x$ in $r(x)$, the $\lambda$-reduction schema is

$$(\lambda x.r(x))(t) = r(t),$$

where $r(t)$ is the result of substituting $t$ for all free occurrences of $x$ in $r(x)$.

The mathematical axioms include the assertions that 0 ($= f_0$) is not a successor and the successor function ($= f_1$) is one-to-one, the defining equations for the primitive recursive function and functional constants $f_2, \ldots, f_p$ ([6], [5]) and the mathematical induction schema extended to $\mathcal{L}(\prec)$. For the countable axiom of choice

$$\text{AC}_{01}. \quad \forall x \exists \alpha A(x, \alpha) \rightarrow \exists \alpha \forall x A(x, \lambda y.\alpha(2^x \cdot 3^y))$$

the $x$ must be free for $\alpha$ in $A(x, \alpha)$.

Finite sequences are coded primitive recursively as in [6]. Let $\langle x_0, \ldots, x_k \rangle = \Pi_0^k p_i^{x_i}$, where $p_i$ is the $i$th prime with $p_0 = 2$, and $(y)_i$ be the exponent of $p_i$ in the prime factorization of $y$. Let $Seq(y)$ abbreviate $\forall i < lh(y)\ (y)_i > 0$, where $lh(y)$ is the number of non-zero exponents in the prime factorization of $y$. If $k \geq 0$ then $\langle x_0 + 1, \ldots, x_k + 1 \rangle$ codes the finite sequence $(x_0, \ldots, x_k)$, and $\langle\,\rangle = 1$ codes the empty sequence. If $Seq(y)$ and $Seq(z)$ then $y * z$ codes the concatenation of the sequences coded by $y$ and $z$.

The finite initial segment of length $n$ of a choice sequence $\alpha$ is coded by $\overline{\alpha}(n)$, where $\overline{\alpha}(0) = 1$ and $\overline{\alpha}(n+1) = \langle \alpha(0)+1, \ldots, \alpha(n)+1 \rangle$. Other useful abbreviations are $\alpha \in w$ for $\overline{\alpha}(lh(w)) = w$, and $w \in \overline{\alpha}$ for $\exists n(w = \overline{\alpha}(n))$. If $Seq(w)$ then $w * \alpha = \beta$ where $\beta \in w$ and $\beta(lh(w) + n) = \alpha(n)$.

### 4.3.4 *Lawless Sequences, Restricted Quantification, and Lawlike Comprehension*

Intuitively, a lawless sequence should not be predictable by any lawlike process, but this negative condition is not enough to satisfy Kreisel's axioms. Instead, call a sequence $\beta$ a *predictor* if $\forall w(Seq(w) \rightarrow Seq(\beta(w)))$, and call a choice sequence $\alpha$ *lawless* if every lawlike predictor correctly predicts $\alpha$ somewhere. Since each prediction affects only finitely many values, this positive condition leaves room for (indeed, guarantees) plenty of chaotic behavior *if there are only countably many lawlike predictors.* From the intuitionistic point of view this assumption needs no justification; but there is no lawlike enumeration of the lawlike sequences.

Formally, let $RLS(\alpha)$ abbreviate

$$\forall b[\forall w(Seq(w) \rightarrow Seq(b(w))) \rightarrow \exists x\, \alpha \in \overline{\alpha}(x) * b(\overline{\alpha}(x))].$$

Note that Troelstra's extension principle fails, since $\forall \alpha(RLS(\alpha) \rightarrow \exists n \alpha(n) = 1)$, but the function assigning to each lawless $\alpha$ the least $n$ such that $\alpha(n) = 1$ cannot be extended continuously to all choice sequences.

Two lawless sequences $\alpha, \beta$ are *independent* if their fair merge $[\alpha, \beta]$ is lawless, and similarly for $\alpha_0, \ldots, \alpha_k$, where $[\alpha_0, \ldots, \alpha_k]((k+1)n + i) = \alpha_i(n)$ for $0 \leq i \leq k$ and all $n$. A *restricted* formula $E$ has arbitrary choice sequence quantifiers only in subformulas of the forms $\forall \alpha_0(RLS([\alpha_0, \ldots, \alpha_k]) \rightarrow A)$ and $\exists \alpha_0(RLS([\alpha_0, \ldots, \alpha_k])\ \&\ A)$, where no arbitrary choice sequence variables but $\alpha_0, \ldots, \alpha_k$ occur free in $A$.

There is a lawlike function-comprehension schema,

$$\mathrm{AC}_{00}^{R}!. \quad \forall x \exists ! y A(x,y) \rightarrow \exists b \forall x A(x, b(x)),$$

where $A(x, y)$ is any restricted $R$-formula and $b$ is free for $y$ in $A(x, y)$. By this axiom, the lawlike sequences are closed under "recursive in."

### 4.3.5 *Axioms for Lawless Sequences*

These are Kreisel's and Troelstra's axioms from [7] and [13], adapted to Kleene's convention for coding continuous functions, with inequality of lawless sequences replaced by independence. There are two *density axioms*:

$$\text{RLS1.}\quad \forall w[Seq(w) \rightarrow \exists\alpha(RLS(\alpha)\ \&\ \alpha \in w)].$$

$$\text{RLS2.}\quad \forall\alpha[RLS(\alpha) \rightarrow \forall w[Seq(w) \rightarrow \exists\beta(RLS([\alpha,\beta])\ \&\ \beta \in w)]].$$

Kreisel's principle of *open data* is stated as follows, on condition that $A(\alpha)$ is restricted and has no other arbitrary choice sequence variables free, and $\beta$ is free for $\alpha$ in $A(\alpha)$:

$$\text{RLS3.}\quad \forall\alpha[RLS(\alpha) \rightarrow (A(\alpha) \rightarrow \exists y\forall\beta(RLS(\beta)\ \&\ \beta \in \overline{\alpha}(y) \rightarrow A(\beta)))].$$

*Effective continuous choice for lawless sequences* is the schema

$$\text{RLS4.}\quad \forall\alpha[RLS(\alpha) \rightarrow \exists bA(\alpha,b)] \rightarrow \exists e\exists b\forall\alpha[RLS(\alpha) \rightarrow$$
$$\exists!ye(\overline{\alpha}(y)) > 0\ \&\ \forall y(e(\overline{\alpha}(y)) > 0 \rightarrow A(\alpha, \lambda x.\, b(\langle e(\overline{\alpha}(y)) \dot{-} 1, x\rangle)))],$$

where $A(\alpha, b)$ is restricted with no arbitrary choice sequence variables but $\alpha$ free, and $e, y, \alpha$ are free for $b$ in $A(\alpha, b)$.

### 4.3.6 *Well-Ordering the Lawlike Sequences*

Axioms W0-4 assert that the lawlike sequences are well-ordered by $\prec$, and W5 weakly specifies the domain of $\prec$:

$$\text{W0.}\quad \alpha = \beta\ \&\ \alpha \prec \gamma \rightarrow \beta \prec \gamma \quad \text{and} \quad \beta = \gamma\ \&\ \alpha \prec \beta \rightarrow \alpha \prec \gamma.$$

$$\text{W1.}\quad \forall a\forall b[a \prec b \rightarrow \neg b \prec a].$$

$$\text{W2.}\quad \forall a\forall b\forall c[a \prec b\ \&\ b \prec c \rightarrow a \prec c].$$

$$\text{W3.}\quad \forall a\forall b[a \prec b \vee a = b \vee b \prec a].$$

$$\text{W4.}\quad \forall a[\forall b(b \prec a \rightarrow A(b)) \rightarrow A(a)] \rightarrow \forall aA(a),$$

where $A(a)$ is any restricted $R$-formula in which $b$ is free for $a$.

$$\text{W5.}\quad \alpha \prec \beta \rightarrow \neg\forall a\forall b\neg(\alpha = a\ \&\ \beta = b).$$

The double negation in W5 is essential. Replacing $\neg\forall a\forall b\neg$ by $\exists a\exists b$ would destroy the realizability model.

### 4.3.7 *Restricted LEM, the Axiom of Closed Data and Lawlike Countable Choice*

The final axiom schema for **RLS**($\prec$) is the *restricted law of the excluded middle*:

$$\text{RLEM.}\quad \forall\alpha(RLS(\alpha) \rightarrow A(\alpha) \vee \neg A(\alpha))$$

for $A(\alpha)$ restricted and with no other arbitrary choice sequence variables free. Lawless sequences exist by RLS1, so $A \vee \neg A$ follows for $A$ without arbitrary choice sequence variables. Thus the logic of the two-sorted subsystem with only number and lawlike sequence variables, omitting W5 and with the $\alpha, \beta, \gamma$ in W0 and the axioms for three-sorted intuitionistic number theory replaced by $a, b, c$ respectively, is classical.

By an easy argument, RLS3 and the restricted LEM entail the following principle of *closed data* with the same restrictions on $A(\alpha)$ as for RLS3:

$$\text{RLS5.} \quad \forall\alpha[RLS(\alpha) \to (\forall y \exists \beta(RLS(\beta)\ \&\ \beta \in \overline{\alpha}(y)\ \&\ A(\beta)) \to A(\alpha))].$$

In an intuitionistic subsystem obtained by omitting RLEM, RLS5 may be taken as an additional axiom schema.

For restricted $R$-formulas $A(x, a)$, the lawlike comprehension schema entails

$$\text{AC}^R_{01}!. \quad \forall x \exists ! a A(x, a) \to \exists b \forall x A(x, \lambda y.b(2^x \cdot 3^y)),$$

with the obvious conditions on the variables. By the well-ordering axioms, with the same conditions, this schema can be strengthened to lawlike countable choice,

$$\text{AC}^R_{01}. \quad \forall x \exists a A(x, a) \to \exists b \forall x A(x, \lambda y.b(2^x \cdot 3^y)).$$

*Lawlike classical analysis* is the subsystem **R** of **RLS**$(\prec)$ obtained by restricting the language to number and lawlike sequence variables as indicated, omitting $\prec$ and its axioms, and replacing RLEM by $A \vee \neg A$ and $\text{AC}^R_{00}!$ by $\text{AC}^R_{01}$.

### 4.3.8 *Brouwer's Bar Theorem and Troelstra's Generalized Continuous Choice*

Two axiom schemas, in addition to those already stated (in the two-sorted language $\mathcal{L}_{\text{FIM}}$ without $\prec$ and with number and arbitrary choice sequence variables only), characterize Kleene and Vesley's full system **FIM** of intuitionistic two-sorted number theory in [6] and [5]. These specifically intuitionistic axiom schemas are Brouwer's "bar theorem" and strong continuous choice, stated now for $\mathcal{L}(\prec)$.

Kleene gave four versions of the bar theorem which are interderivable using $\text{AC}_{00}!$ (a consequence of $\text{AC}_{01}$). The version adopted here assumes a thin bar:

$$\text{BI!}. \quad \forall\alpha\exists!x R(\overline{\alpha}(x))\ \&\ \forall w(Seq(w)\ \&\ R(w) \to A(w))\ \&$$
$$\forall w(Seq(w)\ \&\ \forall n A(w * \langle n+1 \rangle) \to A(w)) \to A(\langle\,\rangle).$$

Kleene's strong version of Brouwer's principle of continuous choice is

$$\text{CC}_{11}. \quad \forall\alpha\exists\beta A(\alpha,\beta) \to \exists\sigma\forall\alpha(\forall n \exists! x \sigma(\langle n+1\rangle * \overline{\alpha}(x)) \neq 0\ \&$$
$$\forall\beta(\forall n \exists x \sigma(\langle n+1\rangle * \overline{\alpha}(x)) = \beta(n)+1 \to A(\alpha,\beta))),$$

which can be abbreviated by $\forall\alpha\exists\beta A(\alpha,\beta) \rightarrow \exists\sigma\forall\alpha(\{\sigma\}[\alpha]\downarrow \;\&\; A(\alpha,\{\sigma\}[\alpha]))$. The necessary restrictions on the bound variables are obvious and will not be stated.

Troelstra [12] found a strengthening of Brouwer's principle which precisely characterizes the function-realizability notion of [6]. Troelstra's *generalized continuous choice* schema can be abbreviated by

$$\mathrm{GC}_{11}.\quad \forall\alpha(A(\alpha) \rightarrow \exists\beta B(\alpha,\beta)) \rightarrow \exists\sigma\forall\alpha(A(\alpha) \rightarrow \{\sigma\}[\alpha]\downarrow \;\&\; B(\alpha,\{\sigma\}[\alpha])),$$

where $A(\alpha)$ must be *almost negative* (i.e. contain no $\vee$, and no $\exists$ except immediately before an equation between terms). The maximal system considered in this paper, with all schemas extended to the language $\mathcal{L}(\prec)$, is

$$\mathbf{FIRM}(\prec) = \mathbf{RLS}(\prec) + \mathrm{BI!} + \mathrm{GC}_{11}.$$

### *4.3.9 Classical and Intuitionistic Analysis as Subsystems of FIRM($\prec$)*

Four subsystems of the maximal system are of independent interest. The main theorem, that **FIRM**($\prec$) is consistent, depends on the assumption that its subsystem **RLS**($\prec$) has a classical model with just the standard integers and all the classical arbitrary choice sequences, but only countably many lawlike sequences.

The subsystem **FIRM** obtained by restricting the language to $\mathcal{L}$ (without $\prec$), dropping the axioms W0–W5, and replacing $\mathrm{AC}^R_{00}!$ by $\mathrm{AC}^R_{01}$ has lawlike classical analysis **R** (described in §4.3.7) and Kleene's intuitionistic analysis **FIM** and basic analysis **B** as two-sorted subsystems in the languages $\mathcal{L}_{\mathbf{R}}$ and $\mathcal{L}_{\mathbf{FIM}}$ $(= \mathcal{L}_{\mathbf{B}})$ respectively, where $\mathcal{L}_{\mathbf{R}}$ is obtained by excluding the arbitrary choice sequence variables from $\mathcal{L}$. For **FIM**, Troelstra's $\mathrm{GC}_{11}$ is replaced by its consequence $\mathrm{CC}_{11}$; **B** is the neutral subsystem of **FIM** with BI! but without continuous choice.

A classical version of the bar theorem for $\mathcal{L}_{\mathbf{R}}$, obtained from BI! by changing $\alpha$ to $a$ and dropping the !, is provable in **R**.[3] Except for the choice of Latin rather than Greek letters for sequence variables, **R** is Kleene's **B** with classical logic.

### *4.3.10 Consistency of FIRM($\prec$)*

This depends on two theorems and an assumption whose consistency follows from the work of Levy [8]. Proofs of these theorems are outlined in Sections 4.4 and 4.5.

**Theorem 1** *Assuming a certain definably well-ordered subset $\mathcal{R}$ of $\omega^\omega$ (determined by iterating definability by restricted R-formulas) is countable,* **RLS**($\prec$) *has a classical model $\mathcal{M}(\prec_{\mathcal{R}})$ with $\mathcal{R}$ as the class of lawlike sequences. The class $\mathcal{RLS}$ of lawless sequences of the model is Baire comeager in $\omega^\omega$, and has classical measure 0.*

**Theorem 2** *Under the same countability assumption,* **FIRM**($\prec$) *is consistent by a classical realizability interpretation. So the classical sequences can consistently be viewed as the lawlike elements of Brouwer's continuum, while the lawless sequences form a dense collection of choice sequences disjoint from the lawlike part.*

## 4.4 Construction of the Classical Model and Proof of Theorem 1

### 4.4.1 *Definability Over* $(A, \prec_A)$ *by a Restricted Formula of* $\mathcal{L}(\prec)$

If $F(a_0, \ldots, a_{k-1})$ is a restricted formula of $\mathcal{L}(\prec)$ of the form $\forall x \exists ! y E(x, y, a_0, \ldots, a_{k-1})$ where $x, y$ are all the distinct number variables free in $E$ and $a_0, \ldots, a_{k-1}$ are all the distinct variables free in $F$, in order of first free occurrence, and if $A \subseteq \omega^\omega$, $\prec_A$ is a binary relation on $A$, $\varphi \in \omega^\omega$ and $\psi_0, \ldots, \psi_{k-1} \in A$, then *$E$ defines $\varphi$ over $(A, \prec_A)$ from $\psi_0, \ldots, \psi_{k-1}$* if and only if, when number variables range over $\omega$, lawlike sequence variables range over $A$, choice sequence variables range over $\omega^\omega$, $\prec$ is interpreted by $\prec_A$, and $a_0, \ldots, a_{k-1}$ by $\psi_0, \ldots, \psi_{k-1}$:

(i) $F$ is true, and

(ii) for all $x, y \in \omega$: $\varphi(x) = y$ if and only if $E(\mathbf{x},\mathbf{y})$ is true.

Let **Def**$(A, \prec_A)$ be the class of all $\varphi \in \omega^\omega$ such that some restricted $R$-formula $E$ of $\mathcal{L}(\prec)$ defines $\varphi$ over $(A, \prec_A)$ from some $\psi_0, \ldots, \psi_{k-1}$ in $A$. Observe that $A \subseteq \mathrm{Def}(A, \prec_A)$, since $a(x) = y$ defines every $\varphi \in A$ over $A$ from itself.

For the intended application, $\prec_A$ will be a well-ordering of $A$ determined inductively with the help of a fixed enumeration $E_0(x, y), E_1(x, y), \ldots$ of all restricted R-formulas of $\mathcal{L}(\prec)$ containing free no-number variables but $x, y$, where $E_0(x, y) \equiv a(x) = y$. In the case that $A$ is non-empty, $\mathrm{Def}(A, \prec_A)$ will be well-ordered by an end extension $\prec_A^*$ of $\prec_A$, defined as follows. For $\varphi, \theta \in \mathrm{Def}(A, \prec_A)$, set $\varphi \prec_A^* \theta$ if and only if $\Delta_A(\varphi) < \Delta_A(\theta)$, where $\Delta_A(\varphi)$ is the smallest tuple $(i, \psi_0, \ldots, \psi_{k-1})$ in the lexicographic ordering $<$ of $\bigcup_{k>o}(\omega \times A^k)$ determined by $<$ on $\omega$ and $\prec_A$ on $A$ such that $E_i$ defines $\varphi$ over $(A, \prec_A)$ from $\psi_0, \ldots, \psi_{k-1}$. Then $\prec_A^*$ well-orders $\mathrm{Def}(A, \prec_A)$; and if $\varphi \in A$ then $\Delta_A(\varphi) = (0, \varphi)$, so $\prec_A$ is an initial segment of $\prec_A^*$. The case $A = \phi$ will be treated separately.

### 4.4.2 *The Classical Model* $\mathcal{M}(\prec_{\mathcal{R}})$

The next step is to define a structure $\mathcal{M}(\prec_{\mathcal{R}})$ for $\mathcal{L}(\prec)$ by iterating the process described in Section 4.4.1. More accurately, $\mathcal{M}(\prec_{\mathcal{R}}) = (\omega, \mathcal{R}, \omega^\omega, f_0, \ldots, f_p, =, \prec_{\mathcal{R}})$, where $f_0, \ldots, f_p$ are the standard interpretations of the primitive recursive functional constants $\mathrm{f}_0, \ldots, \mathrm{f}_p$ respectively, $=$ is equality of natural numbers, and $\mathcal{R}$ and $\prec_{\mathcal{R}}$ are defined as follows. The construction will guarantee that $\prec_{\mathcal{R}}$ is a well-ordering of $\mathcal{R}$.

As above, let $E_0(x, y), E_1(x, y), \ldots$ be a fixed enumeration of all restricted R-formulas in the language $\mathcal{L}(\prec)$ containing free no-number variables but $x$ and $y$, where $E_0(x, y) \equiv a(x) = y$. Begin by defining $\mathcal{R}_0 = \phi$, $\prec_0 = \phi$, and $\mathcal{R}_1 = \mathrm{Def}(\mathcal{R}_0 \prec_0)$.

Since $RLS(\alpha)$ always holds and $u \prec v$ always fails when the lawlike sequence variables range over the empty set and $\prec$ is the empty relation, every restricted $R$-formula of $\mathcal{L}(\prec)$ with only the distinct number variables $x, y$ free is equivalent, by an easy translation, over the structure $(\omega, \phi, \omega^\omega, f_0, \ldots, f_p, =, \phi)$ to a formula of the two-sorted language $\mathcal{L}_{\mathrm{FIM}}$, with only $x$ and $y$ free. So $\mathcal{R}_1$ consists of all analytic sequences.

If $\varphi$ is analytic, let $\Delta_0(\varphi)$ be the least $i \in \omega$ such that $E_i(x,y)$ is a formula of $\mathcal{L}_{\text{FIM}}$ with only $x, y$ free, and $E_i(x,y)$ defines $\varphi$. Then define $\varphi \prec_1 \psi$ if and only if $\varphi, \psi \in \mathcal{R}_1$ and $\Delta_0(\varphi) < \Delta_0(\psi)$. Evidently $\prec_1$ well-orders $\mathcal{R}_1$.

For $\zeta > 0$ define $\mathcal{R}_{\zeta+1} = \text{Def}(\mathcal{R}_\zeta, \prec_\zeta)$ and $\prec_{\zeta+1} = \prec^*_\zeta$. At limit ordinals take unions. By cardinality considerations there is a least ordinal $\eta_0$ (with cardinality $\leq 2^{\aleph_0}$) such that $\mathcal{R}_{\eta_0} = \mathcal{R}_{\eta_0+1}$. Let $\mathcal{R} = \mathcal{R}_{\eta_0}$ and $\prec_{\mathcal{R}} = \prec_{\eta_0}$.

By transfinite induction, $\prec_\zeta$ well-orders $\mathcal{R}_\zeta$ and is an initial segment of $\prec_{\zeta+1}$ for each $\zeta < \eta_0$, so $\prec_{\mathcal{R}}$ well-orders $\mathcal{R}$. Hence $\mathcal{M}(\prec_{\mathcal{R}})$, the natural classical model in which lawlike sequence variables range over the subset $\mathcal{R}$ of $\omega^\omega$, satisfies axioms W0–W5.[4]

Since $\mathcal{M}(\prec_{\mathcal{R}})$ is a classical model, it evidently satisfies RLEM. Since $\mathcal{M}(\prec_{\mathcal{R}})$ satisfies axiom schema $\text{AC}^R_{00}!$ by construction, the following lemmas hold.[5]

**Lemma 1** *If $\mathcal{M}(\prec_{\mathcal{R}})$ satisfies the axiom schema RLS3 of open data for a restricted formula $A(\alpha)$ with only $\alpha$ and a list $\Psi$ of lawlike sequence variables free, then $\mathcal{M}(\prec_{\mathcal{R}})$ satisfies RLS3 for $\neg A(\alpha)$.*

***Proof*** Recall that 0 is not a sequence code and 1 is the code of the empty sequence. Given $A(\alpha)$ with only $\alpha$ and the lawlike variables $\Psi$ free, define restricted $R$-formulas

$$J(x,y) \equiv (Seq(x) \rightarrow Seq(y))\ \&\ (y = 1 \rightarrow \forall\beta(RLS(\beta) \rightarrow (\beta \in x \rightarrow \neg A(\beta))))\ \&$$
$$(y > 1 \rightarrow \forall\beta(RLS(\beta) \rightarrow (\beta \in x * y \rightarrow A(\beta)))),$$

$$K(x,y) \equiv J(x,y)\ \&\ \forall z(z < y \rightarrow \neg J(x,z)).$$

Assume $\mathcal{M}(\prec_{\mathcal{R}})$ satisfies RLS3 for $A(\alpha)$. Then under any assignment $\mathit{\Psi}$ of elements of $\mathcal{R}$ to $\Psi$: $\mathcal{M}(\prec_{\mathcal{R}}) \models_{\mathit{\Psi}} \forall x \exists! y K(x,y)$ and so $K(x,y)$ defines a lawlike predictor $\pi \in \mathcal{R}$ with the property that $\mathcal{M}(\prec_{\mathcal{R}}) \models_{\mathit{\Psi},\pi} \forall x K(x, b(x))$. If $\alpha \in \mathcal{RLS}$ there is a least $n$ at which $\pi$ correctly predicts $\alpha$, so if $x = \overline{\alpha}(n)$ then $\alpha \in x * \pi(x)$ and $\mathcal{M}(\prec_{\mathcal{R}}) \models_{\mathit{\Psi},\pi} K(\mathbf{x}, b(\mathbf{x}))$. If $\mathcal{M}(\prec_{\mathcal{R}}) \models_{\mathit{\Psi},\alpha} \neg A(\alpha)$ then $\pi(x) = 1$; so $\mathcal{M}(\prec_{\mathcal{R}})$ satisfies RLS3 for $\neg A(\alpha)$ also. ■

**Lemma 2** *If $\alpha \in \mathcal{RLS}$, then (i) $w * \alpha \in \mathcal{RLS}$ for every finite sequence code $w$, (ii) $\alpha \circ \varphi \in \mathcal{RLS}$ for every injection $\varphi \in \mathcal{R}$ the characteristic function of whose range is also in $\mathcal{R}$, and (iii) if $\pi \in \mathcal{R}$ is a predictor and $n \in \omega$ then $\pi$ correctly predicts $\alpha$ at some $m \geq n$. Hence, any sequence obtained by changing or omitting finitely many values of any $\alpha \in \mathcal{RLS}$ is also in $\mathcal{RLS}$; if $[\alpha_1, \ldots, \alpha_k] \in \mathcal{RLS}$ then $\alpha_i \in \mathcal{RLS}$ for each $i \leq k$; and every predictor $\pi \in \mathcal{R}$ correctly predicts every $\alpha \in \mathcal{RLS}$ infinitely many times.*

***Proof*** Assume $\alpha \in \mathcal{RLS}$.

(i) Since $\mathcal{M}(\prec_{\mathcal{R}})$ satisfies $\text{AC}^R_{00}!$, for each finite sequence code $w$ and each predictor $\pi \in \mathcal{R}$ there is a predictor $\rho \in \mathcal{R}$ such that $\rho(y) = \pi(w * y)$ for each finite sequence code $y$, and $\rho$ correctly predicts $\alpha$ at some $n$, so $\pi$ correctly predicts $w * \alpha$ at $lh(w) + n$.

(ii) Let $\varphi, \psi, \pi \in \mathcal{R}$ where $\varphi$ is an injection, $\psi$ is the characteristic function of its range, and $\pi$ is a predictor. The problem is to show that $\pi$ correctly predicts $\alpha \circ \varphi$ somewhere. If $y$ codes a finite sequence, let $y \circ \varphi$ be the sequence code of length $n = \max\{i : \varphi(i) < lh(y)\} + 1$ such that for $i < n$: $(y \circ \varphi)_i = (y)_{\varphi(i)}$ if $\varphi(i) < lh(y)$, and $(y \circ \varphi)_i = 1$ otherwise. Let $z = (y \circ \varphi) * \pi(y \circ \varphi)$ and define $\rho(y)$ to be the sequence code of length $m = \max\{\varphi(i) : i < lh(z)\} + 1 \dot{-} lh(y)$ such that for $j < m$: $(\rho(y))_j = (z)_k$, where $\varphi(k) = j + lh(y)$ if there is such a $k$, otherwise $(\rho(y))_j = 1$. If $y$ is not a sequence code set $\rho(y) = 0$.

Then $\rho$ is a predictor and $\rho \in \mathcal{R}$, so $\rho$ correctly predicts $\alpha$ somewhere, so $\alpha \in y * \rho(y)$ for some finite sequence code $y$. To show: $\alpha \circ \varphi \in (y \circ \varphi) * \pi(y \circ \varphi) = z$ for this $y$. With $n, m$ as above, suppose $i < lh(z)$. Then either $\varphi(i) < lh(y)$, so $(\alpha \circ \varphi)(i) + 1 = \alpha(\varphi(i)) + 1 = (y)_{\varphi(i)} = (y \circ \varphi)_i = (z)_i$; or else $\varphi(i) = lh(y) + j$ where $j < m = lh(\rho(y))$, so $lh(y) \leq \varphi(i) < lh(y * \rho(y))$, so $(\alpha \circ \varphi)(i) + 1 = \alpha(\varphi(i)) + 1 = (y * \rho(y))_{\varphi(i)} = (\rho(y))_j = (z)_i$. So $z = \overline{\alpha \circ \varphi}(lh(z))$, so $\pi$ correctly predicts $\alpha \circ \varphi$ at $n$.

(iii) If $\pi$ is a lawlike predictor and $n \in \omega$, define $\rho(y) = \pi(\overline{\alpha}(n) * y)$ for all finite sequence codes $y$, and $\rho(y) = 0$ otherwise. Then $\rho$ is a lawlike predictor, so by (ii) (just established) $\rho$ correctly predicts $\lambda x.\alpha(n + x)$ at some $k$, so $\alpha \in \overline{\alpha}(n + k) * \pi(\overline{\alpha}(n + k))$, so $\pi$ correctly predicts $\alpha$ at $m = n + k$. ■

### 4.4.3 *Outline of the Proof of Theorem 1*

The theorem can be restated as follows.

Assume $\mathcal{R}$ is countable. Then

(a) The structure $\mathcal{M}(\prec_{\mathcal{R}})$ is a classical model of **RLS**($\prec$) with $\mathcal{R}$ as the class of lawlike sequences, and otherwise standard.

(b) The collection $\mathcal{RLS} = \{\alpha \in \omega^\omega \mid \mathcal{M}(\prec_{\mathcal{R}}) \models_\alpha RLS(\alpha)\}$ is disjoint from $\mathcal{R}$ and is Baire comeager in $\omega^\omega$, with classical measure 0.

Part of (a) has already been proved. For (b), the fact that $\mathcal{RLS}$ is disjoint from $\mathcal{R}$ is obvious, since if $\pi \in \mathcal{R}$ then $\sigma = \lambda x.\langle \pi(lh(x)) + 3\rangle \in \mathcal{R}$ is a lawlike predictor which never correctly predicts $\pi$. To prove that $\mathcal{M}(\prec_{\mathcal{R}})$ satisfies RLS1–RLS**4**, and to finish the proof of (b), an enumeration of $\mathcal{R}$ is needed.

Since $\mathcal{R}$ is an ordinal-definable subset of $\omega^\omega$ and $\prec_{\mathcal{R}}$ is an ordinal-definable well-ordering of $\mathcal{R}$, by [8] we may consistently assume that $\eta_0$, and hence $\mathcal{R}$, is countable. For the rest of this section, assume that $\Gamma : \omega \to \mathcal{R}$ is a bijection and $\chi : \omega \times \omega \to \{0, 1\}$ codes a well-ordering of type $\eta_0$ such that for all $n, m \in \omega$:

$$\Gamma(n) \prec_{\mathcal{R}} \Gamma(m) \Leftrightarrow \chi(n, m) = 1.$$

To show that $\mathcal{M}(\prec_{\mathcal{R}})$ satisfies RLS1, define sequences $\{w_n\}_{n\in\omega}$, $\{x_n\}_{n\in\omega}$ as follows:

$$w_0 = \langle\,\rangle = 1,$$
$$w_{n+1} = \begin{cases} w_n * ((\Gamma(n))(w_n)) & \text{if } (\Gamma(n))(w_n) \text{ codes a finite sequence,} \\ w_n & \text{otherwise.} \end{cases}$$
$$x_n = lh(w_n).$$

Since $\mathcal{R}$ contains infinitely many predictors, $\{x_n\}_n$ is cofinal in $\omega$ so there is a unique $\alpha \in \omega^\omega$ such that $\overline{\alpha}(x_n) = w_n$ for each $n$. Evidently this $\alpha \in \mathcal{RLS}$.

To show that $\mathcal{M}(\prec_{\mathcal{R}})$ satisfies RLS2, suppose that $\alpha \in \mathcal{RLS}$ and $w$ codes a finite sequence, and define $\overline{\beta}(lh(w)) = w$. Given any predictor $\pi \in \mathcal{R}$ and any prior choice $\overline{\beta}(n)$ with $n \geq lh(w)$, there are predictors $\varphi, \psi \in \mathcal{R}$ so that for finite sequence codes $y$:

$$\varphi(y) = \pi(\overline{[y * \lambda x.0, \overline{\beta}(n) * \lambda x.0]}(2 \cdot lh(y))),$$
$$\psi(y) = \Pi_{2i<lh(\varphi(y))} p_i^{(\varphi(y))_{2i}}.$$

By Lemma 2 there is a least $m > n$ such that $\alpha \in \overline{\alpha}(m) * \psi(\overline{\alpha}(m))$, and then $\varphi(\overline{\alpha}(m)) = \pi(\overline{[\alpha, \overline{\beta}(n) * \lambda x.0]}(2m))$. If $\overline{\beta}(n)$ is extended by setting $\beta(i) = 0$ for $n \leq i < m$ and $\beta(i) = (\varphi(\overline{\alpha}(m)))_{2i+1} \dot{-} 1$ for $2i + 1 < lh(\varphi(\overline{\alpha}(m)))$, then $\pi$ will correctly predict $[\alpha, \beta]$ at $2m$. Treating each of the countably many predictors $\pi \in \mathcal{R}$ in turn in this manner gradually produces a $\beta$ such that $[\alpha, \beta] \in \mathcal{RLS}$.

The proof that $\mathcal{M}(\prec_{\mathcal{R}})$ satisfies RLS3 is by induction on the logical form of $A(\alpha)$. The cases $s(\alpha) = t(\alpha)$ and $u \prec v$ ($u, v$ $R$-functors) are immediate. The case $u(\alpha) \prec v(\alpha)$ uses W5, so is treated after the inductive cases for $\forall a$, $\forall x$, and &. In addition to establishing the inductive case for $\neg$, since $\mathcal{M}(\prec_{\mathcal{R}})$ is a classical model, Lemma 1 gives the inductive cases for $\vee$, $\rightarrow$ and $\forall$ from those for & and $\exists$, simplifying the proof.

The proof that $\mathcal{M}(\prec_{\mathcal{R}})$ satisfies RLS4 uses the fact that $\mathcal{M}(\prec_{\mathcal{R}})$ satisfies RLS3, $\mathrm{AC}^R_{01}$, and RLEM (cf. [9]). This completes the proof of Theorem 1(a). The remaining parts of the proof of (b) are straightforward (cf. [9] and [10]).

## 4.5 The Γ-Realizability Interpretation

### 4.5.1 *Definitions*

Assume $\mathcal{R}$ is countable, with $\Gamma, \chi$ as in the proof of Theorem 1, and let $\Psi = x_1, \ldots, x_n, \alpha_1, \ldots, \alpha_k, a_1, \ldots, a_m$ be a list of distinct variables. Then a *Γ-interpretation* $\Psi$ of $\Psi$ is any choice of $n$ numbers, $k$ elements of $\omega^\omega$, and $m$ numbers $r_1, \ldots, r_m$; and $\Gamma[\Psi]$ is the corresponding list of $n$ numbers, $k$ elements of $\omega^\omega$, and $m$ elements $\Gamma(r_1), \ldots, \Gamma(r_m)$ of $\mathcal{R}$.

With the same assumptions on $\Gamma$ and $\chi$, the Γ-realizability interpretation of $\mathcal{L}(\prec)$ is defined as follows. For $\pi \in \omega^\omega$, $E$ a formula of $\mathcal{L}(\prec)$ with at most the distinct variables $\Psi$

free, and $\mathit{\Psi}$ a $\Gamma$-interpretation of $\Psi$, define $\pi$ $\Gamma$-*realizes*-$\mathit{\Psi}$ $E$ by induction on the logical form of $E$ using Kleene's coding of recursive partial functionals:[6]

(1) $\pi$ $\Gamma$-realizes-$\mathit{\Psi}$ a prime formula $P$, if $\mathcal{M}(\prec_{\mathcal{R}}) \models_{\Gamma[\mathit{\Psi}]} P$.
(2) $\pi$ $\Gamma$-realizes-$\mathit{\Psi}$ $A \,\&\, B$, if $(\pi)_0$ $\Gamma$-realizes-$\mathit{\Psi}$ A and $(\pi)_1$ $\Gamma$-realizes-$\mathit{\Psi}$ B.
(3) $\pi$ $\Gamma$-realizes-$\mathit{\Psi}$ $A \vee B$, if $(\pi(0))_0 = 0$ and $(\pi)_1$ $\Gamma$-realizes-$\mathit{\Psi}$ $A$, or $(\pi(0))_0 \neq 0$ and $(\pi)_1$ $\Gamma$-realizes-$\mathit{\Psi}$ $B$.
(4) $\pi$ $\Gamma$-realizes-$\mathit{\Psi}$ $A \rightarrow B$, if, if $\sigma$ $\Gamma$-realizes-$\mathit{\Psi}$ A, then $\{\pi\}[\sigma]$ $\Gamma$-realizes-$\mathit{\Psi}$ B.
(5) $\pi$ $\Gamma$-realizes-$\mathit{\Psi}$ $\neg A$, if $\pi$ $\Gamma$-realizes-$\mathit{\Psi}$ $A \rightarrow 1 = 0$.
(6) $\pi$ $\Gamma$-realizes-$\mathit{\Psi}$ $\forall x A(x)$, if for each $x \in \omega$, $\{\pi\}[x]$ $\Gamma$-realizes-$\mathit{\Psi}, x$ $A(x)$.
(7) $\pi$ $\Gamma$-realizes-$\mathit{\Psi}$ $\exists x A(x)$, if $(\pi)_1$ $\Gamma$-realizes-$\mathit{\Psi}, (\pi(0))_0$ $A(x)$.
(8) $\pi$ $\Gamma$-realizes-$\mathit{\Psi}$ $\forall a A(a)$, if for each $r \in \omega$, $\{\pi\}[r]$ $\Gamma$-realizes-$\mathit{\Psi}, r$ $A(a)$.
(9) $\pi$ $\Gamma$-realizes-$\mathit{\Psi}$ $\exists a A(a)$, if $(\pi)_1$ $\Gamma$-realizes-$\mathit{\Psi}, (\pi(0))_0$ $A(a)$.
(10) $\pi$ $\Gamma$-realizes-$\mathit{\Psi}$ $\forall \alpha A(\alpha)$, if for each $\alpha \in \omega^\omega$, $\{\pi\}[\alpha]$ $\Gamma$-realizes-$\mathit{\Psi}, \alpha$ $A(\alpha)$.
(11) $\pi$ $\Gamma$-realizes-$\mathit{\Psi}$ $\exists \alpha A(\alpha)$, if $(\pi)_1$ $\Gamma$-realizes-$\mathit{\Psi}, \{(\pi)_0\}$ $A(\alpha)$.

Here e.g. "$\{\pi\}[\alpha]$ $\Gamma$-realizes-$\mathit{\Psi}, \alpha$ $A(\alpha)$" abbreviates "$\{\pi\}[\alpha]$ is completely defined and $\Gamma$-realizes-$\mathit{\Psi}, \alpha$ $A(\alpha)$."

### 4.5.2 *Outline of the Proof of Theorem 2*

The theorem can be restated as follows:

> Assume $\mathcal{R}$ is countable, with $\Gamma$, $\chi$ as in the proof of Theorem 1. Then to each theorem $E$ of **FIRM**($\prec$) with at most the distinct variables $\Psi = \Psi_0, \Psi_1, \Psi_2$ free, there is a function $\varphi[\mathit{\Psi}]$ which is continuous in $\mathit{\Psi}$ and $\Gamma$-realizes-$\mathit{\Psi}$ $E$ for each $\Gamma$-interpretation $\mathit{\Psi}$ of $\Psi$. Since $0 = 1$ is not $\Gamma$-realizable, **FIRM**($\prec$) and **FIRM** are consistent.

The proof is by induction on a derivation of $E$, using three lemmas. Assume $\Gamma$, $\chi$ are as in the proof of Theorem 1, and let "$E$ is true-$\Gamma[\mathit{\Psi}]$" abbreviate "$\mathcal{M}(\prec_{\mathcal{R}}) \models_{\Gamma[\mathit{\Psi}]} E$."

**Lemma 3** *To each list $\Psi$ of distinct* number and lawlike sequence *variables and each restricted R-formula $A(x,y)$ containing free at most $\Psi, x, y$ where $x, y, a \notin \Psi$, there is a partial function $\xi_A[\mathit{\Psi}]$ so that for each $\Gamma$-interpretation $\mathit{\Psi}$ of $\Psi$: If $\forall x \exists! y A(x,y)$ is true-$\Gamma[\mathit{\Psi}]$ then $\xi_A[\mathit{\Psi}]$ is defined and $\forall x A(x, a(x))$ is true-$\Gamma[\mathit{\Psi}, \xi_A[\mathit{\Psi}]]$.*

**Lemma 4** *To each list $\Psi = x_1, \ldots, x_n, \alpha_1, \ldots, \alpha_k, a_1, \ldots, a_m$ of distinct variables and each almost negative formula $E$ of $\mathcal{L}[\prec]$ containing free only $\Psi$, there is a function $\varepsilon_E[\mathit{\Psi}] = \lambda t.\varepsilon_E(\mathit{\Psi}, t)$ partial recursive in $\Gamma$ so that for each $\Gamma$-interpretation $\mathit{\Psi}$ of $\Psi$:*

*(i) If $E$ is $\Gamma$-realized-$\mathit{\Psi}$ then $E$ is true-$\Gamma[\mathit{\Psi}]$, and*

*(ii) $E$ is true-$\Gamma[\mathit{\Psi}]$ if and only if $\varepsilon_E[\mathit{\Psi}]$ is completely defined and $\Gamma$-realizes-$\mathit{\Psi}$ $E$.*

**Lemma 5** *To each list $\Psi = x_1, \ldots, x_n, \alpha_1, \ldots, \alpha_k, a_1, \ldots, a_m$ and each restricted formula E of $\mathcal{L}(\prec)$ containing free at most $\Psi$, there is a continuous partial function $\zeta_E[\mathit{\Psi}]$ such that for each $\Gamma$-interpretation $\mathit{\Psi}$ of $\Psi$ with $[\alpha_1, \ldots, \alpha_k] \in \mathcal{R}LS$:*

*(i) If E is $\Gamma$-realized-$\mathit{\Psi}$ then E is true-$\Gamma[\mathit{\Psi}]$, and*

*(ii) E is true-$\Gamma[\mathit{\Psi}]$ if and only if $\zeta_E[\mathit{\Psi}]$ is completely defined and $\Gamma$-realizes-$\mathit{\Psi}$ E.*

The proof of Lemma 5 uses a sublemma: for each list $\Psi = \Psi_0, \Psi_1, \Psi_2$ of distinct variables where $\Psi_1 = \alpha_0, \ldots, \alpha_{k-1}$ and $\Psi_0, \Psi_2$ are number and lawlike sequence variables respectively, and each restricted formula $E(\Psi)$ with no other variables free, by Theorem 1 there is a partial continuous functional $\{\tau_E[\mathit{\Psi}_0, \mathit{\Psi}_2]\}(\alpha)$ such that if $\mathit{\Psi}$ is a $\Gamma$-interpretation of $\Psi$ with $\mathit{\Psi}_1 = \alpha_0, \ldots, \alpha_{k-1}$ where $\alpha = [\alpha_0, \ldots, \alpha_{k-1}] \in \mathcal{R}LS$, then $\{\tau_E[\mathit{\Psi}_0, \mathit{\Psi}_2]\}(\alpha)$ is defined and equal to 0 if $E$ is true-$\Gamma[\mathit{\Psi}]$, or 1 if $\neg E$ is true-$\Gamma[\mathit{\Psi}]$.

To complete the proof of the theorem, observe that $\Gamma$-realizability extends the function-realizability interpretation of [6] to $\mathcal{L}(\prec)$, so if $E$ is an axiom of **FIM** or an axiom schema of **FIM** + $\mathrm{GC}_{11}$ extended to $\mathcal{L}(\prec)$, then $E$ is $\Gamma$-realized-$\mathit{\Psi}$ (essentially as in Theorem 10 of [6]) by a $\varphi[\mathit{\Psi}]$ primitive recursive in $\Gamma$. For the additional axioms, schemas, and rules of **FIRM**$(\prec)$ use Theorem 1 with Lemmas 3–5. More details are in [11].

## 4.6 Epilogue

The results show that Brouwer's continuum can consistently contain all the lawlike and lawless sequences. Considered separately, these classes are distinguished by the axioms they satisfy:

(a) Brouwer's arbitrary choice sequences satisfy the bar theorem, countable choice, and continuous choice. Their logic is intuitionistic. Equality of choice sequences is not decidable.

(b) The lawlike sequences satisfy the bar theorem and countable choice, but not continuous choice. Their logic is classical. Equality of lawlike sequences is decidable.

(c) The lawless sequences satisfy (restricted) open data and restricted continuous choice, but not the restricted bar theorem. The restricted logic of independent lawless sequences is classical, but equality of lawless sequences is not decidable.

Kleene observed in [6] that, when considered separately, the recursive sequences do not satisfy the fan theorem, the arithmetical sequences satisfy the fan theorem, but even the hyperarithmetical sequences do not satisfy the bar theorem. In $\mathcal{M}(\prec_{\mathcal{R}})$ all analytic sequences are lawlike, but the three-sorted system **FIRM**$(\prec)$ does not obviously entail $\forall x \exists ! y A(x, y) \rightarrow \exists b \forall x A(a, b(x))$ for all formulas $A(x, y)$ of $\mathcal{L}_{\mathrm{FIM}}$ with only $x, y$ free. This may be an open question.

*Lawless* and *random* are orthogonal concepts. A random sequence of natural numbers should possess certain definable regularity properties (e.g. the percentage of even numbers in its $n$th initial segment should approach 50 as $n$ increases), while a lawless sequence should possess none. Any regularity property definable in $\mathcal{L}$ by a restricted formula can be defeated by a suitable lawlike predictor.

The model considered here has an explicit forcing characterization (cf. [10]). Under the assumption that there are only countably many lawlike sequences, $\alpha$ is lawless if and only if $\overline{\alpha} = \{\overline{\alpha}(n) : n \in \omega\}$ is generic with respect to the class of all dense subsets of $\omega^{<\omega}$, which are classically definable in $\mathcal{L}(\prec)$ by restricted formulas with lawlike parameters.

## Acknowledgement

This is an expanded version of a talk given on March 31, 2013 at the Second Bristol Workshop on the Limitations and Scope of Mathematical Knowledge. I thank Leon Horsten and Philip Welch for organizing the conference, the Templeton Foundation for sponsoring it, and James Appleby for asking the questions which inspired this exposition.

## Notes

1. From the translation of [1] on p. 295 of Volume I of Gödel's collected works.
2. The ! expresses uniqueness: $\exists!\beta A(\beta)$ abbreviates $\exists\beta A(\beta)$ & $\forall\beta\forall\gamma(A(\beta)$ & $A(\gamma) \rightarrow \beta = \gamma)$, and $\exists!xA(x)$ abbreviates $\exists xA(x)$ & $\forall x\forall y(A(x)$ & $A(y) \rightarrow x = y)$.
3. See p. 53 of [6]. The ! is essential for the intuitionistic version.
4. In general, "$\mathcal{M}(\prec_{\mathcal{R}})$ satisfies the axiom" means that $\mathcal{M}(\prec_{\mathcal{R}})$ satisfies the universal closure of the axiom; similarly for axiom schemas such as W4.
5. In general, "$\alpha \in \mathcal{RLS}$" abbreviates "$\mathcal{M}(\prec_{\mathcal{R}}) \models_\alpha RLS(\alpha)$."
6. $\{\pi\}[\alpha] \simeq \lambda x.\,\pi(\langle x+1\rangle * \overline{\alpha}(\mu z\,\pi(\langle x+1\rangle * \overline{\alpha}(z)) > 0)) \dot{-} 1$ and $\{\pi\}[x] \simeq \{\pi\}[\lambda y.\,x]$.

## References


[1] K. Gödel, *Zur intuitionistischen arithmetik und zahlentheorie*, Ergebnisse eines math. Koll. **4**, 34–38, 1933.

[2] K. Gödel, *Some basic theorems on the foundations of mathematics and their implications*, Kurt Gödel. Collected Works. Volume III: Unpublished Essays and Lectures (S. Feferman et al, eds.), pp. 304–323, Oxford, 1995.

[3] K. Gödel, Remarks before the Princeton bicentennial conference on problems in mathematics, 1946, *The Undecidable* (M. Davis, ed.), pp. 84–88, Raven Press, New York, 1965.

[4] S.C. Kleene, *Constructive functions in 'The foundations of intuitionistic mathematics'*, Logic, Methodology and Philosophy of Science III: Proceedings, Amsterdam, 1967 (B. van Rootselaar and J. F. Staal, eds.), pp. 137–144, North-Holland, 1968.

[5] S.C. Kleene, *Formalized recursive functionals and formalized realizability*, Memoirs, no. 89, Amer. Math. Soc., 1969.

[6] S.C. Kleene and R.E. Vesley, *The Foundations of Intuitionistic Mathematics, Especially in Relation to Recursive Functions*, North-Holland, 1965.

[7] G. Kreisel, Lawless sequences of natural numbers, *Comp. Math.* **20**, 222–248, 1968.

[8] A. Levy, *Definability in axiomatic set theory II*, Mathematical Logic and Foundations of Set Theory: Proceedings, Jerusalem, 1968 (Y. Bar-Hillel, ed.), pp. 129–145, North-Holland, 1970.

[9] J.R. Moschovakis, Relative lawlessness in intuitionistic analysis, *Jour. Symb. Logic* **52**, 68–88, 1986.
[10] J.R. Moschovakis, More about relatively lawless sequences, *Jour. Symb. Logic* **59**, 813–829, 1994.
[11] J.R. Moschovakis, A classical view of the intuitionistic continuum, *Ann. Pure and Appl. Logic* **81**, 9–24, 1996.
[12] A.S. Troelstra, *Intuitionistic formal systems*, Metamathematical Investigation of Intuitionistic Arithmetic and Analysis (A. S. Troelstra, ed.), Lecture Notes in Math., Springer-Verlag, 1973.
[13] A.S. Troelstra, *Choice sequences, a chapter of intuitionistic mathematics*, Oxford Logic Guides, Clarendon Press, 1977.

# 5

# A Semantics for In-Principle Provability

T. ACHOURIOTI

## 5.1 Introduction

The notion of absolute provability[1] can be traced back to Gödel's translation of intuitionistic logic into the modal logic **S4**. Gödel remarked that an **S4** operator cannot express provability in a formal system, since the **T**-axiom embodies a form of reflection. However, there is a connection between **S4** and an idealised, non-resource-bounded, notion of proof. For Gödel, the aforementioned translation was a confirmation that intuitionistic provability cannot be restricted to a formal system.

More recently, Shapiro [15] and Reinhardt [14] offered independently an axiomatisation of in-principle provability for mathematics.[2] So-called 'epistemic arithmetic' is the theory of Peano arithmetic expanded by **S4** axioms for an added modal operator with the desired interpretation.[3] While provability can be treated as a predicate, the formalisation of in-principle provability requires a treatment in terms of an operator, due to Montague-type inconsistency results for modal predicates [11], which apply to any theory that has the expressive means to allow for diagonalisation.

Part of the philosophical interest in a formal treatment of in-principle provability derives from the fact that it allows for a formalisation of Gödel's famous disjunctive thesis: either the human mind surpasses any Turing machine or there are absolutely undecidable propositions.[4] An epistemic operator may be employed to express mathematical knowability or the capacities of the human mind that the first disjunct compares to those of Turing machines, while its negation can be used to express absolute *un*decidability and hence the limitations of the human mind. It is important to note that in-principle provability is not a mathematical notion per se and that for this reason a formalisation of it cannot be expected to exhaust its meaning. The hope is, however, that its plausible formal properties are sufficient for answering some of the philosophical questions concerning human mathematical knowability. Indeed, Reinhardt proposed a logic for his modal operator $B$ by means of which he could easily prove Gödel's disjunctive thesis itself [14].

In the philosophical literature, the first disjunct has received far more attention than the second[5]; a great deal has been written on the so-called mechanistic thesis of the mind

*Gödel's Disjunction*. First Edition. Leon Horsten & Philip Welch (eds).

and especially on the Lucas–Penrose arguments meant to deny it [9], [12], [13]. Possible philosophical implications of the formal work presented in this chapter, however, relate only indirectly[6] to the first disjunct, since our focus will mostly be on the relation between in-principle provability and mathematical truth.

It was first Reinhardt's proposal to employ the notion of truth in order to formalise Gödel's disjunction, as well as the argument leading up to it.[7] Truth in Reinhardt's framework appears in two roles. First, as mathematical truth which is represented by a superset of the propositions knowable to the human mind in case the second disjunct is true, and, second, as part of the conceptual apparatus needed to specify the formal properties of absolute provability, in particular, soundness. For the latter, Reinhardt makes use of a typed truth predicate accompanied by Tarski's theory of truth.

Formal treatments of $B$ have been proof theoretic, with the exception of Carlson [3]. Carlson's model theory is close in spirit to a Kripke semantics, and this is precisely the difference with the model theory proposed here. Motivated by intensional phenomena that we argue need to be accounted for by a suitable semantics for the epistemic operator, we consider directed sets of models intended to represent the progression of mathematical theories.[8] We propose a formal analogue for mathematical development as the substrate on which we then seek to develop a semantics for the epistemic operator. In this sense we are closer in spirit to Jeroslow's experimental logic, which allows for change in the interpretation of predicates in the course of mathematical practice [8].

## 5.2 In-Principle Provability and Intensionality

Gödel believed that the distinctiveness of the human mathematical mind when compared to a Turing machine is evident in its ability to come up with new axioms and develop new mathematical theories.

> It turns out that in the systematic establishment of the axioms of mathematics, new axioms, which do not follow from formal logic from those previously established, again and again become evident. [6]

Mathematical statements that, although not formally derivable, can be proved once the theory is sufficiently strengthened by new axioms are only *in-principle* provable. Gödel's second disjunct basically raises the question of whether this process of coming up with new axioms can lead to absolute decidability for number theoretic statements, which we know not to be the case for formal systems due to the incompleteness results.[9]

> It is not at all excluded by the negative results mentioned earlier that nevertheless every clearly posed mathematical yes or no question is solvable in this way. For it is just this becoming evident of more and more new axioms on the basis of the meaning of the primitive notions that the machine cannot imitate. [6]

It is precisely because in-principle provability extends beyond given formal theories that it creates intensional context; the latter pointed out by both Shapiro [15] and Reinhardt

[14]. Reinhardt observes that free variables under the scope of *B* may range over different domains, and that if these domains contain non-canonically generated names or objects then it is not clear what *B* precisely applies to: mathematical terms are subject to change, while for non-canonically generated objects it may not be provable that they exist. To deal with such phenomena, variables would ideally need to 'keep track' of both names and objects that names refer to. In Reinhardt's own words:

> We shall allow free variables for natural numbers within the scope of the operator *B*. In view of the canonical process for generating the natural numbers, there is a natural reading of $BP(n)$; it means that we can prove *P* to hold of the object satisfying the canonical concept for *n*. On the other hand, for general variables *x*, it seems problematic to assign a meaning to $BP(x)$. In particular, if *x* ranges over real numbers or ordinals, it is unclear what could be intended by $BP(x)$. Evidently the variable *x* is to refer to the object of which we have proved *P*; but if we were to refer to *x* in some arcane way we might not be able to prove that *P* holds of the object satisfying such a reference.[10] Consequently the variable *x* in the formula $\theta(x)$ when used in the context $B\theta(x)$, must be keeping track not only of its referent (object referred to) but of an intensional entity as well, which we shall call the reference itself. Because of this dual role for variables, the logic is much simpler in the case where the domain admits something like a canonical set of references.
>
> [14], p. 443.

And a page later:

> Another characteristic of the natural numbers which is not valid for all types of entities is that they necessarily exist. In the case of contingent entities, even those for which we have canonical names, it does not seem plausible to say (for example) that from a proof of $\forall x\theta(x)$ we obtain a proof of $\theta(y)$; we first need to have a proof that *y* exists.
>
> [14], p. 444.

In addition, Shapiro [15] p. 21 observes that if one has a rule allowing arbitrary substitution, then $\phi \rightarrow B\phi$ can be derived, which is obviously wrong for in-principle provability. This is a problem already for natural numbers, showing that intensional phenomena are not restricted to non-canonically generated objects.

Generally speaking, intensional phenomena have to do with the way things are given and problems with identity or substitution that this creates. In mathematics we find examples such as the following: functions computed by a Turing machine can be given by single natural numbers, while there is no algorithm to tell for any two such numbers whether the corresponding functions are extensionally equal. Moreover, the way an object is given may depend on its theoretical environment, for example, if one adopts the axiom of choice then there is a unique least uncountable cardinal $\aleph_1$, but without the axiom of choice one can say that this object splits and can have many different properties according to the substitute of the axiom of choice that one adopts; for example, under the axiom of determinacy, $\aleph_1$ is a measurable cardinal. Finally, it may also happen that objects cease to exist, for instance, the differentials of various orders once commonly used in calculus.

A more methodological consideration is that proofs are usually presented in not fully formalised form and this leaves open the possibility that there are mistakes in the proof or worse that the theorem purportedly proven in fact has a counter-example. Hence, proofs and theorems are fallible in practice. This has been pointed out by Jeroslow by means of his experimental logics, whose basic construct is that of a proof acceptable at time $t$ leaving open the possibility that the proof is not acceptable at a later stage or for that matter at an earlier stage.

It is clear that a theory of in-principle provability must take such intensional features into account. To get a first idea of what one should expect from a logic of $B$ we recall some logical principles that fail in this context. We start with existence. Suppose we have a universal formula $\forall x\phi(x)$ and $a$ a term derived from a definite description that is possibly very complicated. It is then not immediately obvious that one can instantiate the universal quantifier by the formula $\phi(a)$. Another area where standard principles fail is identities under the scope of the epistemic operator. The principle $a = b \Rightarrow B(a = b)$ is no longer available. In fact one should not even expect $B(x = x)$ to be derivable.

It is now informative to mention some of the accommodations that Reinhardt makes in order to deal with intensionality before we proceed with our own proposal. Reinhardt's proposal was to specify the *type* of domain that the variables range over, which are then called 'restricted' to be distinguished from 'unrestricted' variables that range over the universe of discourse. Restricted variables can then either be allowed under the scope of $B$ or not; in the first case they are called *intensional*, in the second case *extensional*. 'Intensional' variables are variables that range, for example, over the natural numbers.[11] A logic for $B$ requires of free variables that they be 'intensional'; in fact it is postulated as a syntactic requirement for the well-formedness of $B\theta$ that free variables in $\theta$ are all 'intensional'.[12] Similar conditions are placed on the quantifier elimination rules. Such rules require an existence proof that the 'reference' has a 'referent'. Since such proofs pertain to mathematics and not logic, Reinhardt regards $\forall nB\exists y(y = n)$ as a mathematical axiom, as opposed to $B\forall n\exists y(y = n)$ which is included in the logic.

## 5.3 Modelling Epistemic Mathematics: A Theory of Descriptions

In this section we present the mathematical structure of inverse systems with limit that we propose to use as a formal analogue of mathematical practice. It is in the context of this structure that we then seek to develop a semantics for $B$.

**Definition 1** *A directed set is a set $\mathbf{T}$ together with an ordering relation $\leq$ such that*

*(1) $\leq$ is a partial order, i.e. transitive, reflexive, anti-symmetric*

*(2) $\leq$ is directed, i.e. for any $s, t \in \mathbf{T}$ there is $r \in \mathbf{T}$ with $s, t \leq r$*

**Definition 2** *Let $\mathbf{T}$ be a directed set. An* inverse system *of models indexed by $\mathbf{T}$ is a family of first-order models $\{\mathcal{M}_s | s \in \mathbf{T}\}$ together with a family of homomorphisms*

*$\mathcal{F} = \{h_{st} | s \geq t, h_{st} : \mathcal{M}_s \longrightarrow \mathcal{M}_t\}$. The mappings in $\mathcal{F}$ must satisfy the coherence requirement that if $s \geq t \geq r$, $h_{tr} \circ h_{st} = h_{sr}$. That $h_{st} : \mathcal{M}_s \longrightarrow \mathcal{M}_t$ is a* homomorphism *means that $\mathcal{M}_s \models R(a_1, a_2, \dots)$ implies $\mathcal{M}_t \models R(h_{st}(a_1), h_{st}(a_2), \dots)$*

*We do not require surjectivity of $h_{st}$.*

**Definition 3** *Let $(T, \{\mathcal{M}_s \mid s \in T\}, \mathcal{F})$ be an inverse system. Let $D_s$ be the domain of $\mathcal{M}_s$. Let $\mathcal{D} \subseteq \Pi_{s \in T} D_s$ the set of all $\xi$ such that for $s \geq t$, $h_{st}(\xi(s)) = \xi(t)$. Define a model $\mathcal{M}$ with domain $\mathcal{D}$ by putting $\mathcal{M} \models R(\xi^1, \xi^2, \dots)$ if for all $s \in T$, $\mathcal{M}_s \models R(\xi_s^1, \xi_s^2, \dots)$. $\mathcal{M}$ is called the* inverse limit *of the given inverse system.*

Similarly to Carlson [3], some idealisations need to be made. We take an inverse system to represent the knowledge of the mathematical community at large and the index set to be an ordering on the stages of mathematical development which is unbounded in resources including time. Directedness serves as a convergence condition: the theories produced at different stages should still be viewed as belonging to the same mathematical universe.[13]

The main reason for opting for a semantics for provability in principle in terms of inverse systems and their limits as opposed to the familiar **S4** Kripke structures is the following: the assumption that domains are constant or increasing does not correspond to mathematical practice, where it may happen, for example, that a class of putative mathematical objects resists all attempts at formalisation and hence is no longer studied, or a putative mathematical object may turn out to be inconsistent. Inverse systems account for such phenomena by allowing for non-surjective bonding mappings between the domains of the models.

Another advantage of using this structure is that it allows for conceptual change and evolution of terms in any direction (since concepts are used to define terms, as in 'even', in the 'smallest even number'). That concepts expand is not in conflict with the usual modal interpretation, which, however, does not account for the possibility that the conceptual apparatus shrinks.

Furthermore, quite often a conjecture becomes a theorem only after strengthening the hypotheses of the conjecture to rule out counterexamples; if one thinks of the hypotheses as defining a single concept that constitutes the antecedent of the conjecture, then this means that the concept is not preserved upwards, to later stages. This justifies our choice of an inverse system, where the homomorphisms are directed downwards, rather than a direct one. Here are the thoughts of a famous mathematician on the matter:

> It seems to me that all concepts of mathematics are initially vague. People use them as if they know what they were talking about for ages until eventually precise definitions for them emerge. The definitions may or may not agree and, usually one finally emerges supreme. This is not always the case. There are still a lot of competing notions of smoothness: continuity, differentiability, continuous differentiability and a whole hierarchy of $C^n$ functions. Still, one might say something like "let C be a smooth curve". The question of whether or not a smooth curve has a tangent everywhere is unanswerable until you specify which notion of smoothness is meant. The answer is "no" for continuity, but "yes" for differentiability and beyond. Does a smooth curve have an arc-length?
>
> [Craig Smorynski, FOM blog, 28 June 2013]

Finally, inverse systems are plausible candidates for modelling progression of mathematical knowledge also because they accommodate certain epistemological constraints, most notably the fallible character of mathematical judgement. For instance, filling the missing steps in mathematical proofs is often the reason for refuting theorems that may have earlier been accepted. Inverse systems embody non-monotonicity since a formula $\phi$ evaluated as true at a model $\mathcal{M}_s$ may be found out to be false at a later stage $\mathcal{M}_t$ with $s < t$.

What this structure in fact provides us with is a theory of presentations or descriptions where we should understand 'presentation' or 'description' as 'possible mathematical object'. For a formal definition, pick an element $d$ in a model $\mathcal{M}_s$ and form the set of all $h_{st}$ applied to $d$ for all $t < s$. This set is the presentation of a possible mathematical object. If $d$ is an element of a thread $\xi$, then the set we just defined is an initial segment of the thread, but if no thread runs through $d$ then $d$-segments do not represent any mathematical object. We will require that 'in-principle provability' is stable along such descriptions that represent mathematical objects. This is because in-principle provability has an idealised and dispositional character: it is not about what is provable now, but what may become provable once enough resources are available. $B$ must, therefore, satisfy a co-recursive equation, informally: $\phi(a)$ (where a is a presentation of a mathematical object) is in-principle provable now, if and only if for all later stages there is a presentation $b$ such that $\phi(b)$ is in-principle provable, where $a$ and $b$ are presentations of the same mathematical object.

## 5.4 Intensional Truth

Adopting a semantic approach affects the notion of truth through soundness: in-principle provability has to be sound. Since soundness must be specified with respect to the interpretation of predicates in $\phi$ of $B\phi$, we are forced to interpret the notion of truth on the same structure as $B$. Such a problem of compatibility between the semantics of in-principle provability and truth does not arise in a proof theoretic treatment.

It is a challenge to find a way to talk of mathematical truth in a context where mathematical objects and concepts are subject to change. Our structure, with variable domains and interpretation of the predicates, calls for some sort of stable state as, for example, the one given by a limit model, or at the very least a notion of stable satisfaction.

While a standard extensional satisfaction predicate relates objects and formulas, we are led to employ an intensional satisfaction predicate which relates presentations to formulas. We saw that presentations are intensional objects and may turn out to be inconsistent. For an extensional satisfaction predicate $\mathbf{T}$, one trivially has $\mathbf{T}(\ulcorner x = x \urcorner, a)$ for arbitrary object $a$. A satisfaction predicate $S$ is intensional if $S(\ulcorner x = x \urcorner, a)$ can be false.

It is generally unpredictable if and when a description turns out to be inconsistent. Therefore, in order to define a stable satisfaction predicate, one must be able to look ahead at all later stages:

$$\mathcal{M}_s \models S(\ulcorner \phi(x) \urcorner, a) \iff \forall t \geq s\, \exists b\, [h_{ts}(b) = a \text{ and } \mathcal{M}_t \models S(\ulcorner \phi(x) \urcorner, b)],$$

where $\phi$ may itself contain $S$.

This co-recursive equation calls for a maximal fixed-point construction which gives the interpretation of stable satisfaction $S$ [1]. We construct a decreasing sequence $\{S^\alpha \mid$

$\alpha$ an ordinal$\}$ such that there exists $\alpha$ with the property that for all $\beta \geq \alpha : \emptyset \neq S^\alpha = S^\beta =:$ $S$ and $S$ satisfies the co-recursive equation. To ensure a monotone decreasing sequence we restrict $S$ to positive primitive (p.p.) formulas, that is, formulas that contain only $\exists$, $\wedge$, $\vee$ and $\perp$.

(1) Define $S^0$ by

$$S^0(\ulcorner \phi(x) \urcorner, a) \Longleftrightarrow \phi(x) \in \mathcal{L}(S) \text{ p.p. and } \exists s\,(a \in \mathcal{M}_s).$$

(2) If $\alpha$ is a limit ordinal, put

$$S^\alpha = \bigcap_{\beta<\alpha} S^\beta.$$

(3) If $\alpha$ is a successor ordinal, i.e. of the form $\beta + 1$, put
for all $t$

$$\mathcal{M}_t \models S^\alpha(\ulcorner \phi(x) \urcorner, a) \Longleftrightarrow \forall s \geq t\, \exists b\, [h_{st}(b) = a \text{ and } \mathcal{M}_s \models \phi(S^\beta)(b)],$$

where '$\phi(S^\beta)$' means that the predicate $S$ in $\phi$ is interpreted as $S^\beta$.

The homorphisms ensure downwards stability of satisfaction, since $S$, thus defined, is provably an admissible predicate in the inverse system [1]. Note that the ***T***-schema fails for intensional $S$: we always have $a = a$, but not $S(\ulcorner x = x \urcorner, a)$.

Furthermore, only that part of mathematical truth on the limit $\mathcal{N}$ is accessible to us (in the sense of in-principle provability) which satisfies for all inverse systems with limit: if for all $s$ $\mathcal{M}_s \models \phi(\xi_s, ...)$, then $\mathcal{N} \models \phi(\xi, ...)$; this is because there is no independent access to the limit available. The interpolation theorem ([2], see [1] for an improved version of the proof) shows that formulas preserved from the inverse system to the limit must be equivalent to a conjunction of geometric formulas, i.e. a geometric theory; therefore, in what follows the set of true statements will consist of only geometric implications. This does not constitute a real limitation for the reason that any first-order theory can be translated into a geometric theory (see [10] for a recent version of this translation).

It is shown in [1] that $S$ as defined above can consistently extend to closed geometric implications as long as they are $S$-guarded: a geometric implication is guarded if all variables in the antecedent bound by the universal quantifier are under the scope of $S$. So the predicate $S$ turns out to play a double role: it serves as a guard when it occurs in the antecedent of geometric implications, in which case it singles out elements that are not truncated; and it also applies to the whole geometric implication, in which case it signals truth.

One can easily see that this intensional satisfaction predicate expresses a non-deflationary notion of truth.

## 5.5 Towards Axioms for In-Principle Provability

In our first attempt at formulating axioms for $B$ we start out by listing certain plausible properties, first of all, that it is a modal operator that must be interpretable on inverse systems: hence it has some features of an **S4** operator, since the ordering of the index set is reflexive and transitive. We also assume that $B$ is sound with respect to the type-free satisfaction predicate $S$, and that it satisfies a co-recursive equation of the form: '$\phi$ is in-principle provable now iff it is in-principle provable at all later stages'. We then show that these properties lead to some counterintuitive results.

(1) $B$ must satisfy a co-recursive equation due to its dispositional character, e.g. one or more of the following:
  (i) $B\phi(a)$ holds at $\mathcal{M}_s$ implies $\forall t > s \exists b\, h_{ts}(b) = a$ and $B\phi(b)$ holds at $\mathcal{M}_t$
  (ii) $\forall t > s \exists b\, h_{ts}(b) = a$ and $B\phi(b)$ holds at $\mathcal{M}_t$ implies $B\phi(a)$ holds at $\mathcal{M}_s$
  (iii) $B\phi(a)$ holds at $\mathcal{M}_s$ iff $\forall t > s \exists b\, h_{ts}(b) = a$ and $\phi(b)$ holds at $\mathcal{M}_t$
(2) soundness w.r.t. stable satisfaction: for all $\phi$ in the domain of $S$, $B\forall x(B\phi(x) \rightarrow S(\ulcorner\phi(x)\urcorner, x))$.
(3) $B$ satisfies the **T, 4, K** axioms.
(4) $B$ may take as argument formulas that involve $S$.

Property 1(i) says that in-principle provability is monotonic; it is a version of the **4** axiom in modal logic. Property 1(ii) initially seems plausible: if $\phi$ is in-principle provable at all possible successors $t > s$, then $\phi$ seems in-principle provable at $s$, as in 'proof by cases'. But perhaps we need to know that the left-hand side of 1(ii) is in-principle provable, not just true. The situation is similar to the intuitionistic interpretation of the $\rightarrow$, which is a function that transforms any proof of the antecedent into a proof of the consequent. The question there naturally arises: is this function provable itself? Opinions are divided.[14] Hence, 1(ii) is perhaps questionable as it stands.

Observe that by soundness (property 2), 1(i) implies the left to right direction of 1(iii), but the converse direction is even more questionable than 1(ii), since the hypothesis does not mention in-principle provability at all, and hence implicitly expresses that what is stably true is provable.

Property 2 is a kind of reflection principle for in-principle provability, where truth is represented by the limit. Its intended meaning is: if $B\phi$ holds at $s$, then $\phi$ holds at the limit. To turn this into a formula that can be interpreted at any stage $s$, we reformulate it as $B\forall x(B\phi(x) \rightarrow S(\ulcorner\phi(x)\urcorner, x))$, with $\phi$ now restricted to the domain of $S$. This formulation also works if the inverse limit is empty; but in all cases we have to allow that $B$ takes as argument formulas that involve $S$ (property 4). Note that property 2 implies that $B(x = x)$ is not derivable; hence necessitation must fail at least for open formulas, although it can be adopted for sentences.

Property 3 lists some **S4** axioms. **T** follows from soundness (property 2), at least for formulas in the domain of $S$ (given one side of the $T$-schema). This means that an instance of

T of the form $BB\phi \rightarrow B\phi$ will have to be derived from property 1. We will investigate below whether soundness can be extended, especially when considered in combination with the converse Barcan formula. Lastly, **4** follows from property 1(i).

We now prove that under these assumptions $B = S$.

**Theorem 1**

(a) *Assume for S only the schema:* $S(\ulcorner\phi(x)\urcorner, x) \rightarrow \phi(x)$. *An operator B satisfying soundness w.r.t. S and the axioms* **K, 4** *also satisfies:* $B\phi(x) \leftrightarrow BS(\ulcorner\phi(x)\urcorner, x)$.

(b) *Now assume S satisfies the co-recursive equation. If, in addition, the co-recursive equation 1(iii) for B holds, we have* $BS(\ulcorner\phi(x)\urcorner, x) \leftrightarrow S(\ulcorner\phi(x)\urcorner, x)$.

(c) *From (a) and (b) we obtain* $B\phi(x) \leftrightarrow S(\ulcorner\phi(x)\urcorner, x)$.

*Proof*

(a) By soundness, for $\phi$ in the domain of $S$, $B\forall x(B\phi(x) \rightarrow S(\ulcorner\phi(x)\urcorner, x))$, which by the **K** property of $B$ (together with converse Barcan) implies

$$\forall x(BB\phi(x) \rightarrow BS(\ulcorner\phi(x)\urcorner, x)),$$

from which we get by half of the ***T***-schema for $S$ and the **K, 4** axioms for $B$,

$$BB\phi(x) \rightarrow BS(\ulcorner\phi(x)\urcorner, x) \rightarrow B\phi(x) \rightarrow BB\phi(x),$$

which entails

$$BS(\ulcorner\phi(x)\urcorner, x) \leftrightarrow B\phi(x).$$

(b) It remains to be shown that

$$BS(\ulcorner\phi(x)\urcorner, x) \leftrightarrow S(\ulcorner\phi(x)\urcorner, x).$$

The left-to-right direction follows from the left-to-right direction of the ***T***-schema for $S$. For the converse direction, observe that the right-hand side of the equivalence satisfies the co-recursive equation for $S$,

$$\mathcal{M}_s \models S(\ulcorner\phi(x)\urcorner, x)[a] \Leftrightarrow \forall t \geq s \exists b\, h_{ts}(b) = a \;\&\; \mathcal{M}_t \models S(\ulcorner\phi(x)\urcorner, x)[b],$$

and we may apply the co-recursive definition 1(iii) of $B$ to the right-hand side of the co-recursive equation for $S$ to get

$$\forall t \geq s \exists b\, h_{ts}(b) = a\, \mathcal{M}_t \models S(\ulcorner\phi(x)\urcorner, x)[b] \Leftrightarrow \mathcal{M}_s \models BS(\ulcorner\phi(x)\urcorner, x)[a],$$

whence

$$\mathcal{M}_s \models S(\ulcorner\phi(x)\urcorner, x)[a] \Leftrightarrow \mathcal{M}_s \models BS(\ulcorner\phi(x)\urcorner, x)[a]. \qquad \blacksquare$$

To summarise, Theorem 1 gives three results on the relationship between $B$ and $S$, varying the assumptions on $S$ and $B$. In part (a) of the theorem, one only assumes **K** and **4** as properties of $B$, $S(\ulcorner\phi(x)\urcorner, x) \rightarrow \phi(x)$ and soundness of $B$ w.r.t. $S$; this yields $B = BS$. In part (b) one assumes co-recursive equations for both $B$ and $S$, which gives $S = BS$. Hence the assumptions taken together give $B = S$. This provides an immediate solution to Gödel's second disjunct: if in-principle provability and truth are equivalent, there are no absolutely undecidable propositions. However, a closer look reveals that the result is implausibly strong; for example, it implies that $B$ is preserved downwards, which means that if $\phi$ is in-principle provable now, then it has always been. The reason turns out to be a formulation of a co-recursive equation for $B$ that allows one to conclude in-principle provability of $\phi$ on the basis of truth of $\phi$ on sufficiently many models (property 1(iii)). This leaves us with 1(i–ii), although it is not easy to see how 1(i–ii) can be realised without realising 1(iii). Hence, a new principle must be sought and Theorem 2 will give a solution.

To conclude this section, we show using diagonalisation that under mild soundness conditions, $B$ cannot satisfy the converse Barcan formula $B\forall x\phi(x) \to \forall x B\phi(x)$. This is to be expected since converse Barcan forces constant or increasing domains which is not true of inverse systems; our domains are not constant or increasing. This is useful to point out also because the converse Barcan is used essentially in the previous proof.

**Lemma 1** *Let $S$ satisfy the schema $\forall x(S(\ulcorner\phi(x)\urcorner, x) \to \phi(x))$. Let $B$ satisfy necessitation for sentences,* **T** *and converse Barcan; then $B$ cannot be sound w.r.t. $S$ for all formulas.*

***Proof*** Let $\psi \leftrightarrow \neg S(\ulcorner\psi(x)\urcorner, x)$ be a provable fixed point. Since $\forall x(S(\ulcorner\phi(x)\urcorner, x) \to \phi(x))$, it follows that $\forall x \neg S(\ulcorner\psi(x)\urcorner, x)$ is provable; hence $\forall x\psi(x)$ is provable. By necessitation for sentences $B\forall x\psi(x)$ is provable. On the other hand, by **T**, $B\psi(x) \to \psi(x)$, and by soundness w.r.t. $S$,

$$B\psi(x) \to S(\ulcorner\psi(x)\urcorner, x),$$

from which we obtain $\forall x \neg B\psi(x)$.

By the converse Barcan formula applied to $B\forall x\psi(x)$ we get $\forall x B\psi(x)$ which together with $\forall x(B\psi(x) \to \bot)$ yields $\bot$.

A new start is required, one that connects $B$ closer to provability. ■

## 5.6 Intensional Semantics for 'It is In-Principle Provable that'

$B$ as axiomatised in the previous section satisfies some of our desiderata. In particular, $S$ has the right intensionality properties and one advantage of the identity $B = S$ is that $B$ inherits these properties. For example, if $x = x$ implied $B(x = x)$, then by soundness it would follow that $x = x$ implies $S(\ulcorner x = x\urcorner, y)$, which we know not to be the case.

However, the identity of $B$ and $S$ comes at a cost: $B$ no longer has a recognisable connection to provability. In this section, therefore, we attempt a new definition of $B$, one that corresponds to the intuition that $B\phi$ implies that at some stage—not necessarily the stage in which $B\phi$ is asserted—axioms will be found from which $\phi$ follows. Think of Con(PA) as an example of such a statement $\phi$. This statement is not provable in the given formal system, PA. But there are stronger systems from which it is formally provable, which is why it is regarded as 'in-principle provable'. Statements which cannot be proven in any formal system are Gödel's 'absolutely *un*decidable propositions', the very opposite of in-principle provability.

In order to recover a theory that will allow us to establish the connection of $B$ to provability, we take the second disjunct of Gödel's disjunction as our starting point. If Gödel's disjunct is true, so that there are no absolutely undecidable propositions, then on the intended interpretation of the operator $B$ as 'it is in-principle provable that' (and a classical reading of mathematical truth), we must have that for all $\phi$, $B\phi \vee B\neg\phi$, whence $B$ determines a complete theory. If, on the other hand, there are absolutely undecidable propositions, then $B$ determines a set of sentences that is no longer complete.

The idea that whenever $B\phi$ is asserted, axioms will eventually be found from which $\phi$ follows, suggests the following construction for a Kripke-style semantics for $B$, namely, that one starts from a consistent geometric theory and constructs an inverse system of finite parts of that theory, the finite parts being all one has access to at given stages of the development of mathematical theories.[15] Given a first-order theory $\mathcal{T}$[16], let $\mathcal{N}$ be a saturated model of $\mathcal{T}$, and $\{\mathcal{M}_s \mid s \subseteq \mathcal{T} \text{ finite}\}$ an inverse system with limit $\mathcal{N}$. We aim to interpret $B$ on this inverse system in such a way that $B\phi$ is true on an indexed model iff $\phi$ can be derived from (a finite part of) $\mathcal{T}$.

The following is a slight generalisation of the construction used in the proof of the interpolation theorem found in [1]:

**Lemma 2** *Let $\mathcal{T}$ be a geometric first-order theory, $\Sigma = \{s \subseteq \mathcal{T} \mid s \text{ finite}\}$. Let $\mathcal{N}$ be a saturated model of $\mathcal{T}$, then there exists an inverse system $\{\mathcal{M}_s \mid s \in \Sigma\}$ such that*

(1) $\mathcal{M}_s \models s$
(2) $\lim_{\rightarrow} \mathcal{M}_s = \mathcal{N}$
(3) *geometric implications are preserved to the limit.*

**Definition 4** *An inverse system with limit satisfying the properties of Lemma 2 will be called* $\mathcal{T}$-canonical.

### 5.6.1 *Dynamical Proofs*

Our aim is to provide a semantics of $B$ for $\mathcal{T}$-canonical structures such that it distinguishes actual from potential provability. We first need a notion of provability tailored for geometric theories; to this end we employ the so-called 'dynamical proofs'. In a dynamical proof one takes a geometric theory[17] as defining a consequence relation holding between two sets of facts. In particular, geometric implications are viewed as inference rules that transform finite sets of atomic sentences ('facts') into other such sets; an advantage of using this notion of proof is that it avoids trivialities such as $A(b) \vdash A(b)$.

An example, taken from Coquand [4], illustrates the idea. The theory is:[18]

(1) $P(x) \wedge U(x) \rightarrow Q(x) \vee \exists y R(x, y)$
(2) $P(x) \wedge Q(x) \rightarrow \bot$
(3) $P(x) \wedge R(x, y) \rightarrow S(x)$
(4) $P(x) \wedge T(x) \rightarrow U(x)$
(5) $U(x) \wedge S(x) \rightarrow V(x) \vee Q(x)$

Next we show how to derive $V(a_0)$ from $P(a_0), \boldsymbol{T}(a_0)$:

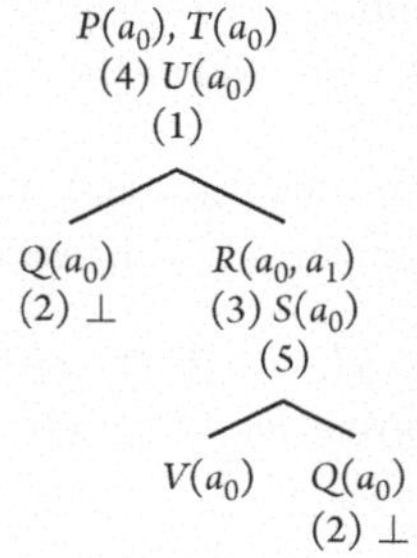


Some comments on this derivation. The geometric theory defines the consequence relation; hence the geometric implications occurring in it are taken to be inference rules. Disjunctions lead to branching of the tree, as we see in (1) and (5). The existential quantifier in formula (1) introduces a new term in the proof, here $a_1$, which appears in the right branch of (1). Lastly, a fact is derivable if it appears on every branch not marked by $\bot$, which leaves $V(a_0)$. If $X$ is a collection of facts whose terms are collected in $I$, $F$ a fact with terms in $I$, and $\boldsymbol{T}$ a geometric theory, then there exists a dynamical proof of $F$ from $X$ if and only if $\boldsymbol{T}, X \vdash F$ in intuitionistic logic. We may extend the concept of dynamical proof slightly by allowing $\exists$ introduction on the leaves of the tree, and we say that an existential formula is derivable from a set of facts if the formula occurs on all leaves not marked by $\bot$. The reader will be able to do the case of $\vee$.

It will be clear how a dynamical proof of a geometric implication from a geometric theory proceeds: if $\boldsymbol{T}$ is the geometric theory and $\forall\bar{x}(\tau(x) \rightarrow \theta(x))$ the geometric implication ($\tau$ is a conjunction of atomic formulas, and for simplicity take $\theta$ an existentially quantified conjunction $\theta'$ of atomic formulas; we will interpret $\theta'$ as a set), choose new terms not occurring in either $\boldsymbol{T}$ or $\forall\bar{x}(\tau(\bar{x}) \rightarrow \theta(\bar{x}))$, plug these terms into $\tau$, and construct a dynamical proof tree with the sets $\theta'$ at the leaves. There may occur terms in $\theta'$ not in $\tau$; these have to be quantified existentially. Introduce any other existential quantifiers on $\theta'$ as required by $\theta$. The result is an intuitionistic derivation of $\forall\bar{x}(\tau(\bar{x}) \rightarrow \theta(\bar{x}))$ from $\boldsymbol{T}$. Conversely, if there is an intuitionistic derivation of $\forall\bar{x}(\tau(\bar{x}) \rightarrow \theta(\bar{x}))$ from $\boldsymbol{T}$, then there exists a dynamical proof in the sense just sketched.

For our purposes the advantage of dynamical proofs is that facts can only issue in other facts if the former can be acted upon by a geometric inference rule; hence there exists a clear distinction between given facts and facts to be proved. If $s$ is a finite set of geometric implications, and $\Delta, \Gamma$ finite sets of facts, we use $\Delta \Vdash_s \Gamma$ to mean that $\Gamma$ is dynamically derivable from $\Delta$ using the rules in $s$.

### 5.6.2 *Bringing Provability Back into 'In-Principle Provability'*

In this section, we propose a modal interpretation of $B$ in terms of potential dynamical provability, which is given in Definition 6. We restrict applicability of $B$ to positive primitive formulas and closed geometric implications. This is because we are primarily interested in the relation between in-principle provability and mathematical truth as represented by S. Definition 6 needs a preliminary definition of covering.[19] We then go on to list properties that follow from this definition, one of which will be the negation of the earlier result of the identity of B and S.

Given a model $\mathcal{M}_s$, let $\Delta_s$ be its positive (atomic) diagram, i.e. the set of all atomic sentences true on $(\mathcal{M}_s, \bar{a})$, the expansion of $\mathcal{M}_s$ with new constants.

**Definition 5** *Let s be an index and $a \in \mathcal{M}_s$ such that for all $t > s$ there exists b with $h_{ts}(b) = a$. A covering of $(s, a)$ is a set $C_{s,a}$ satisfying: for all $t \geq s$ there exists $s \leq t \leq r$, $b \in \mathcal{M}_r$ with $(r, b) \in C_{s,a}$ and $h_{rs}(b) = a$.*

**Lemma 3** *Suppose $C_{s,a}$ is a covering of $(s, a)$ and for each $(r, b) \in C_{s,a}$ there exists a covering $C_{r,b}$ of $(r, b)$. Then*

$$\bigcup\{C_{r,b} \mid (r, b) \in C_{s,a}\}$$

*is a covering of $(s, a)$.*

***Proof*** Choose $t \geq s$ and $(r, b) \in C_{s,a}$ with $s \leq t \leq r$, $h_{rs}(b) = a$. Since $t \leq r$ and there exists a covering of $(r, b)$, there is $t \leq r \leq p$, $c \in \mathcal{M}_p$ with $h_{pr}(c) = b$. But then $h_{ps}(c) = a$. ■

**Definition 6** *If $\theta(x)$ is a positive primitive formula without occurrences of B or S, we define $\mathcal{M}_s \models B\theta(a)$ iff there exists a covering $C_{s,a}$ such that for all $(r, b) \in C_{s,a}$ there exists finite $\Delta \subseteq \Delta_r$ such that $\Delta \Vdash_r \{\theta(b)\}$. Under the same hypothesis on $\theta$, $\mathcal{M}_s \models BB\theta(a)$ iff there exists a covering $C_{s,a}$ such that for all $(r, b) \in C_{s,a}$ there exists a covering $C_{r,b}$ of $(r, b)$ such that for all $(p, c) \in C_{r,b}$: there exists finite $\Delta \subseteq \Delta_p$ $\Delta \Vdash_p \{\theta(c)\}$.*

What follows is an intermediate step for the next definition:
If $\phi = \forall x(\theta(x) \rightarrow \tau(x))$ is geometric, let $\phi^g$ be its B-guarded counterpart $\forall x(B\theta(x) \rightarrow \tau(x))$. Then, by Tarski's definition of truth on a model, we obtain

$$\mathcal{M}_s \models \forall x(B\theta(x) \rightarrow \tau(x)) \Leftrightarrow \forall a(\mathcal{M}_s \models B\theta(a) \Rightarrow \mathcal{M}_s \models \tau(a)),$$

which is well defined by the preceding definition.

**Definition 7** $\mathcal{M}_s \models B\phi^g$ *iff there exists a covering $C_{s,a}$ (a arbitrary) such that*

$$\forall (r, b) \in C_{s,a} \mathcal{M}_r \models \forall x(B\theta(x) \rightarrow \tau(x)).$$

The elements in the covering do not play a role in evaluating sentences (as, for example, a formula expressing Con(PA)) but they are needed for possibly open positive primitive formulas. In the case of sentences the role of guardedness is to prevent that all branches truncate. This means that if it is not the case that for all $t > s$ there exists $b$ with $h_{ts}(b) = a$, then $\mathcal{M}_s \not\models B\theta(a)$.

For illustration purposes we may apply these definitions to the so-called proof-by-cases, where we regard a set of cases to be a single covering. In such proofs an object $a$ may require objects $b$ with $h_{rs}(b) = a$ for its explicit construction. The in-principle proof may, therefore, operate on terms that cannot be evaluated when the in-principle proof is announced. This marks a clear contrast of 'in-principle provability' to actual provability.

Theorem 2 lists some properties and explores soundness for $B$.

**Theorem 2** *Let $\theta$ range over positive primitive formulas, $\phi$ over geometric implications, $\mathcal{T}$ over geometric theories. Then we have on $\mathcal{T}$-canonical structures:*

(1) (a) $B\theta(a) \to \theta(a)$ *if $\theta$ is in the signature of all indexed models*
(b) $BB\theta(a) \to B\theta(a)$ *generally*

(2) $B\theta(a) \to BB\theta(a)$

(3) (a) $B\theta(a) \to S(\ulcorner\theta(x)\urcorner, a)$ *if $\theta$ is in the signature of all indexed models*
(b) $B\theta(a) \to BS(\ulcorner\theta(x)\urcorner, a)$ *generally*

(4) $\mathcal{M}_s \models B\theta(a)$ *iff for all $t > s$ there exists $b$ with $h_{ts}(b) = a$ and $\mathcal{M}_t \models B\theta(b)$*

(5) $\not\models B(x = x)$; *hence necessitation is not valid for open formulas, although it is valid for sentences*

(6) $B\phi^g \to \phi^g$ *if $\phi$ is in the signature of all indexed models*

(7) $\mathcal{M}_s \models B\phi^g \implies \forall t \geq s \mathcal{M}_t \models B\phi^g$

(8) $B\phi^g \to BB\phi^g$

***Proof***

(1(a)) Assume $\mathcal{M}_s \models B\theta(a)$. For some $(r, b) \in C_{s,a}$, $r \geq s$, $b \in \mathcal{M}_r$ and $h_{rs}(b) = a$: there exists finite $\Delta \subseteq \Delta_r$ such that $\Delta \Vdash_r \{\theta(b)\}$, whence $\mathcal{M}_r \models \theta(b)$; since $\theta$ is in the signature of $\mathcal{M}_s$, by downwards preservation we obtain $\mathcal{M}_s \models \theta(a)$.

(1(b)) This follows from Lemma 3.

(2) Assume $\mathcal{M}_s \models B\theta(a)$, then $\mathcal{M}_s \models BB\theta(a)$ because we may take $C_{r,b} = \{(r, b)\}$ in Lemma 3.

(3(a)) Assume $\mathcal{M}_s \models B\theta(a)$, then by the same argument as for 1(a), for all $t$ there is $b$ with $h_{ts}(b) = a$ and $\mathcal{M}_t \models \theta(b)$. From this it follows that $\mathcal{M}_s \models S(\ulcorner\theta(x)\urcorner, a)$. Similar for 3(b) using Lemma 3.

(4) From right to left, apply Lemma 3. From left to right: choose $t > s$, $u \geq t$ and $(r, c) \in C_{s,a}$ with $r \geq u$, $h_{rs}(c) = a$ and there exists finite $\Delta \subseteq \Delta_r$ such that $\Delta \Vdash_r \{\theta(c)\}$. Put $h_{rt}(c) = b$, then the covering $C_{s,a}$ can be trimmed to obtain a covering $C_{t,b}$ that proves $\mathcal{M}_t \models B\theta(b)$.

(5) Consider $\mathcal{M}_s \models Bx = x[a]$ for $a$ such that not for all $t > s$ there exists $b$ with $h_{ts}(b) = a$.

(6, 7, 8) follow easily. ∎

Corollary 1 gives further insight into the intensional character of $B$, which is no longer found to be identical to $S$.

**Corollary 1**

(1) $\models B\forall x(x = x)$, *but* $\not\models \forall x B(x = x)$

(2) *Extensionality for B fails:* $\models \forall x(\top \leftrightarrow x = x)$, *but* $\not\models \forall x(B\top \leftrightarrow B(x = x)$

(3) *For S we have* $\models \forall x(S(\ulcorner \top \urcorner, x) \leftrightarrow S(\ulcorner x = x \urcorner, x))$; *hence* $B \neq S$

Corollary 1 holds anywhere where $B$ and $S$ are defined. Note that according to (1) converse Barcan fails. The notion of extensionality employed in (2) is the one of substitution in the context of modal operators: $\phi \leftrightarrow \psi$ is true iff $B\phi \leftrightarrow B\psi$ is true. The reason for the non-identity of $B$ and $S$, as can be seen in (2) and (3), is that $B$ sometimes suppresses the variable, as in $BT$, while $S$ always carries the variable with it. This is responsible for the intensional behaviour of $B$ that goes beyond the intensionality of $S$.

### 5.6.3 *B and Theory* $\mathcal{T}$

We may now formulate an easy result showing that $B$ represents a consistent geometric theory $\mathcal{T}$ on $\mathcal{T}$-canonical structures, which was our intuitive motivation:

**Corollary 2** *Let* $\mathcal{T}$ *be a geometric theory, and let* $\phi$ *range over geometric implications.*

(*i*) *If for all* $\mathcal{T}$*-canonical structures there exists s such that* $\mathcal{M}_s \models B\phi^g$, *then* $\mathbf{T} \vdash \phi$

(*ii*) *if* $\mathbf{T} \vdash \phi$, *then for every* $\mathcal{T}$*-canonical structure there exists s such that* $\mathcal{M}_s \models B\phi^g$

***Proof***

(i) It suffices to show that $\mathcal{N} \models \mathcal{T}$ implies $\mathcal{N} \models \phi$ for saturated $\mathcal{N}$. Let $\mathcal{N}, \{\mathcal{M}_s \mid s \in \Sigma\}$ be $\mathcal{T}$-canonical. Assume $\mathcal{N} \models \mathcal{T}$ and let $s$ be such that $\mathcal{M}_s \models B\phi^g$. We have to show $\mathcal{N} \models \phi$. Let $\phi = \forall x(\theta(x) \to \tau(x))$, then $\phi^g = \forall x(B\theta(x) \to \tau(x))$. By definition, $\mathcal{M}_s \models B\phi^g$ iff there exists a covering $C_{s,a}$ ($a$ arbitrary) such that

$$\forall (r, b) \in C_{s,a}\mathcal{M}_r \models \forall x(B\theta(x) \to \tau(x)).$$

Let $\mathcal{N} \models \theta(\xi)$. Put $d := \xi_r$ and define a covering $C_{r,d} := \{(u, c) \mid u > r, c = \xi_u\}$. $C_{r,d}$ is a witness for $\mathcal{M}_r \models B\theta(\xi_r)$, whence $\mathcal{M}_r \models \tau(\xi_r)$ and $\mathcal{N} \models \tau(\xi)$. Since $\mathcal{N}$ was arbitrary, we have $\mathcal{T} \models \phi$, whence $\mathcal{T} \vdash \phi$.

(ii) Let $\phi = \forall x(\theta(x) \to \tau(x))$ and assume $\mathcal{T} \models \phi$, then for some finite $s \subseteq \mathbf{T}$, $s \vdash \phi$. It follows that $s \vdash \phi$ also intuitionistically, and since dynamical derivability is complete for intuitionistic logic, $\Vdash_s \phi$, and hence for $t \geq s$, $\Vdash_t \phi$. Since $\phi$ is in the signature of all $\mathcal{M}_t$ for $t \geq s$, we have for all $a \in \mathcal{M}_t$,

$$\mathcal{M}_t \models B\theta(a) \to \theta(a),$$

hence for all $t \geq s$

$$\mathcal{M}_t \models \forall x(\theta(x) \to \tau(x)),$$

from which it follows that

$$\mathcal{M}_s \models B\phi^g.$$

■

We should emphasise that soundness for $B$ does not hold generally on arbitrary indexed models $\mathcal{M}_s$. This is because the signatures may not match in the sense that the $B\phi$ may be evaluated to be true on a model $\mathcal{M}_s$, even though the concepts in $\phi$ are not part of the signature of the model which is required for $S$. It is of course the case that $S\phi$ will be found true once the signature accommodates these concepts, which we know will be the case at some model $\mathcal{M}_t$ with $t > s$. One may be misled into thinking that due to downwards preservation of $S$, $S\phi$ must also be true at $\mathcal{M}_s$, but this is not so. We cannot demand downwards preservation of predicates if these are not available in the signature of the models. This observation highlights another difference between $B$ and $S$, namely, that the interpretation of $B$ is not compositional, unlike that of $S$. Soundness holds on the limit (which also means that the properties of $B$ go beyond **S4**), and therefore the formulas $\phi$ that are in-principle provable form a consistent theory, true on the limit; conversely, a theory true on the limit gives rise to an in-principle provability operator, which is the content of Corollary 2.

Lastly, it is informative to state a completeness theorem for the proposed semantics of $B$, although we omit the proof in this chapter: Suppose we have a consistent geometric theory, then there is a limit model with inverse system, i.e. a $\mathcal{T}$-canonical structure, such that the limit makes true the geometric theory and for each $\phi$ in the theory, $B\phi$ holds at some $\mathcal{M}_s$.

## 5.7 Conclusion

Reinhardt used an operator for in-principle provability in order to formalise the consequences of Gödel's incompleteness theorems. Unlike our first attempt, the semantics that we've proposed for $B$ does not lend definite support to either the second disjunct of Gödel's disjunctive thesis, or its negation. We hope, however, to have made advances on the logic in which to carry out the philosophical discussion surrounding Gödel's disjunctive thesis.

It was noticed early on by both Reinhardt and Shapiro that in-principle provability exhibits a number of intensional features, which we believe separate it from standard quantified **S4**, that would otherwise be a good candidate logic for in-principle provability. Our aim has been to provide a semantics for in-principle provability that takes into account these intensional features while keeping the relation between in-principle provability and actual provability.

Properties 1(a) and 6 in Theorem 2 correspond to the **T**-axiom which embodies a reflection principle that is generally not sound for formal derivability. In [15] we find an attempt to interpret an **S4** *a priori* knowability operator as provability in a formal system which is shown not to work (p.14). It is indeed an essential characteristic of in-principle provability that it is non-formalisable in the following sense: B$\phi$ should not mean that there exists a formal proof of $\phi$ in a given formal system. But this is also in line with our definition of in-principle provability which, however, uses the machinery of dynamical proofs. The reason is that in constructing inverse systems with limit we start from an arbitrary theory $\mathcal{T}$, as opposed to restricting ourselves to theories that are recursively enumerable. This means that the set of finite subsets of the theory is not arithmetically definable; hence there is no way to determine what the formal system is going to be. So we can say that non-formalisability is captured by the notion of covering, which is not related to a particular formal system.

There are interesting commonalities between Carlson [3] and the present proposal. Most notably, Carlson's operator $K$ is studied in terms of knowledge operators relativised to different stages (hence his use of 'stratification'). Moreover, the construction of $K$ is carried out in such a way that for any sentence $\sigma$, $\sigma$ is in theory $\mathcal{T}$ iff $K\sigma$ is in $\mathcal{T}$. We saw that this is also true of $B$ on the limit for guarded geometric implications. A main difference, however, is that, although Carlson rejects the Barcan formula, he maintains converse Barcan which forces increasing domains (p. 8).

The basic principle of our semantics is that mathematics grows both with respect to the entities it constructs and the theorems it proves, where 'growth' is understood in a non-monotonic manner. This means that the modal structures that provide an interpretation to the in-principle provability operator cannot be of the standard quantified **S4** kind, but are given by more general inverse systems. While there is, thus, significant difference between our approach and that of Shapiro and Carlson, there is also strong commonality in that our epistemic operator is also meant to formalise what Carlson calls a theory of knowledge, i.e. mathematical knowledge.

## Notes

1. The type of mathematical knowledge meant here goes under the names of *a priori* knowability, in-principle, or absolute or intuitive provability. These names are used as synonymous in the literature although they may emphasise different aspects of what the operator represents, e.g. the *a priori* character as opposed to a restriction to using mathematical concepts only, or a kind of provability that is distinct from formal derivability. We will be using the term 'in-principle provability' occasionally alternating with 'absolute provability'. The reason is that our final proposal of a semantics for the modal operator can be seen as an attempt to elucidate the notion of provability involved.
2. The motivation driving this work was not the same: while Shapiro's primary aim was to develop a system that is strong enough to contain both classical and intuitionistic logic, Reinhardt's motivation was to investigate the philosophical implications surrounding Gödel's disjunctive thesis.
3. With the additional adjustment that the induction schema allows for formulas that contain the modal operator.
4. There exist different formulations of Gödel's disjunctive thesis in the literature. The one mostly cited comes from Wang [17] p. 324: 'Either the human mind surpasses all machines (to be more precise: it can decide more number theoretic questions than any machine) or else there exist number theoretical questions undecidable for the human mind.'
5. A version of the second disjunct is proven in [7].
6. The disjunction is inclusive; one of the disjuncts says something of the other only if it is proven to be false.

7. Recently, Koellner has also explored the role and use of a truth predicate in the study of the absolute knowability operator (see present volume).
8. We believe that the structure we propose could accommodate Feferman's transfinite progressions of theories [5], yet we do not undertake this task here.
9. Gödel's own view was on the affirmative.
10. Not to be confused with the fregean use of the term.
11. It is somewhat confusing that Reinhardt calls 'intensional' the types of free variables that are allowed under the scope of *B*, precisely because they do not create intensional context. To avoid confusion—especially since we make frequent use of the term in its usual meaning—we will signal Reinhardt's use by quotation marks.
12. Reinhardt also extends the 'intensional' contexts from what is under the scope of *B* to the arguments of predicates; a relation *R* specifies intensional or extensional argument place-holders.
13. The construction can be adjusted to accommodate incompatible theories at different stages of the mathematical development, In that case, directedness ensures that mathematics converges to a consistent mathematical universe on the limit. However, accommodating incompatible theories is not of concern in the literature due to the emphasis that is placed on the soundness condition for *B*.
14. According to Kreisel, yes; according to Troelstra, no.
15. One has access to finite parts of a theory in the following standard sense: (i) any actual proof uses only finitely many axioms; (ii) theories like PA are schematic and become infinite when you consider all instances of the schemata; the resulting theory is recursive and hence has a finite description. Theories that cannot be described recursively are not considered, since one cannot decide which formulas are axioms.
16. All theories will be countable.
17. We assume the geometric implications in the theory have antecedents consisting of conjunctions of atomic formulas only.
18. We omit the universal quantifiers.
19. The notion of covering here is inspired by Coquand [4]. A covering corresponds to a Grothendieck topology. See [16] where this is spelled out.

# References

[1] T. Achourioti. Intensional truth: a study beyond deflationism. (Unpublished doctoral thesis.)

[2] T. Achourioti and M. van Lambalgen. A formalisation of Kant's transcendental logic. *Review of Symbolic Logic*, 4 (02):254–289, 2011.

[3] T.J. Carlson. Knowledge, machines, and the consistency of Reinhardt's strong mechanistic thesis. *Annals of Pure and Applied Logic*, **105** (1–3):51–81 (2000), 1999.

[4] T. Coquand. A completeness proof for geometrical logic. See http://www.cse.chalmers.se/~coquand/site.pdf

[5] S. Feferman. Transfinite recursive progressions of formal theories. *Journal of Symbolic Logic*, 27:259–316, 1962.

[6] K. Gödel. The modern development of the foundations of mathematics in the light of philosophy. In *Collected Works*, pp. 374–375. Oxford University Press, 1961.

[7] L. Horsten. Provability in principle and controversial constructivistic principles. *Journal of Philosophical Logic*, **26**:635–660, 1997.

[8] R.G. Jeroslow. Experimental logics and $\Delta^0_2$-theories. *Journal of Philosophical Logic*, **4**:253–267, 1975.
[9] J. Lucas. Minds, machines and Gödel. *Philosophy*, **36**:120–124, 1961.
[10] G. Mints. Classical and intuitionistic geometric logic. *Proceedings of PhML 2012, St. Petersburgh*, pp. 112–117, 2012.
[11] R. Montague. Syntactical treatment of modality, with corollaries on reflection principle and finite axiomatizability. *Acta Philosophica Fennica*, **16**:153–167, 1963.
[12] R. Penrose. *The Emperor's New Mind*. Oxford University Press, 1989.
[13] R. Penrose. *Shadows of the Mind*. Oxford University Press, 1994.
[14] W.N. Reinhardt. Epistemic theories and the interpretation of Gödel's incompleteness theorems. *Journal of Philosophical Logic*, **15** (4):427–474, 1986.
[15] S. Shapiro. Epistemic and intuitionistic arithmetic. In S. Shapiro, ed., *Intensional Mathematics*, pp. 51–82. North-Holland, 1985.
[16] A.S. Troelstra and D. Van Dalen. *Constructivism in Mathematics: An Introduction*, volume 2. Elsevier, 1988.
[17] H. Wang. *From Mathematics to Philosophy*. Humanities Press, 1974.

# PART II
## Mind and Machines

# 6

## Collapsing Knowledge and Epistemic Church's Thesis

TIMOTHY J. CARLSON

*Epistemic Church's Thesis*, or *ECT*, is a schema formalizing a strong version of Church's Thesis within the theory of Epistemic Arithmetic, *EA*, which says

> For any function I can compute,
> I can find a Turing machine which computes the same function.

*ECT* was shown to be consistent with *EA* by R. Flagg [3]. We will show the collapsing technique of [2] can be used to provide an alternative proof of this result by showing that there is a knowing machine which knows *EA* + *ECT*. This argument can be combined with that of [2] to show that there is a knowing machine which knows *EA* + *SMT* + *ECT*, where *SMT* is Reinhardt's Strong Mechanistic Thesis. Further elaboration of the arguments establishes the independence of *SMT* and *ECT*. These results will appear elsewhere.

### 6.1 Introduction

Assume $\mathcal{L}$ is a first-order language with the equality symbol. We fix an enumeration of a countable set of variables $v_0, v_1, \ldots$ which we assume are included in the variables of every first-order language.

For a set $\mathcal{K}$ disjoint from the symbols of $\mathcal{L}$, $\mathcal{L}^{\mathcal{K}}$ is the language obtained from $\mathcal{L}$ by adding the elements of $\mathcal{K}$ as modal operators. Formulas are defined inductively as usual, with the additional clause that $K\varphi$ is a formula whenever $K \in \mathcal{K}$ and $\varphi$ is a formula. We adopt the convention that atomic formulas are of the form $Pt_1 \ldots t_n$, where $P$ is an $n$-place predicate symbol and $t_1, \ldots, t_n$ are terms. We often write certain binary predicate symbols, e.g. =, informally using infix notation. In that case, we sometimes include outer parentheses. Similarly, for binary operation symbols that are traditionally written using infix, we often include outer parentheses. On the other hand, we informally drop parentheses in formulas when doing so improves readability.

*Gödel's Disjunction*. First Edition. Leon Horsten & Philip Welch (eds).

Formulas of the form $K\varphi$, where $K \in \mathcal{K}$, will be referred to as *epistemic* formulas. Each $K \in \mathcal{K}$ can be thought of as representing knowledge of some entity. $K\varphi$ may be read as $K$ *knows* $\varphi$.

When $\mathcal{K} = \{K\}$, we often write $\mathcal{L}^K$ for $\mathcal{L}^{\mathcal{K}}$.

When $\varphi$ is a formula and $x$ is a variable, we define an occurrence of $x$ in $\varphi$ to be *free* in $\varphi$ in the usual way, i.e. it must not be in the scope of any quantification of $x$. An occurrence of a variable is *bound* if it is not free. Two formulas are *alphabetic variants* if they differ only in the choice of bound variables. We will identify alphabetic variants.

When $x_1, \ldots, x_n$ are distinct variables, we will use $(x_1, \ldots, x_n | y_1, \ldots, y_n)$ to denote the function which maps $x_i$ to $y_i$ for each $i = 1, \ldots, n$. If $\varphi$ is a formula and each $y_i$ is a term, then $\varphi(x_1, \ldots, x_n | y_1, \ldots, y_n)$ is the result of simultaneously replacing each free occurrence of $x_i$ by $y_i$ in $\varphi$ for $i = 1, \ldots, n$. We define $t(x_1, \ldots, x_n | y_1, \ldots, y_n)$ analogously when $t$ is a term. When $x_i = v_i$ for $i = 0, \ldots, n$ we may write $\varphi(t_0, \ldots, t_n)$ and $u(t_0, \ldots, t_n)$ for $\varphi(x_0, \ldots, x_n | t_0, \ldots, t_n)$ and $u(x_0, \ldots, x_n | t_0, \ldots, t_n)$ respectively. Since we are identifying alphabetic variants, all substitutions are without unintended quantification of variables.

An *assignment* is a function whose domain is a set of variables. If $s_1$ and $s_2$ are assignments, we define $s_1 s_2$ to be the assignment whose domain is the union of the domains of $s_1$ and $s_2$ and which sends any $x$ in the domain of $s_2$ to $s_2(x)$ and any $x$ in the domain of $s_1$ but not in the domain of $s_2$ to $s_1(x)$. For example, $s(x|a)$ is the function $\tilde{s}$ whose domain is $dom(s) \cup \{x\}$, with the property that $\tilde{s}(x) = a$ and $\tilde{s}(z) = s(z)$ whenever $z$ is in the domain of $s$ and is distinct from $x$.

A *structure* for $\mathcal{L}^{\mathcal{K}}$ consists of a structure for $\mathcal{L}$ extended to give interpretations to the epistemic formulas $K\varphi$ by assigning to $K\varphi$ a subset of the collection of all functions mapping the free variables of $K\varphi$ into the universe of the structure. These interpretations are required to satisfy the following *coherence condition.*

($\star$) If $x_1, \ldots, x_n$ lists the free variables of $\varphi$ without repetitions, $y_1, \ldots, y_n$ is a list of variables (possibly with repetitions) and $K \in \mathcal{K}$, then the interpretation of $K\varphi(x_1, \ldots, x_n | y_1, \ldots, y_n)$ consists of all functions $\sigma$ with domain $\{y_1, \ldots, y_n\}$ so that the function which sends $x_j$ to $\sigma(y_j)$ for each $j$ is in the interpretation of $K\varphi$.

We remark that one of the conditions in the definition of structure in [2] could be omitted since we are identifying alphabetic variants.

For $\mathcal{A}$ a structure for $\mathcal{L}^{\mathcal{K}}$, $t$ term and $s$ an assignment whose domain includes the variables in $t$, we use $\bar{s}^{\mathcal{A}}(t)$ to denote the value of $t$ in $\mathcal{A}$ under the assignment $s$. When $\mathcal{A}$ is clear from context, we will drop the superscript $\mathcal{A}$ and simply write $\bar{s}(t)$.

For $\mathcal{A}$ a structure for $\mathcal{L}^{\mathcal{K}}$, $\varphi$ a formula and $s$ an assignment in the universe of $\mathcal{A}$ whose domain includes the free variables of $\varphi$, define $\mathcal{A} \models \varphi[s]$ inductively as usual, with the additional clause that

$$\mathcal{A} \models K\psi[s]$$

iff

$s$ extends an element of the interpretation of $K\psi$.

For $\Sigma$ a set of formulas of $\mathcal{L}^{\mathcal{K}}$ and $\varphi$ a formula of $\mathcal{L}^{\mathcal{K}}$, we define $\Sigma \models \varphi$ iff for every structure $\mathcal{A}$ for $\mathcal{L}^{\mathcal{K}}$ and every $s$ which assigns elements of the universe of $\mathcal{A}$ to the variables, if $\mathcal{A} \models \sigma[s]$ for each $\sigma$ in $\Sigma$ then $\mathcal{A} \models \varphi[s]$.

As in [2], we will omit the details of defining analogues of other notions of first-order logic in this context and will use the same notation that is used for first-order logic when the meaning is clear.

The coherence condition in the definition of the notion of a structure for $\mathcal{L}^{\mathcal{K}}$ implies that

$$\mathcal{A} \models \varphi(x_1,\ldots,x_n|y_1,\ldots,y_n)[s] \text{ iff } \mathcal{A} \models \varphi[s(x_1,\ldots,x_n|s(y_1),\ldots,s(y_n))]$$

whenever $y_1,\ldots,y_n$ are variables. Therefore,

$$\mathcal{A} \models (x_1 = y_1 \wedge \cdots \wedge x_n = y_n \rightarrow (\varphi \leftrightarrow \varphi(x_1,\ldots,x_n|y_1,\ldots,y_n)))\,[s]$$

for all $s$ and all $\mathcal{A}$ which respect equality, i.e. interpret the equality symbol as equality. In general, one could interpret the equality symbol in a structure as an arbitrary congruence over the reduct to $\mathcal{L}$. The equivalence above does not generalize to the case where $y_1,\ldots,y_n$ are arbitrary terms.

Henceforth, the logic we use when discussing formulas of $\mathcal{L}^{\mathcal{K}}$ is the base logic.

S. Shapiro [7] and W. Reinhardt [4, 5, 6] independently developed an epistemic version of Peano Arithmetic. Shapiro called his theory *Epistemic Arithmetic*, Abbreviated as *EA*.

The *language of Peano Arithmetic*, $\mathcal{L}_{PA}$, is the first-order language with non-logical symbols $\mathbf{0}$,$\mathbf{S}$, $+$ and $\cdot$ as usual.

The *basic arithmetic axioms* are the following sentences of $\mathcal{L}_{PA}$.

- Sentences saying $\mathbf{0}$ is not in the range of $\mathbf{S}$ and $\mathbf{S}$ is one-to-one:

$$\forall x(x \neq \mathbf{0})$$

and

$$\forall x \forall y(\mathbf{S}x = \mathbf{S}y \rightarrow x = y)$$

- The usual defining formulas for addition:

$$\forall x(x + \mathbf{0} = x)$$

and

$$\forall x \forall y(x + \mathbf{S}y = \mathbf{S}(x + y))$$

- The usual defining formulas for multiplication:

$$\forall x(x \cdot \mathbf{0} = \mathbf{0})$$

and

$$\forall x \forall y (x \cdot \mathbf{S}y = x \cdot y + x)$$

The *standard model of arithmetic* is the structure for $\mathcal{L}_{PA}$, denoted by $\mathcal{N}$, with universe $\omega$ and which interprets $\mathbf{0}$, $\mathbf{S}$, $+$ and $\cdot$ as the number zero and the standard operations of successor, addition, and multiplication.

For $\mathcal{K}$ a set of modal operators, the *Peano axioms* for $\mathcal{L}^{\mathcal{K}}_{PA}$ consist of the basic arithmetic axioms along with the *induction schema* for $\mathcal{L}^{\mathcal{K}}_{PA}$ made up of the universal closures of all formulas of the form

$$\varphi(x|\mathbf{0}) \wedge \forall x(\varphi \rightarrow \varphi(x|\mathbf{S}x)) \rightarrow \forall x \varphi$$

We will also need the following schemas.

$E_1$: Universal closures of formulas of the form

$$K\varphi$$

where $\varphi$ is valid (in the base logic)

$E_2$: Universal closures of formulas of the form

$$K(\varphi \rightarrow \psi) \rightarrow K\varphi \rightarrow K\psi$$

$E_3$: Universal closures of formulas of the form

$$K\varphi \rightarrow \varphi$$

$E_4$: Universal closures of formulas of the form

$$K\varphi \rightarrow KK\varphi$$

As shown in [2], *EA* in the language $\mathcal{L}^{K}_{PA}$ is the theory in the base logic axiomatized by $\Sigma \cup K\Sigma$, where $\Sigma$ consists of all sentences $\sigma$ such that $\sigma$ is either a Peano axiom or an element of some $E_i$.

*ECT* is the schema consisting of the universal closures of formulas of the form

$$K\forall x \exists y K\varphi \rightarrow \exists e K \forall x \exists y (E(e, x, y) \wedge \varphi)$$

where $e$ does not occur free in $\varphi$ and $E$ is a fixed formula of $\mathcal{L}_{PA}$ with free variables $v_0$, $v_1$ and $v_2$ such that, letting $\mathcal{N}$ be the standard model of arithmetic,

$$\mathcal{N} \models E(e, x, y)\ [e, x, y | a, m, n]$$

iff on input $m$, the $a^{th}$ Turing machine halts and outputs $n$. For convenience, we will write $\{t_1\}(t_2) \simeq t_3$ for $E(t_1, t_2, t_3)$ when $t_1$, $t_2$, and $t_3$ are terms.

## 6.2 Knowing Entities and Syntactic Encoding

When we build structures for $\mathcal{L}^{\mathcal{K}}$, we will represent the interpretations of epistemic formulas syntactically by specifying a set of formulas $T_K$ for each $K \in \mathcal{K}$ which codes $K$'s knowledge.

Suppose $\mathcal{M}$ is a structure for $\mathcal{L}$ and let $M$ denote the universe of $\mathcal{M}$.

**Definition 6.2.1** *Assume $t_a$ is a term of $\mathcal{L}$ for $a \in M$. For $\varphi$ a formula and $s$ an assignment in $M$, define $\varphi^s$ to be the result of simultaneously substituting $t_{s(x)}$ for $x$ in $\varphi$ for all variables $x$.*

Notice we have suppressed the dependence of $\varphi^s$ on $t_a$ $(a \in M)$.

Assume for each $a \in M$ we have chosen a ground term $t_a$ to represent $a$ (a *ground term* is a term without occurrences of variables). Also, assume $T_K$ is a set of sentences of $\mathcal{L}^{\mathcal{K}}$ for each $K \in \mathcal{K}$. We can expand $\mathcal{M}$ to a structure $\mathcal{A}$ for $\mathcal{L}^{\mathcal{K}}$ so that

$$(\star\star) \qquad \mathcal{A} \models K\varphi\,[s] \text{ iff } \varphi^s \in T_K$$

**Definition 6.2.2** *Assume $\mathcal{A}$ to be an expansion of $\mathcal{M}$ to $\mathcal{L}^{\mathcal{K}}$ and $T_K$ is a set of sentences of $\mathcal{L}^{\mathcal{K}}$ for $K \in \mathcal{K}$. If $t_a$ $(a \in M)$ is a family of ground terms, we say $T_K$ $(K \in \mathcal{K})$* encodes *$\mathcal{A}$ with respect to $t_a$ $(a \in M)$ if $(\star\star)$ holds for all formulas $\varphi$ of $\mathcal{L}^{\mathcal{K}}$. We say $T_K$ $(K \in \mathcal{K})$* encodes *$\mathcal{A}$ if $T_K$ $(K \in \mathcal{K})$ encodes $\mathcal{A}$ with respect to some family $t_a$ $(a \in M)$.*

Notice $\varphi^s$ is always a sentence and if $\varphi$ is a sentence itself then $\varphi^s = \varphi$. This implies that if $T_K$ $(K \in \mathcal{K})$ encodes $\mathcal{A}$ with respect to $t_a$ $(a \in M)$ then

$$\begin{aligned} T_K &= \{\varphi^s \mid \mathcal{A} \models K\varphi[s]\} \\ &= \{\varphi \mid \varphi \text{ is a sentence and } \mathcal{A} \models K\varphi\} \end{aligned}$$

**Definition 6.2.3** *A family $t_a$ $(a \in M)$ of ground terms is a* proper assignment of terms *for $\mathcal{M}$ if $t_a^{\mathcal{M}} = a$ for all $a \in M$.*

**Lemma 6.2.4** *Assume $T_K$ $(K \in \mathcal{K})$ encodes $\mathcal{A}$ with respect to $t_a$ $(a \in M)$ where $t_a$ $(a \in M)$ is proper. For any formula $\varphi$ and assignment $s$*

$$\mathcal{A} \models \varphi[s] \text{ iff } \mathcal{A} \models \varphi^s$$

***Proof*** Straightforward induction. ■

The lemma shows that we cannot expect every expansion of $\mathcal{M}$ to be encoded using $t_a$ $(a \in M)$. For example, assume $\mathcal{A} \models \forall x\, K(x = x)$. This implies $\mathcal{A} \models \forall x \exists y K(x = y)$. By the lemma, $\mathcal{A} \models \exists y K(t_a = y)$ for all $a \in M$, which may be interpreted as saying $K$ knows how to verify a computation of $t_a$.

The lemma also implies that $\mathcal{A} \models \varphi^s$ whenever $\varphi$ is valid. This is not true in general, e.g. $\forall x KPx \rightarrow KPx$ is valid but $\forall x KPx \rightarrow KPt$ is not unless $t$ is a variable.

For $\Sigma$ a set of formulas of $\mathcal{L}^{\mathcal{K}}$, we write $K\Sigma$ for the set of formulas $K\sigma$ where $\sigma \in \Sigma$.

**Lemma 6.2.5** *Assume $T$ encodes $\mathcal{A}$ with respect to $t_a$ $(a \in M)$ where $t_a$ $(a \in M)$ is proper.*

1. $\mathcal{A} \models E_1$ *iff* $\varphi^s \in T$ *for all valid* $\varphi$ *and all* $s$.
2. $\mathcal{A} \models E_2$ *iff* $T$ *is closed under modus ponens.*
3. $\mathcal{A} \models E_3$ *iff* $\mathcal{A} \models T$.
4. $\mathcal{A} \models E_4$ *iff* $KT \subseteq T$.
5. *For* $\Sigma$ *a set of sentences,* $\mathcal{A} \models K\Sigma$ *iff* $\Sigma \subseteq T$.

***Proof*** Straightforward by Lemma 6.2.4. ■

**Definition 6.2.6** *Assume $T$ is a set of sentences of $\mathcal{L}^K$. $T$ is a* knowing entity *over $\mathcal{M}$ iff there is an expansion of $\mathcal{M}$ encoded by $T$ which satisfies $\Sigma \cup K\Sigma$ where $\Sigma = E_1 \cup E_2 \cup E_3 \cup E_4$. $T$ is a* knowing machine *over $\mathcal{M}$ if $T$ is a computably enumerable knowing entity over $\mathcal{M}$.*

Notice that if $T$ is a knowing entity over $\mathcal{M}$ then $T$ is a theory in $\mathcal{L}^K$. This follows from parts 1 and 2 of Lemma 6.2.5. Also, $T$ must be $K$-closed by part 4: if $\sigma \in T$ then $K\sigma \in T$.

We remark that if $T$ is a knowing entity over $\mathcal{M}$ then $T$ encodes a unique expansion $\mathcal{A}$ of $\mathcal{M}$ that is a model of $\Sigma \cup K\Sigma$ where $\Sigma$ is the union of the $E_i$.

We also remark that in certain cases, e.g. when not every element of $M$ is the value of a ground term, the requirement that encoding terms be ground terms can be dropped.

These issues will be discussed further elsewhere.

## 6.3 Hierarchies and Stratification

As in Section 6.2, fix a structure $\mathcal{M}$ for a first-order language $\mathcal{L}$.

In order to produce and verify that certain theories in $\mathcal{L}^K$ are knowing machines, we first produce a stratified version of the theory by adding modal operators $K_\alpha$ for each element $\alpha$ of a linear ordering $\mathcal{Q} = (Q, \leq^{\mathcal{Q}})$. The elements of $Q$ can be thought of as moments in time and $K_\alpha \varphi$ can be thought of as saying that $\varphi$ is known, or is knowable, at time $\alpha$.

Fix $\mathcal{Q}$ for the rest of this section. We will write $<$ and $\leq$ for $<^{\mathcal{Q}}$ and $\leq^{\mathcal{Q}}$, respectively, when no confusion is likely.

In Section 6.4, we will see how the operators $K_\alpha$ $(\alpha \in Q)$ can sometimes be collapsed into a single operator.

Later, we will want to assume that $\mathcal{Q}$ is a well-ordering, but that will not be necessary for this section.

**Definition 6.3.1** *The* stratified *formulas of $\mathcal{L}^{\{K_\alpha \,|\, \alpha \in Q\}}$ are the formulas generated from the atomic formulas of $\mathcal{L}$ as usual with the modified condition that $K_\alpha \varphi$ is a formula whenever $\varphi$ is a formula such that $\beta < \alpha$ for all $\beta$ such that $K_\beta$ occurs in $\varphi$. $\mathcal{L}^{\mathcal{Q}}$ is the fragment of $\mathcal{L}^{\{K_\alpha \,|\, \alpha \in Q\}}$ consisting of the stratified formulas.*

Notice the stratified formulas of $\mathcal{L}^{\{K_\alpha \,|\, \alpha \in Q\}}$ with respect to $\mathcal{Q}$ are those formulas such that $\beta < \alpha$ whenever $K_\beta$ is in the scope of $K_\alpha$.

We will identify structures for $\mathcal{L}^Q$ with structures for $\mathcal{L}^{\{K_\alpha \mid \alpha \in Q\}}$ in which the interpretation of $K_\alpha\varphi$ is empty whenever $K_\alpha\varphi$ is not stratified. For $X \subseteq Q$ we will write $\mathcal{L}^X$ for $\mathcal{L}^{Q \upharpoonright X}$. When $\alpha \in Q$, $\mathcal{L}^\alpha$ will denote $\mathcal{L}^X$ when $X$ is the set of $\beta$ with $\beta < \alpha$. Moreover, we will abuse notation and say $T_\alpha$ ($\alpha \in Q$) encodes $\mathcal{A}$ with respect to $t_a$ ($a \in M$) when $T_K$ ($K \in \{K_\alpha \mid \alpha \in Q\}$) encodes $\mathcal{A}$ with respect to $t_a$ ($a \in M$).

The following schemas are stratified analogues of the schemas $E_i$ for $i = 1, 2, 3, 4$. Assume $\alpha \in Q$.

$SE_1(\alpha)$: Universal closures of stratified formulas of the form

$$K_\alpha\varphi$$

where $\varphi$ is valid.

$SE_2(\alpha)$: Universal closures of stratified formulas of the form

$$K_\alpha(\varphi \to \psi) \to K_\alpha\varphi \to K_\alpha\psi$$

$SE_3(\alpha)$: Universal closures of stratified formulas of the form

$$K_\alpha\varphi \to \varphi$$

$SE_4(\alpha)$: Universal closures of stratified formulas of the form

$$K_\beta\varphi \to K_\alpha K_\beta\varphi$$

For $i = 1, 2, 3, 4$ and $Y \subseteq Q$, $SE_i(Y)$ is the collection of all sentences of $\mathcal{L}^Y$ in $SE_i(\alpha)$ for some $\alpha$. We will write $SE_i$ for $SE_i(Q)$.

The following lemma is an analogue of Lemma 6.2.5.

**Lemma 6.3.2** *Assume $T_\alpha$ ($\alpha \in Q$) encodes $\mathcal{A}$ with respect to $t_a$ ($a \in M$) where $t_a$ ($a \in M$) is proper and $T_\alpha$ is a set of sentences in $\mathcal{L}^\alpha$ for $\alpha \in Q$.*

1. *$\mathcal{A} \models SE_1(\alpha)$ iff $\varphi^s \in T_\alpha$ for all valid $\varphi$ of $\mathcal{L}^\alpha$ and all $s$.*
2. *$\mathcal{A} \models SE_2(\alpha)$ iff $T_\alpha$ is closed under modus ponens.*
3. *$\mathcal{A} \models SE_3(\alpha)$ iff $\mathcal{A} \models T_\alpha$.*
4. *$\mathcal{A} \models SE_4(\alpha)$ iff $K_\beta T_\beta \subseteq T_\alpha$ for all $\beta < \alpha$.*
5. *For $\Sigma$ a set of sentences in $\mathcal{L}^\alpha$, $\mathcal{A} \models K_\alpha\Sigma$ iff $\Sigma \subseteq T_\alpha$.*

***Proof*** Straightforward by Lemma 6.2.4. ■

The following lemma applies to part 4 of Lemma 6.3.2.

**Lemma 6.3.3** *Assume $\Sigma(\alpha)$ is a set of sentences in $\mathcal{L}^\alpha$ and $T_\alpha$ is the theory in $\mathcal{L}^\alpha$ axiomatized by $\Sigma(\alpha)$ for $\alpha \in Q$. If $\beta < \alpha$ and $SE_1(\beta) \cup SE_2(\beta) \cup K_\beta\Sigma(\beta) \subseteq \Sigma(\alpha)$ then $K_\beta T_\beta \subseteq T_\alpha$.*

***Proof*** Since $\Sigma(\alpha)$ contains $SE_1(\beta)$ and $SE_2(\beta)$, the collection of sentences $\sigma$ of $\mathcal{L}^\beta$ such that $\Sigma(\alpha) \models K_\beta\sigma$ is closed under logical implication. Since $K_\beta\Sigma(\beta) \subseteq \Sigma(\alpha)$, this implies $\Sigma(\alpha) \models K_\beta T_\beta$. Notice $SE_1(\beta)$ and $SE_2(\beta)$ are only needed for sentences. ■

## 6.4 Collapsing

Assume $\mathcal{Q} = (Q, \leq)$ is a linear ordering and $\mathcal{M}$ is a structure for the first-order language $\mathcal{L}$ with universe $M$. In this section, we describe how some structures for $\mathcal{L}^\mathcal{Q}$ can be collapsed to a structure for $\mathcal{L}^K$.

**Definition 6.4.1** *Assume $\varphi$ is a formula of $\mathcal{L}^\mathcal{Q}$. The* collapse *of $\varphi$, $\varphi^-$, is the formula of $\mathcal{L}^K$ obtained from $\varphi$ by replacing each occurrence of any of the $K_\alpha$ in $\varphi$ by $K$.*

**Lemma 6.4.2** *Assume $Q$ is infinite. If $\varphi$ is a formula of $\mathcal{L}^K$ then there exists a formula $\tilde{\varphi}$ of $\mathcal{L}^\mathcal{Q}$ such that $\tilde{\varphi}^- = \varphi$. Moreover, if $\varphi$ is valid then $\tilde{\varphi}$ can be chosen to be valid.*

***Proof*** To ensure $\tilde{\varphi}$ is valid if $\varphi$ is, we want to replace each occurrence of $K$ in $\phi$ by some $K_\alpha$ in a uniform way so that substitution is preserved. When $\mathcal{Q}$ has an infinite increasing sequence, one way of doing this is to define a formula $\tilde{\varphi}$ by induction on the complexity of $\varphi$ so that whenever $\varphi$ has the form $K\psi$ then $\tilde{\varphi}$ is $K_\alpha\tilde{\psi}$ where $\alpha$ is the least element of the sequence with the property that $K_\alpha$ does not occur in $\tilde{\psi}$. Since $\mathcal{Q}$ may not have an infinite increasing sequence, we will modify this approach by considering the collection $\Phi$ of subformulas of $\varphi$. Let $Q'$ be a subset of $Q$ of size the cardinality of $\Phi$. We can now carry out the construction for the formulas in $\Phi$, replacing $Q$ by $Q'$. One easily checks that for $\varphi_1, \varphi_2 \in \Phi$, if $\varphi_2 = \varphi_1(x_1, x_2, \ldots, x_m|y_1, y_2, \ldots, y_m)$ then $\tilde{\varphi_2} = \tilde{\varphi_1}(x_1, x_2, \ldots, x_m|y_1, y_2, \ldots, y_m)$.

Assume $\varphi$ is valid. In order to show that $\tilde{\varphi}$ is valid, suppose $\mathcal{A}$ is a structure for $\mathcal{L}^\mathcal{Q}$. By the properties of the function $\psi \mapsto \tilde{\psi}$, we can find a structure $\mathcal{B}$ for $\mathcal{L}^K$ such that the interpretation of $\psi$ in $\mathcal{B}$ is the same as the interpretation of $\tilde{\psi}$ in $\mathcal{A}$ whenever $\psi$ is a subformula of $\varphi$ of the form $K\theta$. By induction on the complexity of formulas $\psi \in \Phi$, $\mathcal{A} \models \tilde{\psi}[s]$ iff $\mathcal{B} \models \psi[s]$ for all assignments $s$. Therefore, $\mathcal{A} \models \tilde{\phi}[s]$ for all $s$. ■

**Definition 6.4.3** *Assume $\mathcal{A}$ is a structure for $\mathcal{L}^\mathcal{Q}$ which extends $\mathcal{M}$ and $C \subseteq Q$. We say $\mathcal{A}$* collapses *on $C$ provided the interpretations of $\varphi$ and $\psi$ are identical for all epistemic formulas $\varphi$ and $\psi$ of $\mathcal{L}^C$ with $\varphi^- = \psi^-$. If $\mathcal{A}$ collapses on $C$ and $C$ is infinite, the* collapse *of $\mathcal{A}$ on $C$ is the structure $\mathcal{A}^-$ for $\mathcal{L}^K$ which extends $\mathcal{M}$ with the property that the interpretation of $\varphi^-$ in $\mathcal{A}^-$ is identical to the interpretation of $\varphi$ in $\mathcal{A}$ for any epistemic formula $\varphi$ of $\mathcal{L}^C$.*

**Theorem 6.4.4** *Assume $\mathcal{A}$ collapses on $C$ where $C \subseteq Q$ is infinite and let $\mathcal{A}^-$ be the collapse of $\mathcal{A}$ on $C$. For all formulas $\varphi$ of $\mathcal{L}^C$*

$$\mathcal{A} \models \varphi[s] \text{ iff } \mathcal{A}^- \models \varphi^-[s]$$

*for all assignments $s$ in $M$.*

***Proof*** Straightforward. ■

**Definition 6.4.5** *Assume $\Sigma$ is a set of formulas of $\mathcal{L}^{\mathcal{Q}}$ and $X \subseteq \mathcal{Q}$. The $K_X$-*closure *of $\Sigma$ is the smallest collection of formulas $\Delta$ of $\mathcal{L}^{\mathcal{Q}}$ containing $\Sigma$ such that $K_\alpha\varphi \in \Delta$ whenever $\varphi \in \Delta$, $\alpha \in X$ and $K_\alpha\varphi$ is stratified. We say $\Sigma$ is $K_X$-*closed *if it equals its $K_X$-*closure.

**Theorem 6.4.6** *Let $\Sigma$ be the $K_{\mathcal{Q}}$-closure of $SE_1 \cup SE_2 \cup \Gamma$ where $\Gamma$ is the collection of universal closures of formulas of the form*

$$K_\beta\varphi \leftrightarrow K_\alpha\varphi$$

*with $\alpha < \beta$. For all formulas $\varphi$ and $\psi$ of $\mathcal{L}^{\mathcal{Q}}$, if $\varphi^- = \psi^-$ then $\Sigma \models \varphi \leftrightarrow \psi$.*

***Proof*** Clearly, it suffices to establish the theorem for finite $\mathcal{Q}$. We do so by induction on cardinality.

Assume $n \in \omega$ and the theorem holds for all $\mathcal{Q}$ with $card(\mathcal{Q}) < n$.

The statement of the theorem follows easily from the case where $\varphi$ and $\psi$ are epistemic formulas. So, suppose $\varphi = K_\alpha\varphi_0$ and $\psi = K_\beta\psi_0$ are formulas of $\mathcal{L}^{\mathcal{Q}}$ such that $\varphi^- = \psi^-$ or, equivalently, $\varphi_0^- = \psi_0^-$.

By symmetry, we may assume $\alpha \leq \beta$. Let $\mathcal{Q}'$ be the restriction of $\mathcal{Q}$ to the set of $\gamma$ with $\gamma < \beta$. Let $\Gamma'$ be the collection of formulas of $\mathcal{L}^{\mathcal{Q}'}$ which are in $\Gamma$ and let $\Sigma'$ be the $\mathcal{L}^{\mathcal{Q}'}$-closure of $SE_1(\mathcal{Q}') \cup SE_2(\mathcal{Q}') \cup \Gamma'$. By the induction hypothesis, $\Sigma' \models \varphi_0 \leftrightarrow \psi_0$.

Since $SE_1(\beta) \cup SE_2(\beta) \subseteq \Sigma$, the set of formulas $\theta$ such that $\Sigma \models K_\beta\theta$ is a subset of $\mathcal{L}^\beta$ that is closed under logical implication. Since $\Sigma$ is $K_{\mathcal{Q}}$-closed, $K_\beta\Sigma' \subseteq \Sigma$. Since $\Sigma' \models \varphi_0 \leftrightarrow \psi_0$, $\Sigma \models K_\beta(\varphi_0 \leftrightarrow \psi_0)$. Appealing to $SE_1(\beta)$ and $SE_2(\beta)$ again, we see $\Sigma \models K_\beta\varphi_0 \leftrightarrow K_\beta\psi_0$. If $\alpha = \beta$, this is the desired conclusion. So, assume $\alpha < \beta$. Since the universal closure of $K_\beta\varphi_0 \leftrightarrow K_\alpha\varphi_0$ is in $\Gamma$, $\Sigma \models K_\beta\varphi_0 \leftrightarrow K_\alpha\psi_0$. ■

## 6.5 A Computable Collapsing Relation

In this section we discuss a computable relation on the ordinal $\varepsilon_0 \cdot \omega$, $\preceq_1$ of [1], which is useful in constructing stratified structures which collapse. The definition of $\preceq_1$ is based on the notion of $\Sigma_1$-elementary substructure. Since the definition of a $\Sigma_1$ formula can vary, we specify that a $\Sigma_1$ formula is a quantifier-free formula prefixed by a string of existential quantifications. If $\mathcal{A}$ and $\mathcal{B}$ are structures for the same languages we will write $\mathcal{A} \preceq_{\Sigma_1} \mathcal{B}$ to indicate that $\mathcal{A}$ is a $\Sigma_1$-elementary substructure of $\mathcal{B}$.

**Definition 6.5.1** *Define the binary relation $\preceq_1$ on the ordinals inductively so that*

$$\alpha \preceq_1 \beta \text{ iff } (\alpha, \leq, \preceq_1) \text{ is a } \Sigma_1\text{-elementary substructure of } (\beta, \leq, \preceq_1).$$

*The induction is on $\beta$ simultaneously for all $\alpha$. We will write $\alpha \preceq_1 \infty$ when $\alpha \preceq_1 \beta$ for all $\beta \geq \alpha$.*

Notice that for any ordinal $\alpha$ the collection of all $\beta$ such that $\alpha \preceq_1 \beta$ is an interval of ordinals and either $\alpha \preceq_1 \infty$ or there is a largest $\beta$ such that $\alpha \preceq_1 \beta$.

The construction in Section 6.6 will require properties of $\preceq_1$ described in the following two theorems. The proofs hinge on the main result of [1] and will appear elsewhere.

**Definition 6.5.2** *The* pointwise partial ordering *of the collection of finite sets of ordinals is the partial ordering $\preceq_p$ where $X \preceq_p Y$ iff $card(X) = card(Y)$ and the $i^{th}$ element of X is less than or equal to the $i^{th}$ element of Y for each $i < card(X)$. If A is a collection of finite sets of ordinals and $X \in A$ has the property that $X \preceq_p Y$ for all $Y \in A$ then X is the* pointwise minimum *element of A.*

**Theorem 6.5.3** *If X is a finite set of ordinals then the collection of Y with $X \cong Y$ (as substructures of $(ORD, \leq, \preceq_1)$) has a pointwise minimum element. Moreover, if $\tilde{X}$ is the pointwise minimum of such Y and the largest element of $\tilde{X}$ is $\tilde{\mu}$, then there is an embedding of $[0, \tilde{\mu}]$ into ORD which maps X onto $\tilde{X}$.*

We will want to identify ordinals below $\varepsilon_0 \cdot \omega$ with notations using a fully reduced Cantor normal form as follows. Every ordinal below $\varepsilon_0$ can be written uniquely in Cantor normal form as

$$\omega^{\alpha_1} + \cdots + \omega^{\alpha_n},$$

where $\varepsilon_0 > \alpha_1 \geq \cdots \geq \alpha_n$ (we interpret the empty sum as 0). If we similarly expand each $\alpha_i$ and so on, we obtain a term for $\alpha$ which uses only exponentiation to base $\omega$, addition, and 0. If we identify $\alpha$ with this term we obtain a notation system for ordinals below $\varepsilon_0$. We extend this system to ordinals below $\varepsilon_0 \cdot \omega$ by using terms

$$\varepsilon_0 + \cdots + \varepsilon_0 + \omega^{\alpha_1} + \cdots + \omega^{\alpha_n}$$

with $\varepsilon_0 > \alpha_1 \geq \cdots \geq \alpha_n$ (where each $\alpha_i$ is now a term). The projections of addition, multiplication, and exponentiation to base $\omega$ to this set of notations are well known to be computable.

**Theorem 6.5.4**

1. *If $\alpha \preceq_1 \beta < \varepsilon_0 \cdot \omega$ then $(\alpha, \leq, \preceq_1)$ is a $\Sigma_1$-elementary substructure of $(\beta, \leq, \preceq_1)$. In particular, if $\alpha \preceq_1 \beta$ then $\alpha \leq \beta$.*
2. *If $0 < n \leq m < \omega$ then $\varepsilon_0 \cdot n \preceq_1 \varepsilon_0 \cdot m$.*
3. *$\preceq_1$ restricted to $\varepsilon_0 \cdot \omega$ is computable.*
4. *The function which maps a finite subset X of $\varepsilon_0 \cdot \omega$ to the pointwise minimum among the isomorphic copies of X is computable.*

## 6.6 A Machine That Knows *EA* + *ECT*

To build a knowing machine that knows *EA* + *ECT*, we will introduce a hierarchy $H(\alpha)$ ($\alpha < \varepsilon_0 \cdot \omega$) which will generate a family of theories encoding an expansion of the standard model of arithmetic, $\mathcal{N}$, to a structure for $\mathcal{L}_{PA}^{\varepsilon_0 \cdot \omega}$. This structure will collapse on the set consisting of positive multiples of $\varepsilon_0$. The sentences known by *K* in the collapsed structure will be a knowing machine that knows *EA* + *ECT*.

For $n \in \omega$, let $\overline{n}$ be the *numeral* for $n$ i.e. the term $\mathbf{SS} \cdots \mathbf{S0}$ with $n$ occurrences of $\mathbf{S}$. We will use the terms $\overline{n}$ to encode the expansion of $\mathcal{N}$ to $\mathcal{L}_{PA}^{\varepsilon_0 \cdot \omega}$. In particular, whenever $\varphi$ is a formula of either $\mathcal{L}_{PA}^{\varepsilon_0 \cdot \omega}$ or $\mathcal{L}_{PA}^{K}$ and $s$ is an assignment in $\omega$, we let $\varphi^s$ be the formula obtained by simultaneously substituting $\overline{s(x)}$ for $x$ in $\varphi$ for all variables $x$.

**Definition 6.6.1** *Suppose X and Y are subsets of $\varepsilon_0 \cdot \omega$ of the same order type. Let $h : X \to Y$ be the unique order preserving bijection of X and Y. If $\varphi$ is a formula of $\mathcal{L}^X$ we define the formula $\varphi^{(X,Y)}$ of $\mathcal{L}^Y$ to be the result of replacing any occurrence of $K_\alpha$ in $\varphi$ by $K_{h(\alpha)}$.*

When $\Phi$ is a set of formulas of $\mathcal{L}^X$ we will write $\Phi^{(X,Y)}$ for the collection of formulas $\varphi^{(X,Y)}$ where $\varphi \in \Phi$.

**Lemma 6.6.2** *Suppose X and Y are subsets of $\varepsilon_0 \cdot \omega$ of the same order type. If $\Phi$ is a set of formulas in $\mathcal{L}^X$ and $\varphi$ is a formula of $\mathcal{L}^X$ then*

$$\Phi \models \varphi \;\; \textit{iff} \;\; \Phi^{(X,Y)} \models \varphi^{(X,Y)}$$

***Proof*** Straightforward. ■

Before defining $H(\alpha)$, we describe a set of formulas $H_0$ and an operation $H_1$ defined on the finite subsets of $\varepsilon_0 \cdot \omega$.

$H_0$ consists of all sentences of $\mathcal{L}_{PA}^{\varepsilon_0 \cdot \omega}$ which are in one of the following collections.

- The collection of sentences of the form $\varphi^s$ where $\varphi$ is a valid sentence of $\mathcal{L}_{PA}^{\varepsilon_0 \cdot \omega}$.
- $SE_i(\alpha)$ for $i = 1, 2, 3, 4$ and $\alpha < \varepsilon_0 \cdot \omega$.
- The Peano axioms for $\mathcal{L}_{PA}^{\varepsilon_0 \cdot \omega}$.
- The universal closures of formulas of the form

$$K_\gamma \varphi \leftrightarrow K_\beta \varphi$$

  where $\beta \preceq_1 \gamma$.
- The universal closures of formulas of the form

$$K_\gamma \forall x \exists y K_\beta \varphi \to \exists z K_\delta \forall x \exists y (\{z\}(x) \simeq y \wedge \varphi)$$

  where $\gamma < \delta$ and $z$ is not free in $\varphi$.

For $X \subseteq \varepsilon_0 \cdot \omega$, we will write $H_0(X)$ for the set of sentences of $\mathcal{L}_{PA}^{X}$ which are in $H_0$.

We can now use the recursion theorem to define a function $H_1$ from the collection of finite subsets of $\varepsilon_0 \cdot \omega$ into the power set of the collection of sentences of $\mathcal{L}_{PA}^{\varepsilon_0 \cdot \omega}$ which satisfies the following inductive description and has the property that the relation on $\sigma$ and $X$ determined by $\sigma \in H_1(X)$ is computably enumerable.

$H_1(X)$ is the union of the following three collections of sentences.

1. $H_0(X)$.
2. The set of sentences of the form

$$K_\beta \sigma$$

   where there are $Y \subseteq X$ and $\beta \in X$ such that $Y < \beta$, $\sigma$ is a sentence of $\mathcal{L}_{PA}^{Y}$ and $H_1(Y) \models \sigma$.
3. The set of sentences of the form

$$\forall x \exists y(\{\overline{e}\}(x) \simeq y \wedge \varphi)$$

   where there exists a bounded subset $Y$ of $X$ and $\beta \in Y$ such that letting $\tilde{Y}$ be the pointwise minimum of the isomorphic copies of $Y$ and letting $\tilde{\beta}$ correspond to $\beta$ under the isomorphism of $\tilde{Y}$ and $Y$ the following conditions hold:
   (i) $\varphi$ is a formula of $\mathcal{L}_{PA}^{Y \cap \beta}$.
   (ii) $H_1(Y) \models \forall x \exists y K_\beta \varphi$.
   (iii) $e$ is the code for the Turing machine which on input $n$ searches for the first proof of a sentence of the form $\varphi^{(Y,\tilde{Y})}(x, y|\overline{n}, \overline{m})$ from $\bigcup\{H_1(Z) \mid Z$ *is a finite subset of* $\tilde{\beta}\}$ and outputs $m$.

Some comments on the inductive description of $H_1$ are in order. To say that $Y$ is a bounded subset of $X$ means that $Y$ is a subset of $X$ such that $Y < \gamma$ for some $\gamma \in X$. Also, the description of the code $e$ needs to be clarified. We fix a computable way of going from a code for a Turing machine which enumerates a theory to a code for a Turing machine which searches through proofs of the given theory as described. Since we are using the recursion theorem to construct $H_1$, we can effectively produce a code for $H_1(Z)$ from $Z$ and the code for $H_1$.

**Lemma 6.6.3** *For finite subsets $X$ and $Y$ of $\varepsilon_0 \cdot \omega$, if $X \subset Y$ then $H_1(X) \subseteq H_1(Y)$.*

***Proof*** Straightforward. ■

We now define $H(X)$ for an arbitrary subset $X$ of $\varepsilon_0 \cdot \omega$ (not only for $X$ an ordinal) to be the union of all $H_1(Y)$ where $Y$ ranges over the finite subsets of $X$. Notice $H(X)$ is a set of sentences of $\mathcal{L}_{PA}^{X}$.

By Lemma 6.6.3, $H$ extends $H_1$.

Finally, we define $\mathcal{N}^H$, the expansion of $\mathcal{N}$ to $\mathcal{L}_{PA}^{\varepsilon_0 \cdot \omega}$ mentioned in the first paragraph of this section, so that

$$\mathcal{N}^H \models K_\alpha \varphi[s] \qquad \text{iff} \qquad H(\alpha) \models \varphi^s$$

whenever $\alpha < \varepsilon_0 \cdot \omega$ and $K_\alpha \varphi$ is stratified.

Lemma 6.2.4 implies

$$\mathcal{N}^H \models \varphi[s] \qquad \text{iff} \qquad \mathcal{N}^H \models \varphi^s$$

for all formulas $\varphi$ of $\mathcal{L}_{PA}^{\varepsilon_0 \cdot \omega}$ and all assignments $s$ in $\omega$. This fact will be used repeatedly throughout the rest of this section.

**Lemma 6.6.4** *For all subsets $X$ and $Y$ of $\varepsilon_0 \cdot \omega$, if $X \cong Y$ (as substructures of $(\varepsilon_0 \cdot \omega, \leq, \preceq_1)$) then $H(Y) = H(X)^{(X,Y)}$.*

***Proof*** It suffices to show that for all finite subsets $X$ and $Y$ of $\varepsilon_0 \cdot \omega$, if $X \cong Y$ then $H_1(Y) = H_1(X)^{(X,Y)}$. This can be accomplished by a straightforward induction on the cardinality of $X$. ■

We want to establish that $\mathcal{N} \models H(\varepsilon_0 \cdot \omega)$. We begin by showing that $\mathcal{N}^H$ is a model of $H_0(\varepsilon_0 \cdot \omega) - SE_3$ and, next, of the sentences arising from the second clause in the definition of $H_1$. Afterwards, the truth of the sentences in $SE_3$ and those arising from the third clause in the definition of $H_1$ will be established by induction.

**Lemma 6.6.5** $\mathcal{N}^H \models H_0(\varepsilon_0 \cdot \omega) - SE_3$.

***Proof*** The following claims establish that all sentences in $H_0(\varepsilon_0 \cdot \omega) - SE_3$ are true in $\mathcal{N}^H$.

**Claim 1.** If $\varphi$ is a valid formula of $\mathcal{L}_{PA}^{\varepsilon_0 \cdot \omega}$ then $\mathcal{N}^H \models \varphi^s$ for all $s$.

Claim 1 is immediate from the fact that $\mathcal{N}^H \models \varphi^s$ iff $\mathcal{N}^H \models \varphi[s]$.

**Claim 2.** $\mathcal{N}^H \models SE_1$.

Suppose $\sigma \in SE_1$. For some $\alpha$ and valid $\varphi$, $\sigma$ is a universal closure of $K_\alpha \varphi$. Assume $s$ is an assignment in $\omega$. Since $\varphi^s \in H_0(\alpha) \subseteq H(\alpha)$, $\mathcal{N}^H \models K_\alpha \varphi[s]$.

**Claim 3.** $\mathcal{N}^H \models SE_2$.

Assume $\sigma \in SE_2$. There are $\alpha$, $\varphi$, and $\psi$ such that $\sigma$ is a universal closure of $K_\alpha(\varphi \to \psi) \to K_\alpha \varphi \to K_\alpha \psi$. Assume $s$ is an assignment in $\omega$ such that $\mathcal{N}^H \models K_\alpha(\varphi \to \psi)[s]$ and $\mathcal{N}^H \models K_\alpha \varphi[s]$. By definition of $\mathcal{N}^H$, $H(\alpha) \models (\varphi \to \psi)^s$ and $H(\alpha) \models \varphi^s$. Since $(\varphi \to \psi)^s = \varphi^s \to \psi^s$, $H(\alpha) \models \psi^s$. By definition, $\mathcal{N}^H \models K_\alpha \psi[s]$.

**Claim 4.** $\mathcal{N}^H \models SE_4$.

Assume $\sigma \in SE_4$. There are $\alpha$, $\beta$, and $\varphi$ such that $\sigma$ is a universal closure of $K_\beta \varphi \to K_\alpha K_\beta \varphi$. Assume $s$ is an assignment in $\omega$ such that $\mathcal{N}^H \models K_\beta \varphi[s]$. By the definition of $\mathcal{N}^H$, $H(\beta) \models \varphi^s$. Therefore, there exists finite $Y \subseteq \beta$ such that $\varphi^s \in \mathcal{L}_{PA}^{Y}$ and $H_1(Y) \models \varphi^s$. By the second clause in the definition of $H_1$, $K_\beta \varphi^s \in H_1(Y \cup \{\beta\})$. Therefore, $K_\beta \varphi^s \in H(\alpha)$. By definition, $\mathcal{N}^H \models K_\alpha K_\beta \varphi[s]$.

**Claim 5.** $\mathcal{N}^H$ is a model of the Peano axioms for $\mathcal{L}_{PA}^{\varepsilon_0 \cdot \omega}$.

Since $\mathcal{N}^H$ expands the standard model of arithmetic, $\mathcal{N}^H$ is a model of the basic arithmetic axoms.

Suppose $\sigma$ is in the induction schema for $\mathcal{L}_{PA}^{\varepsilon_0\cdot\omega}$. For some formula $\varphi$ and variable $x$, $\sigma$ is a universal closure of $\varphi(x|\mathbf{0}) \wedge \forall x(\varphi \rightarrow \varphi(x|\mathbf{S}x)) \rightarrow \forall x\varphi$. Assume $s$ is an assignment in $\omega$ such that $\mathcal{N}^H \models \varphi(x|\mathbf{0})[s]$ and $\mathcal{N}^H \models \forall x(\varphi \rightarrow \varphi(x|\mathbf{S}x))[s]$. Noticing that $\varphi(x|\mathbf{0})^s = \varphi^{s(x|0)}$ and $(\varphi \rightarrow \varphi(x|\mathbf{S}x))^{s(x|n)} = \varphi^{s(x|n)} \rightarrow \varphi^{s(x|n+1)}$, we see that $\mathcal{N}^H \models \varphi^{s(x|0)}$ and $\mathcal{N}^H \models \varphi^{s(x|n)} \rightarrow \varphi^{s(x|n+1)}$ for all $n \in \omega$. By induction, $\mathcal{N}^H \models \varphi^{s(x|n)}$ for all $n \in \omega$. Therefore, $\mathcal{N}^H \models \varphi[s(x|n)]$ for all $n \in \omega$ or, equivalently, $\mathcal{N}^H \models \forall x\varphi[s]$.

**Claim 6.** If $\alpha \preceq_1 \beta$ and $\varphi$ is a formula of $\mathcal{L}_{PA}^{\alpha}$ then $\mathcal{N}^H$ satisfies any universal closure of $K_\beta\varphi \leftrightarrow K_\alpha\varphi$.

Assume $s$ is an assignment.

Since $\alpha \leq \beta$, $\mathcal{N}^H \models K_\alpha\varphi \rightarrow K_\beta\varphi[s]$ follows easily from the definition of $\mathcal{N}^H$ and the monotonicity of $H$.

To see that $\mathcal{N}^H \models K_\beta\varphi \rightarrow K_\alpha\varphi[s]$, assume $\mathcal{N}^H \models K_\beta\varphi[s]$. By the definition of $\mathcal{N}^H$, $H(\beta) \models \varphi^s$ implying there is finite $Z \subseteq \beta$ such that $\varphi \in \mathcal{L}_{PA}^{Z}$ and $H_1(Z) \models \varphi^s$. Write $Z = X \cup Y$ where $X \subseteq \alpha$ and $Y \subseteq [\alpha, \beta)$. Since $\alpha \preceq_1 \beta$, there exists $\tilde{Y} \subseteq \alpha$ such that $X < \tilde{Y}$ and $X \cup \tilde{Y} \cong X \cup Y$. Notice that $\varphi$ is a formula of $\mathcal{L}_{PA}^{X}$. Therefore, $(\varphi^s)^{(X\cup Y, X\cup\tilde{Y})} = \varphi^s$. By Lemma 6.6.4, $H_1(X \cup \tilde{Y}) = H_1(X \cup Y)^{(X\cup Y, X\cup\tilde{Y})}$. Since $H_1(X \cup Y) \models \varphi^s$, Lemma 6.6.2 implies $H_1(X \cup \tilde{Y}) \models \varphi^s$. Therefore, $H(\alpha) \models \varphi^s$ implying $\mathcal{N}^H \models K_\alpha\varphi[s]$.

**Claim 7.** Assume $\beta < \gamma < \delta < \varepsilon_0 \cdot \omega$, $\varphi$ is a formula of $\mathcal{L}_{PA}^{\beta}$ and $z$ is not free in $\varphi$. $\mathcal{N}^H$ satisfies any universal closure of

$$K_\gamma \forall x \exists y K_\beta \varphi \rightarrow \exists z K_\delta \forall x \exists y(\{z\}(x) \simeq y \wedge \varphi)$$

Suppose $s$ is an assignment and $\mathcal{N}^H \models K_\gamma\forall x\exists yK_\beta\varphi[s]$. By the definition of $\mathcal{N}^H$, $H(\gamma) \models \forall x\exists yK_\beta\psi$ where $\psi$ is the result of simultaneously replacing all free occurrences of each variable $w$ which is different from $x$ and $y$ by $\overline{s(w)}$. There is a finite subset $Y$ of $\gamma$ such that $\forall x\exists yK_\beta\psi$ is a formula of $\mathcal{L}_{PA}^{Y}$ and $H_1(Y) \models \forall x\exists yK_\beta\psi$. The third collection of formulas comprising $H_1(Y \cup \{\gamma\})$ in the inductive definition of $H_1$ contains a formula $\forall x\exists y(\{\overline{e}\}(x) \simeq y \wedge \psi)$. Since $Y \cup \{\gamma\} \subseteq \delta$, $\mathcal{N}^H \models K_\delta\forall x\exists y(\{\overline{e}\}(x) \simeq y \wedge \psi)$. Equivalently, $\mathcal{N}^H \models K_\delta\forall x\exists y(\{z\}(x) \simeq y \wedge \varphi)[s(z|e)]$. Therefore, $\mathcal{N}^H \models \exists zK_\delta\forall x\exists y(\{z\}(x) \simeq y \wedge \varphi)[s]$. ∎

The following lemma implies that all sentences in $H(\varepsilon_0 \cdot \omega)$ which arise from the second collection in the definition of $H_1$ are true in $\mathcal{N}^H$.

**Lemma 6.6.6** *Assume $\beta < \gamma < \varepsilon_0 \cdot \omega$. If $\sigma$ is a sentence of $\mathcal{L}_{PA}^{\beta}$ and $H(\beta) \models \sigma$ then $\mathcal{N}^H \models K_\beta\sigma$ and $K_\beta\sigma \in H(\gamma)$.*

***Proof*** Assume $\sigma$ is a sentence of $\mathcal{L}_{PA}^{\beta}$ and $H(\beta) \models \sigma$.

The definition of $\mathcal{N}^H$ immediately implies $\mathcal{N}^H \models K_\beta\sigma$.

There exists finite $Y \subseteq \beta$ such that $H_1(\beta) \models \sigma$. Considering the second collection of sentences from the definition of $H_1$, we see $K_\beta\sigma \in H_1(Y \cup \{\beta\})$. Since $Y \cup \{\beta\} \subseteq \gamma$, $K_\beta\sigma \in H(\gamma)$. ∎

**Lemma 6.6.7** $\mathcal{N}^H \models H(\varepsilon_0 \cdot \omega)$.

***Proof*** We will show that $\mathcal{N}^H \models H(\alpha)$ by induction on $\alpha \leq \varepsilon_0 \cdot \omega$.

Assume $\alpha \leq \varepsilon_0 \cdot \omega$ and $\mathcal{N}^H \models H(\beta)$ for $\beta < \alpha$. Suppose $\sigma \in H(\alpha)$. We need to show $\mathcal{N}^H \models \sigma$. By the previous two lemmas we are left with the two cases below.

**Case 1.** $\sigma \in SE_3$.

There exist $\beta$ and a formula $\varphi$ of $\mathcal{L}_{PA}^X$ such that $\sigma$ is a universal closure of $K_\beta\varphi \rightarrow \varphi$. Since $\sigma \in H(\alpha)$, $\beta < \alpha$. Suppose $s$ is an assignment such that $\mathcal{N}^H \models K_\beta\varphi[s]$. By definition, $H(\beta) \models \varphi^s$. The induction hypothesis implies $\mathcal{N}^H \models \varphi^s$. Therefore, $\mathcal{N}^H \models \varphi[s]$.

**Case 2.** There is a finite subset $X$ of $\alpha$ such that $\sigma$ is in the third collection in the definition of $H_1(X)$.

In this case, there exist $e \in \omega$ and a stratified formula $\varphi$ such that $\sigma$ is

$$\forall x \exists y (\{\overline{e}\}(x) \simeq y \wedge \varphi)$$

and there exists a bounded subset $Y$ of $X$ and $\beta \in Y$ where, letting $\tilde{Y}$ be the pointwise minimum of the isomorphic copies of $Y$ and letting $\tilde{\beta}$ correspond to $\beta$ under the isomorphism of $\tilde{Y}$ and $Y$, the following conditions hold:

- $\varphi$ is a formula of $\mathcal{L}_{PA}^{Y \cap \beta}$.
- $H_1(Y) \models \forall x \exists y K_\beta \varphi$.
- $e$ is the code for the Turing machine which on input $n$ searches for the first proof of a sentence of the form $\varphi^{(Y,\tilde{Y})}(x, y|\overline{n}, \overline{m})$ from $\bigcup\{H_1(Z) \mid Z \textit{ is a finite subset of } \tilde{\beta}\}$ and outputs $m$.

Choose $\gamma \in X$ such that $Y < \gamma$. Since $X \subseteq \alpha$, $\gamma < \alpha$. There will be no further need for $X$.

Since $H_1(Y) \models \forall x \exists y K_\beta \varphi$, we see that $H_1(\tilde{Y}) \models \forall x \exists y K_{\tilde{\beta}} \varphi^{(Y,\tilde{Y})}$ by Lemma 6.6.2. Using the fact that $\tilde{Y}$ is a bounded subset of $\alpha$, the induction hypothesis implies that $\mathcal{N}^H \models \forall x \exists y K_{\tilde{\beta}} \varphi^{(Y,\tilde{Y})}$. Equivalently, for all $n \in \omega$ there exists $m \in \omega$ such that $H(\tilde{\beta}) \models \varphi^{(Y,\tilde{Y})}(x, y|\overline{n}, \overline{m})$. Therefore, the Turing machine with code $e$ computes a total function.

Fix $n \in \omega$ and let $m = \{e\}(n)$. We need to show $\mathcal{N}^H \models \varphi(x, y|\overline{n}, \overline{m})$. By the choice of $e$, $H(\tilde{\beta}) \models \varphi^{(Y,\tilde{Y})}(x, y|\overline{n}, \overline{m})$. Let $\tilde{\mu}$ be the largest element of $\tilde{Y}$. By Theorem 6.5.3, there is an embedding $h : [0, \tilde{\mu}] \rightarrow ORD$ which maps $\tilde{Y}$ onto $Y$. Let $Z$ be $\{h(\xi) \mid \xi \leq \tilde{\mu}\}$. Since $h$ is order preserving and $Y \subset \gamma$, $Z \subseteq \gamma$. By Lemma 6.6.4, $H(\tilde{\mu} + 1)^{(\tilde{\mu}+1,Z)} \subseteq H(\gamma)$. Note that $(\varphi^{(Y,\tilde{Y})})^{(\tilde{\mu}+1,Z)} = \varphi$. By Lemma 6.6.2, $H(\tilde{\mu} + 1)^{(\tilde{\mu}+1,Z)} \models \varphi(x, y|\overline{n}, \overline{m})$. Therefore, $H(\gamma) \models \varphi(x, y|\overline{n}, \overline{m})$. By the induction hypothesis, $\mathcal{N}^H \models \varphi(x, y|\overline{n}, \overline{m})$. ∎

**Lemma 6.6.8** *$\mathcal{N}^H$ collapses on the set of positive multiples of $\varepsilon_0$.*

***Proof*** Let C be the set of ordinals $\varepsilon_0 \cdot n$ where $0 < n < \omega$. By Theorem 6.5.4, $\gamma \preceq_1 \gamma'$ whenever $\gamma, \gamma' \in C$ and $\gamma \leq \gamma'$.

Let $\Gamma$ be the set of universal closures of stratified formulas $K_\beta\varphi \leftrightarrow K_\alpha\varphi$ where $\alpha < \beta$. For $Y \subseteq \varepsilon_0 \cdot \omega$, let $\Gamma(Y)$ be the set of sentences of $\mathcal{L}^Y_{PA}$ in $\Gamma$. By the definition of $H_0$, $SE_1(Y), SE_2(Y) \subseteq H_0(Y)$ for all $Y$. Moreover, $\Gamma(Y) \subseteq H_0(Y)$ whenever $Y \subseteq C$.

Recalling the second collection in the definition of $H_1(Y)$, a straightforward induction shows that for all finite $Y \subseteq C$, the $K_Y$-closure of $SE_1(Y) \cup SE_2(Y) \cup \Gamma(Y)$ is contained in $H_1(Y)$. This implies that $H(C)$ contains the $K_C$-closure of $SE_1(C) \cup SE_2(C) \cup \Gamma(C)$. By Theorem 6.4.6, $H(C) \models \varphi \leftrightarrow \psi$ whenever $\varphi$ and $\psi$ are formulas of $\mathcal{L}^C_{PA}$ with $\varphi^- = \psi^-$. Since the previous lemma implies $\mathcal{N}^H \models H(C)$, we conclude that $\mathcal{N}^H$ collapses on C. ∎

**Theorem 6.6.9** *There is a knowing machine over the standard model of arithmetic that knows EA + ECT.*

***Proof*** Let C be the collection of ordinals $\varepsilon_0 \cdot n$ for $0 < n < \omega$. By Lemma 6.6.8, $\mathcal{N}^H$ collapses on C. Let $\mathcal{A}$ be the collapse of $\mathcal{N}^H$ on C and let $T$ be the set of sentences $\sigma$ of $\mathcal{L}^K$ such that $\mathcal{A} \models K\sigma$.

By Theorem 6.4.4,

$$\mathcal{N}^H \models \varphi[s] \text{ iff } \mathcal{A} \models \varphi^-[s]$$

for all formulas $\varphi$ of $\mathcal{L}^C_{PA}$. We will make frequent use of this fact as well as Lemma 6.4.2 which implies that for every formula $\varphi$ of $\mathcal{L}^K$, there is a formula $\tilde{\varphi}$ of $\mathcal{L}^C_{PA}$ such that $\tilde{\varphi}^- = \varphi$ and $\tilde{\varphi}$ is valid if $\varphi$ is.

**Claim 1.** $T$ encodes $\mathcal{A}$ with respect to $\overline{n}$ $(n \in \omega)$.

Notice that for $\psi$ a formula of $\mathcal{L}^{\varepsilon_0 \cdot \omega}_{PA}$ and $s$ an assignment, $(\psi^s)^- = (\psi^-)^s$.

Assume $\varphi$ is a formula of $\mathcal{L}^K_{PA}$ and $s$ is an assignment. Let $\tilde{\varphi}$ be a formula of $\mathcal{L}^C_{PA}$ such that $\tilde{\varphi}^- = \varphi$ and choose $\gamma \in C$ such that $\tilde{\sigma}$ is a formula of $\mathcal{L}^\gamma_{PA}$.

$$\begin{aligned} \mathcal{A} \models K\varphi[s] \quad &\text{iff} \quad \mathcal{N}^H \models K_\gamma\tilde{\varphi}[s] \\ &\text{iff} \quad \mathcal{N}^H \models K_\gamma\tilde{\varphi}^s \\ &\text{iff} \quad \mathcal{A} \models K\varphi^s \\ &\text{iff} \quad \varphi^s \in T \end{aligned}$$

**Claim 2.** $\mathcal{A} \models T$.

Assume $\sigma \in T$ or, equivalently, $\mathcal{A} \models K\sigma$. Let $\tilde{\sigma}$ be a sentence of $\mathcal{L}^C_{PA}$ such that $\tilde{\sigma}^- = \sigma$ and choose $\gamma \in C$ such that $\tilde{\sigma}$ is a sentence of $\mathcal{L}^\gamma_{PA}$. Since $\mathcal{A} \models K\sigma$, $\mathcal{N}^H \models K_\gamma\tilde{\sigma}$. Since $K_\gamma\tilde{\sigma} \to \tilde{\sigma} \in SE_3 \subseteq H(\varepsilon_0 \cdot \omega)$, Lemma 6.6.8 implies $\mathcal{N}^H \models \tilde{\sigma}$. Therefore, $\mathcal{A} \models \sigma$.

**Claim 3.** The following collections of sentences are contained in $T$.

- $E_i$ for $i = 1, 2, 3, 4$.
- The Peano axioms for $\mathcal{L}^K$.
- The sentences in the schema $ECT$.

Suppose $\sigma$ is one of these collections. We claim there is $\tilde{\sigma} \in H_0(C)$ such that $\tilde{\sigma}^- = \sigma$. The proofs for the various families are similar and mostly straightforward, so we will only discuss two cases.

First, assume $\sigma \in E_1$. There is a valid formula $\varphi$ of $\mathcal{L}^K_{PA}$ such that $\sigma$ is a universal closure of $K\varphi$. Choose a valid formula $\tilde{\varphi}$ of $\mathcal{L}^C_{PA}$ such that $\tilde{\varphi}^- = \varphi$. Let $\gamma$ be an element of $C$ such that $K_\gamma \tilde{\varphi}$ is stratified. Clearly, there is a universal closure $\tilde{\sigma}$ of $K_\gamma \tilde{\varphi}$ such that $\tilde{\sigma}^- = \sigma$. We have $\tilde{\sigma} \in SE_1(C) \subseteq H_0(C)$.

Next, suppose $\sigma \in ECT$ i.e. $\sigma$ is a universal closure of a formula of the form

$$K\forall x \exists y K\varphi \rightarrow \exists z K \forall x \exists y(\{z\}(x) \simeq y \wedge \varphi)$$

where $z$ is not free in $\varphi$. Choose a formula $\tilde{\varphi}$ of $\mathcal{L}^C_{PA}$ such that $\tilde{\varphi}^- = \varphi$. Let $\beta < \gamma < \delta$ be in $C$ such that $\tilde{\varphi}$ is a formula of $\mathcal{L}^\beta_{PA}$. Clearly, there is a universal closure $\tilde{\sigma}$ of

$$K_\gamma \forall x \exists y K_\beta \tilde{\varphi} \rightarrow \exists z K_\delta \forall x \exists y(\{z\}(x) \simeq y \wedge \tilde{\varphi})$$

such that $\tilde{\sigma}^- = \sigma$. We see $\tilde{\sigma}$ is a sentence of $\mathcal{L}^C_{PA}$ in the last of the five collections which comprise $H_0$.

The remaining cases are similar.

Now, suppose we have $\tilde{\sigma} \in H_0(C)$ such that $\tilde{\sigma}^- = \sigma$. Choose $\gamma \in C$ so that $\tilde{\sigma}$ is a formula of $\mathcal{L}^\gamma_{PA}$. By Lemma 6.6.6, $K_\gamma \tilde{\sigma}$ is in $H(\varepsilon_0 \cdot \omega)$. By Lemma 6.6.7, $\mathcal{N}^H \models K_\gamma \tilde{\sigma}$ implying $\mathcal{A} \models K\sigma$. Therefore, $\sigma \in T$.

**Claim 4.** $T$ is $K$-closed i.e. $KT \subseteq T$.

Assume $\sigma \in T$. Our assumption implies $\mathcal{A} \models K\sigma$. Since $E_4 \subseteq T$ by Claim 3, Claim 2 implies $\mathcal{A} \models K\sigma \rightarrow KK\sigma$. Therefore, $\mathcal{A} \models KK\sigma$ implying $K\sigma \in T$.

**Claim 5.** $T$ is computably enumerable.

Since the relation $\theta \in H_1(X)$ on $\theta$ and $X$ is computably enumerable, so is the relation $\theta \in H(\gamma)$ on $\theta$ and $\gamma$.

Suppose $\sigma$ is a sentence of $\mathcal{L}^K_{PA}$. By the definition of $T$, we see that $\sigma \in T$ iff there exists a sentence $\tilde{\sigma}$ of $\mathcal{L}^C_{PA}$ and $\gamma \in C$ such that $\tilde{\sigma}^- = \sigma$, $K_\gamma \tilde{\sigma}$ is stratified and $\mathcal{N}^H \models K_\gamma \tilde{\sigma}$. Since $\mathcal{N}^H \models K_\gamma \tilde{\sigma}$ is equivalent to $H(\gamma) \models \tilde{\sigma}$, we conclude that $T$ is computably enumerable.

By Claims 2, 3, and 4, $\mathcal{A}$ is a model of $\Sigma \cup K\Sigma$ where $\Sigma$ is the union of the $E_i$ for $i = 1, 2, 3, 4$. Combined with Claim 1, this implies $T$ is a knowing entity. By Claim 5, $T$ is a knowing machine.

By Claims 3 and 4, $T$ contains the axioms of $EA$ as well as $ECT$. Since every knowing entity is a theory, $EA$ is a subset of $T$. ■

## 6.7 Remarks

The heirarchy of $H(\alpha)$ in this chapter has analogues in the proofs that the theories *EA* and *EA* + *SMT* are knowing machines in [2]. In each of the latter proofs, a set $\Sigma$ of sentences of $\mathcal{L}_{PA}^{\varepsilon_0\cdot\omega}$ is constructed directly and one defines the expansion $\mathcal{N}^{\Sigma}$ so that

$$\mathcal{N}^{\Sigma} \models K_\alpha\varphi[s] \text{ iff } \Sigma \cap \mathcal{L}_{PA}^{\alpha} \models \varphi^s$$

So, $\Sigma \cap \mathcal{L}_{PA}^{\alpha}$ is the analogue of $H(\alpha)$ and $\mathcal{N}^{\Sigma}$ is the analogue of $\mathcal{N}^{H}$. While the construction in this chapter is apparently more complex than those in [2], there is a subtle point which comes up in [2] which we avoid here. This is due to the fact that in [2] our task is to show that specific theories, *EA* and *EA* + *SMT*, are knowing machines while here we have only showed that there exists a knowing machine which contains the theory *EA* + *ECT*.

The proofs in all cases show that the resulting knowing machine is the collapse of $H(\varepsilon_0 \cdot \omega) \cap \mathcal{L}_{PA}^{C}$ (where *C*, as previously, is the collection of positive multiples of $\varepsilon_0$). Since we want the encoded structures to satisfy $E_1$, the machines must know $\varphi^s$ for every valid formula of $\mathcal{L}_{PA}^{K}$ and every *s*. In turn, this implies that $\psi^s \in H(\varepsilon_0 \cdot \omega)$ for every valid formula $\psi$ of $\mathcal{L}_{PA}^{C}$ and every *s*. This is not an issue in the present chapter: we simply add these formulas to $H(\varepsilon_0 \cdot \omega)$ through the definition of $H_0$. In [2], we ultimately must argue that the two theories contain all formulas $\varphi^s$ for $\varphi$ valid. There are several approaches to handling this issue.

The approach chosen in [2] was to to define $\Sigma$ so that the collapse of $H(\varepsilon_0 \cdot \omega) \cap \mathcal{L}_{PA}^{C}$ was easily seen to be the required theory. We then had to establish that $H(\varepsilon_0 \cdot \omega)$ logically implied the required sentences $\psi^s$.

A second approach, the approach taken here, is to include the sentences $\psi^s$ in $\Sigma$. The resulting machine would then be axiomatized by the desired theory along with the formulas $\varphi^s$. One could then argue that the additional sentences are redundant since their collapses follow from *EA*.

A third approach would be to avoid ground terms when encoding structures and simply to use variables. In this way, the formulas $\psi^s$ will also be valid and there is no reason to include them in $H(\varepsilon_0 \cdot \omega)$.

The possibility of using variables when encoding structures leads to another matter worth discussion: the form of the induction schema in *EA*. Since we are defining theories in terms of the base logic, one might expect the induction axioms in $\mathcal{L}_{PA}^{K}$ to be true in any expansion of the standard model of arithmetic to $\mathcal{L}_{PA}^{K}$. This is false for the form that we have used (taken from [7]). Moreover, if we used variables for encoding structures then a more natural form of induction axiom would be universal closures of formulas

$$\forall x(x = 0 \to \varphi) \land \forall x \forall y(\varphi \land y = \mathbf{S}x \to \varphi(x|y)) \to \forall x\varphi$$

Induction axioms of this form are obviously true in every expansion of $\mathcal{N}$ to $\mathcal{L}_{PA}^{K}$.

Finally, a word concerning the choice of formula representing $\{e\}(n) \simeq m$. We could choose a formula so that *EA* implies

$$\forall x \forall y \forall z(\{z\}(x) \simeq y \leftrightarrow K(\{z\}(x) \simeq y))$$

and

$$\forall x \forall y_1 \forall y_2 \forall z(\{z\}(x) \simeq y_1 \wedge \{z\}(x) \simeq y_2 \rightarrow y_1 = y_2)$$

In this case, $EA + ECT$ would imply the universal closures of formulas of the form

$$K\forall x \exists y K\varphi \rightarrow \exists z K \forall x \exists y K(\{z\}(x) \simeq y \wedge \varphi)$$

where $z$ is not free in $\varphi$. Even if we do not choose such a formula to represent $\{e\}(n) \simeq n$, the stronger version of $ECT$ above would be known by the resulting machine if we expanded $H$ somewhat.

## References

[1] Timothy J. Carlson, Ordinal arithmetic and $\Sigma_1$-elementarity, *Arch. Math. Logic* **38**, 449–460, 1999.

[2] Timothy J. Carlson, Knowledge, machines and Reinhardt's strong mechanistic thesis, *Ann. Pure Appl. Logic* **105**, 51–82, 2000.

[3] Robert C. Flagg, Church's thesis is consistent with epistemic arithmetic, In: S. Shapiro (ed.), *Intensional Mathematics*, pp. 121–172, North-Holland, Amsterdam, 1988.

[4] William N. Reinhardt, The consistency of a variant of church's thesis with an axiomatic theory of an epistemic notion, Revista Colombiana de Matem'aticas, Special volume for the Proceedings of the 5th Latin American Symposium on Mathematical Logic, 1981, **19**, pp. 177–200, 1985.

[5] William N. Reinhardt, Absolute versions of the incompleteness theorems, *Nous* **19**, pp. 317–346, 1985.

[6] William N. Reinhardt, Epistemic theories and the interpretation of Gödel's incompleteness theorems, *J. Philos. Logic* **15**, pp. 427–474, 1986.

[7] Stewart Shapiro, Epistemic and intuitionistic arithmetic, In: S. Shapiro (ed.), *Intensional Mathematics*, North-Holland, Amsterdam, pp. 11-46, 1988.

# 7

# Gödel's Disjunction

PETER KOELLNER

Gödel's disjunction concerns two central philosophical claims. The first is the claim that "the mind cannot be mechanized" (or, more precisely, that "the outputs of the idealized human mind cannot be simulated by an idealized finite machine"). The second is the claim that "the mathematical world is independent of human reason, insofar as there are mathematical truths that lie outside the scope of human reason" (or, equivalently, that "there are absolutely undecidable statements"). The disjunction states that at least one of these claims must hold, that is, either "the mind cannot be mechanized" or "mathematical truth outstrips human reason."

Now, it is rarely the case in philosophy that significant claims are actually established beyond a shadow of a doubt, especially when those claims concern such large matters as the nature of mind and mathematical realism. But Gödel—who was very cautious in his claims about what he could actually establish—thought that the above disjunctive conclusion was a consequence of his incompleteness theorems. Indeed he went so far as to call it a "mathematically established fact" ([13], p. 310).

If the disjunction is indeed a "mathematically established fact," then the question arises: Which is it? Is it the case that the mind cannot be mechanized? Or is it the case that mathematical truth outstrips human reason? Or is it perhaps both?

Gödel himself was convinced that the first disjunct was true and the second was false—that is, he was convinced that the mind could not be mechanized and that human reason was sufficiently powerful to capture all mathematical truths.[1] But although he was convinced of these stronger statements he did not believe that he was in a position to establish either. He thought only that he could establish the disjunction and, although he had strong convictions about each disjunct, he did not think that he could prove or refute either.

He did, however, think that one day we would be in a position to prove the first disjunct. What was missing, as he saw it, was an adequate resolution of the paradoxes involving self-applicable concepts like the concept of truth. And, according to Wang's reports on their conversations, he thought that if we had this missing ingredient, then we would be able to prove the first disjunct:

> If one could clear up the intensional paradoxes somehow, one would get a clear proof that mind is not machine.
>
> ([47], p. 187).

*Gödel's Disjunction*. First Edition. Leon Horsten & Philip Welch (eds).

But he did not think that we had yet arrived at an adequate resolution of the paradoxes. And, lacking such a resolution, he felt that the most he could claim to have established was the disjunctive conclusion.

Others, who have discussed these matters since Gödel, have claimed something stronger. They have claimed that the incompleteness theorems actually imply the first disjunct, that is, they have claimed that the incompleteness theorems show that the mind cannot be mechanized. There has been a massive amount of literature on the subject, mainly centering on the arguments of Lucas and Penrose.[2]

In this chapter I would like to examine and assess these arguments. But my approach will be somewhat different. One problem with the discussion in the literature as it currently stands is that the background assumptions on the underlying concepts—most importantly, the concepts of *truth, absolute provability*, and *idealized human knowability*—are seldom fully articulated and, as a consequence, it is difficult to assess the cogency of the arguments. One of my goals is to sharpen the discussion by making the background assumptions explicit. Once we do this we will be able to pull the entire debate into a framework where we can actually establish definitive results, results of the form: "If the principles governing the fundamental concepts are such and such, then there is no hope of proving or refuting the first disjunct."

We will see that once we do this, there is a natural framework governing the concepts of truth and absolute provability and that in this framework one can give a rigorous proof of Gödel's disjunction, thereby vindicating, in some sense at least, his claim that the disjunction is a "mathematically established fact." We will also see that (for reasons anticipated by Gödel) the arguments for the first disjunct are based on an oversight and, indeed, that results of Reinhardt and Carlson show that there can be *no* argument for the first disjunct in this framework. Now, the framework in question involves treating the concept of truth as *typed* (in that the truth predicate is applied only to sentences that do not themselves involve the truth predicate). But, as Gödel noted, an adequate formalization of the concept of truth involves treating it as *type-free* (in that the truth predicate *can* apply to sentences involving the truth predicate). And, as we have seen, he held out hope that once we discovered an adequate type-free theory of truth we would be in a position to prove the first disjunct. Interestingly, Penrose's new argument involves treating truth as type-free. In order to reconstruct his argument I devise a system for truth and absolute provability where truth is treated as type-free. And at this point an interesting twist emerges. For when one enters the realm of type-free truth one runs the risk of encountering statements that are indeterminate, like the paradoxical liar sentence (which asserts of itself that it is not true, and so involves an application of the truth predicate to a sentence involving the truth predicate). In particular, one runs the risk that the very statement one is attempting to prove—say, the first disjunct—is actually (provably) indeterminate. It turns out that in this setting there are actually two versions of the disjunction and its disjuncts. The first, fully general versions end up being (provably) indeterminate. Thus, in this case, one cannot even get started since the statement one is attempting to prove—say, the first disjunct—is seen to be neither true nor false. But the second, restricted versions end up being (provably) determinate, and so, in this case there is at least an initial prospect of success. However, in that case we will see that although the disjunction itself is provable, neither disjunct is provable or refutable in the framework. Thus, as in the previous setting, not only does the Penrose argument for the first disjunct fail, but there can be *no* argument for the first disjunct within this framework.

This chapter is part of a two-part series, the second being [19]. Both can be read independently of one another. But while in the companion paper the discussion is largely philosophical and the theorems are stated but not proved, here the philosophical discussion is kept to a minimum and the focus is on proving the main theorems. The companion paper also contains skeptical remarks on the very questions of whether "the mind can be mechanized" and whether "there are absolutely undecidable statements," but here I will set these concerns aside and simply assume, for the sake of argument, that these questions are unproblematic and are understood in the idealized way that the participants in the debate undertand them. The conclusion of both papers is that there is good reason to believe that the questions of whether "the mind can be mechanized" and whether "there are absolutely undecidable statements" are either problematic to begin with or, when suitably regimented, very likely themselves cases of absolutely undecidable statements.[3]

## 7.1 The Disjunction

Let us begin with an informal discussion of the disjunction and the conclusions that Gödel arrived at concerning it.

The disjunction concerns the concepts of *relative provability, absolute provability,* and *truth,* as well as the concepts of an *idealized finite machine* and an *idealized human mind,* which, as we shall see are closely related to the concepts of relative provability and absolute provability. We will begin by focusing on the first three concepts.

Let '$F$' stand for an arbitrary formal system with the feature that each sentence of $F$ is true and the rules of $F$ are truth preserving; let '$K$' stand for the set of all sentences which are absolutely provable; and let '$T$' stand for the set of true sentences.[4] It should be noted that $K$ and $T$ are fixed throughout, while $F$ is a variable term intended to capture the notion of relative provability, the relativity being to the system in question. We shall restrict $F$ such that $F \subseteq K$ and we shall assume that $K$ has the basic feature that $K \subseteq T$. Thus our initial assumptions guarantee that

$$F \subseteq K \subseteq T.$$

The question then arises as to whether we can draw more substantive conclusions concerning the relationship between $F$, $K$, and $T$. For example, are any of the above inclusions proper?

### 7.1.1 *Relative Provability and Truth*

The first substantive conclusion is provided by the incompleteness theorems, and it concerns the relationship between $F$ and $T$. For the incompleteness theorems tell us that for any such system $F$ there are true statements that are outside of the scope of $F$. In the words of Gödel:

> [N]o well-defined system [$F$] of correct axioms can comprise all [of] objective mathematics [$T$], since the proposition which states the consistency of the system is true, but not demonstrable in the system.
>
> ([13], p. 309).

In other words, for any $F$ we have

$$F \subsetneq T.$$

This is a clear and definite statement since the concepts involved—relative provability and arithmetical truth—are clear and definite. And the result is a pure mathematical result.

In contrast to the notions of $F$ and $T$ it is far from clear that the notion $K$ (absolute provability) is clear and definite. Nevertheless, even without having a clear and definite sense of this notion, we can conclude that *whatever K might be* we must have that for any $F$

$$\text{either } F \subsetneq K \quad \text{or} \quad K \subsetneq T.$$

That is, either *absolute provability* outstrips *relative provability* (with respect to any $F$), or *truth* outstrips *absolute provability*.

But to arrive at a more definite conclusion—for example, concerning which inclusion is proper or whether both inclusions are proper—we require a better grip on the nature of $K$.

### 7.1.2 *Absolute Provability*

Although the incompleteness theorems show that certain statements are undecidable *relative* to particular systems (and hence that $F \subsetneq T$), it is unclear whether any of these statements are *absolutely* undecidable ($K \subsetneq T$). Gödel certainly thought that his theorems did *not* show this:

> [These statements are] not at all absolutely undecidable; rather, one can always pass to "higher" systems in which the sentence in question is decidable. (Some sentences, of course, nevertheless remain undecidable.) In particular, for example, it turns out that analysis is a system higher in this sense than number theory, and the axiom system of set theory is higher still than analysis.
>
> ([11], p. 35).

Here the concept of absolute provability seems to be understood as "what is provable from some set of well-justified axioms." But instead of trying to give a substantive analysis of the concept of absolute provability we will work rather with its structural properties.

The concept of absolute provability will be here understood in a highly idealized sense. For example, we shall assume that it satisfies the following principles (where we have used the symbol '$K$' as an operator):

(1) $K\varphi$, where $\varphi$ is a first-order logical validity.

(2) $(K(\varphi \to \psi) \wedge K\varphi) \to K\psi$.

(3) $K\varphi \rightarrow \varphi$.

(4) $K\varphi \rightarrow KK\varphi$.

The first of these principles—known as *logical omniscience*—reveals that $K$ is understood in a highly idealized sense since some (indeed most) of the logical validities are too long for an actual agent to comprehend; to be specific, some (indeed most) have more symbols than there are fundamental particles in the observable universe.[5]

These idealizing assumptions might seem like grand assumptions to make. But we should stress that all of the idealizing assumptions we make are on behalf of our opponent—who wishes to maintain that $K$ outstrips any $F$. Since our goal is to show that the arguments of our opponent do not establish the conclusion, the stronger the assumptions we make on his behalf, the easier his task, and hence the stronger our limitative results concerning the failure of his arguments.

With these remarks on the concept of absolute provability in place let us return to the previous quote of Gödel. This phenomenon—whereby we capture missing arithmetical truths by ascending to higher types—provides us with some hint that perhaps absolute provability (and not just truth) outstrips *all* forms of relative provability. But contrary to appearances it does not quite establish this. For perhaps there is a "master system," $F^*$, such that relative provability with regard to $F^*$ coincides with absolute provability. What we can conclude is merely that *if* there is such a master system, then we could never know (in the sense of being able to absolutely prove) that all of its axioms were true. This is precisely what Gödel had in mind when he wrote the following about his incompleteness theorems:

> For, *it makes it impossible that someone should set up a certain well-defined system of axioms and rules and consistently make the following assertion about it*: *All of these axioms and rules I perceive (with mathematical certitude) to be correct, and moreover I believe that they contain all of mathematics.* If someone makes such a statement he contradicts himself. For if he perceives the axioms under consideration to be correct, he also perceives (with the same certainty) that they are consistent. Hence he has a mathematical insight not derivable from his axioms.
>
> ([13], p. 309).

In other words, although the incompleteness theorems enable us to conclude that $F \subsetneq T$, we cannot conclude, in parallel fashion, that $F \subsetneq K$. But we *can* draw the conditional conclusion to the effect that *if* the correctness of $F$ is absolutely provable (that is, in $K$) *then* $F \subsetneq K$.

But can we draw the stronger, non-conditional conclusion? The trouble is that for all we have shown, we have not ruled out the possibility that there *is* an $F$ that coincides with $K$. Gödel was well aware of this possibility:

> However, as to subjective mathematics, it is not precluded that there should exist a finite rule producing all its evident axioms. However, if such a rule exists, we with our human understanding could certainly never know it to be such, that is, we could never know with mathematical certainty that all propositions it produces are correct.[12]

> 12 For this (or the consequence concerning the consistency of the axioms) would constitute a mathematical insight not derivable from the axioms [and] rules under consideration, contrary to the assumption.
>
> ([13], p. 309).

In other words (for all we have shown) it may indeed be the case that there is such an $F$ such that $F = K$—we have only shown that if there is such an $F$ then it must be "hidden" in the sense that we cannot absolutely prove that it has this feature. We can only arrive at a conditional conclusion, namely,

$$\textit{if } \text{there is an } F \text{ such that } F = K, \textit{ then } K \subsetneq T$$

(which, in turn, implies that there are $\varphi$ in $T$ such that neither $\varphi$ nor $\neg\varphi$ is in $K$). In the words of Gödel:

> [I]f the human mind were equivalent to a finite machine, then objective mathematics not only would be incompletable in the sense of not being contained in any well-defined axiomatic system, but moreover there would exist *absolutely* unsolvable diophantine problems of the type described above, where the epithet "absolutely" means that they would be undecidable, not just within some particular axiomatic system, but by *any* mathematical proof that the human mind can conceive.
>
> ([13], p. 310).

Reformulating this in disjunctive form we have that

$$\textit{either } \text{for all } F, F \subsetneq K \textit{ or } K \subsetneq T$$

(where, again, the second alternative implies that there are $\varphi$ in $T$ such that neither $\varphi$ nor $\neg\varphi$ is in $K$). In the words of Gödel:

> So the following disjunctive conclusion is inevitable: *Either mathematics is incompletable in this sense, that its evident axioms can never be comprised in a finite rule, that is to say, the human mind (even within the realm of pure mathematics) infinitely surpasses the powers of any finite machine, or else there exist absolutely unsolvable diophantine problems of the type specified* (where the case that both terms of the disjunction are true is not excluded, so that there are, strictly speaking, three alternatives).
>
> ([13], p. 310).

In other words, either absolute provability outstrips all forms of relative provability or there are absolutely undecidable sentences of arithmetic.

### 7.1.3 *Idealized Finite Machines and Idealized Human Minds*

In the above discussion the concepts of *relative provability, absolute provability,* and *truth* have taken center-stage and the concepts of an *idealized finite machine* and an *idealized human mind* were set to the side. But these latter concepts do appear in the quotations from

Gödel and they are central in the standard formulations of the first disjunct. So let us say something about them.

As we noted, the concept of an idealized finite machine closely corresponds to the concept of relative provability, and the concept of an idealized human mind closely corresponds to the concept of absolute provability.

The first correspondence is well established and easy to explain. In speaking of relative provability we mean provability *with respect to a recursively axiomatizable formal system*. This latter concept—that of being a recursively axiomatizable formal system—is a precise mathematical concept. It is one of a large cluster of precise, formal concepts that were all shown to be extensionally equivalent and thought by many (and conjectured by Church) to capture the informal concept of *being computable* (or *being effective* or *being algorithmically decidable*). Now, Turing produced one such concept—that of a Turing machine—which he showed was equivalent to the others. Moreover, this concept had the advantage that it seemed, in addition, to provide an adequate conceptual analysis of the informal concept of computability (by an idealized finite machine). In what follows, we shall accept this analysis and take the concept of computable by an *idealized finite machine* to be coextensive with the concept of *relative provability* (with respect to a recursively axiomatizable formal system).[6]

The second correspondence is more difficult since neither the concept of an idealized human mind nor the concept of absolute provability is precise. Indeed I think that both concepts are problematic, for reasons I explain in the companion paper [19]. But I will set those concerns aside here. What is important for our purposes is that the idealizing assumptions made on the concept of an idealized human mind are parallel to those made on the concept of absolute provability. And so, for our purposes, it will be safe to assume (following Gödel) that these two concepts are co-extensive.

### 7.1.4 *Summary*

We can summarize this discussion as follows: Gödel thought he was in a position to establish the disjunctive conclusion but he did not think that he was in a position to establish the first disjunct—to the effect that for any $F$, $F \subsetneq K$—since he was aware that he could not rule out the possibility that there was indeed a master system $F^*$ such that $F^* = K$; he could, rather, only show that if there were such a master system then it must be "hidden" in the sense that we could never be in a position to absolutely prove that *it* produced only truths. Others—most notably Lucas and Penrose—have claimed something much stronger, namely, that they could establish the first disjunct.

In the remainder of this chapter we will place these arguments in sharper relief by being explicit about the assumptions on the concepts of *relative provability*, *absolute provability*, and *truth* (and, given the above correspondence, the concepts of *provable by an idealized finite machine* and *knowable by an idealized human mind*). This will enable us to pull the entire discussion into a precise setting, where we can establish mathematical results that bear on these questions. We shall see that the disjunction is indeed provable in the appropriate systems (thereby vindicating Gödel's claim that it is a "mathematically established fact") and that there are serious obstacles facing any attempt to establish either disjunct.

## 7.2 Notation

Let us now turn to a formal discussion of the disjunction. We begin with notation.

Our base language is the language of arithmetic, $L_{PA}$, consisting of the usual *logical symbols* (connectives, quantifiers, and equality) and the following *non-logical symbols*: The constant symbol '0', the unary function symbol '$S$', the binary operation symbols '+' and '×', and the binary relation symbol '$\leqslant$'. The *terms* are generated by starting with '0' and the variables and iteratively applying '$S$', '+', and '×'. The *canonical numerals* are '0', '$S0$', '$SS0$', .... For perspicuity we shall use '$\bar{0}$', '$\bar{1}$', '$\bar{2}$', ... as shorthand for the canonical numerals. This language will be extended to the language of epistemic arithmetic, $L_{EA}$, by adding the symbol '$K$' and then to the language of epistemic arithmetic with truth, $L_{EA_T}$, by adding both the symbols '$K$' and '$T$'. Here '$K$' is treated as a symbol for an intensional operator that takes propositions as arguments and is intended to express *absolute provability*, and the symbol '$T$' is a predicate intended to express *truth*. Later, when we consider our final system, DTK, the language $L_{DTK}$ will have the same symbols, only that there '$K$' will be treated as a predicate which takes terms as arguments.

We fix in advance a coding of the expressions of each language and we use '$\ulcorner\varphi\urcorner$' to indicate the Gödel number of '$\varphi$'. It is convenient to have notation for the arithmetical operations on Gödel numbers that mirrors the syntactic operations on the corresponding formulas. To this end we shall employ the "dot notation" of Feferman, whereby the arithmetical operation on Gödel numbers is denoted by placing a dot under the associated syntactic symbol. For example, $\underset{\cdot}{\neg}$ is the operation such that for any Gödel number $\ulcorner\varphi\urcorner$ of a sentence we have $\underset{\cdot}{\neg}\ulcorner\varphi\urcorner = \ulcorner\neg\varphi\urcorner$.

We need one more piece of notation. Notice that although it makes sense to write '$\text{Prov}_{PA}(\ulcorner\varphi\urcorner)$', it does not strictly speaking make sense to write '$(\forall x)\,\text{Prov}_{PA}(\ulcorner\varphi(x)\urcorner)$' since '$x$' ranges over natural *numbers* but the intension here is to say that every *numeral* substitution instance of '$\varphi(x)$' is provable in PA. This is where we use the notation '$\dot{-}$' with the understanding that '$(\forall x)\,\text{Prov}_{PA}(\ulcorner\varphi(\dot{\bar{x}})\urcorner)$' means that for every natural number $x$, if you take the canonical numeral for $x$, substitute it for the dot in '$\varphi(\cdot)$', then the Gödel number of the resulting expression is in the range of the arithmetical relation '$\text{Prov}_{PA}$' (which is an arithmetical representation of the syntactic statement that '$\varphi(SS\ldots 0)$' is provable in PA for all canonical numerals '$SS\ldots 0$').

For a given language $L$, let 'Var($x$)', 'At-Sent($x$)', and 'Sent($x$)' be the arithmetical formulas indicating that $x$ is the Gödel number of a variable, that $x$ is the Gödel number of an atomic (quantifier-free) sentence of $L$, and that $x$ is the Gödel number of a sentence of $L$, respectively. (Here $L$ is one of the languages mentioned above. Strictly speaking we should indicate '$L$' in the notation. But we will omit this since the relevant language will always be clear from context.) If, in the context of a given language, we wish to restrict to Gödel numbers of a fragment of the language this will be indicated by a subscript; for example, if $L_{PA}$ is a sublanguage of $L$, then '$\text{Sent}_{L_{PA}}(x)$' is the arithmetical formula indicating that $x$ is the Gödel number of a sentence in the language of PA.

Finally, if $x$ is the Gödel number of a formula, $z$ is the Gödel number of a variable, and $y$ is a natural number, then $x(\dot{\bar{y}}/z))$ is the Gödel number of the formula obtained by substituting the canonical numeral for $y$ for (the variable numbered by) $z$ in (the expression numbered by) $x$.

## 7.3 Arithmetic

We now introduce the system, PA, which will lie at the heart of all of the systems we shall consider.

To begin with assume that we have fixed a sound and complete system for logic, whose only rule is modus ponens. All of our systems are taken to include this system of logic. Now, work in the language of arithmetic, $L_{\mathrm{PA}}$. A quantifier is *bounded* if it is the form '$\exists x \leqslant t$' or '$\forall x \leqslant t$', where $t$ is a term not involving $x$. A formula is a *bounded formula* (denoted $\Delta^0_0$) if all of its quantifiers are bounded. For $n \geqslant 0$ the classes of formulas $\Sigma^0_n$ and $\Pi^0_n$ are defined as follows: $\Sigma^0_0 = \Pi^0_0 = \Delta^0_0$. $\Sigma^0_{n+1}$ is the set of formulas of the form $\exists\vec{x}\varphi$, where $\varphi$ is a $\Pi^0_n$-formula and $\vec{x}$ is a (possibly empty) list of variables. $\Pi^0_{n+1}$ is the set of formulas of the form $\forall\vec{x}\varphi$, where $\varphi$ is a $\Sigma^0_n$-formula and $\vec{x}$ is a (possibly empty) list of variables. The classes $\Sigma^0_n$ and $\Pi^0_n$ for all $n < \omega$ constitute the *arithmetical hierarchy of formulas* and provide a stratification of $L_{\mathrm{PA}}$ in terms of quantifier complexity.

**Definition 7.3.1** *Q is the theory in the language of arithmetic consisting of the following axioms:*

(1) $S(x) \neq 0$

(2) $S(x) = S(y) \rightarrow x = y$

(3) $x \neq 0 \rightarrow (\exists y)(x = S(y))$

(4) $x + 0 = x$

(5) $x + S(y) = S(x + y)$

(6) $x \times 0 = 0$

(7) $x \times S(y) = (x \times y) + x$

(8) $x \leqslant y \leftrightarrow (\exists z)(x + z = y)$

*This is a very weak theory. To strengthen Q one adds induction principles.*

**Definition 7.3.2** *Let $\Gamma$ be a collection of formulas in the language of arithmetic. $\Gamma$-induction is the scheme consisting of all instances of*

$$(\varphi(0) \wedge (\forall x)(\varphi(x) \rightarrow \varphi(S(x)))) \rightarrow (\forall x)\varphi(x)$$

*for $\varphi \in \Gamma$. $I\Sigma_n$ is the theory Q + $\Sigma^0_n$-induction. PA is the theory Q + $\bigcup_{n<\omega} \Sigma^0_n$-induction.*

## 7.4 Incompleteness

Let us say that a formal system $T$ is *$\Sigma^0_1$-sound* iff for every $\Sigma^0_1$-statement $\varphi$, if $T$ proves $\varphi$ then $\varphi$ is true. A formal system $T$ is *$\Sigma^0_1$-complete* if it proves every true $\Sigma^0_1$-statement. It is not hard to show that the system PA (in fact, even Q) is $\Sigma^0_1$-complete. This raises the question of whether PA is $\Pi^0_1$-complete, that is, whether it can capture the truths at the next level of the arithmetical hierarchy—the $\Pi^0_1$-truths. The first incompleteness theorem answers this question negatively.

**Theorem 7.4.1 (Gödel)** *Assume that* PA *is* $\Sigma^0_1$*-sound. Then there is a* $\Pi^0_1$*-statement such that* PA *can neither prove* $\varphi$ *nor* $\neg\varphi$.

In other words, assuming that PA is $\Sigma^0_1$-sound, the statement $\varphi$ is *independent* of PA. Rosser improved this theorem by showing that there is a different $\Pi^0_1$-statement which could be shown to be independent of PA on the weaker assumption that PA is consistent.

The second incompleteness theorem shows that PA cannot prove its own consistency:

**Theorem 7.4.2 (Gödel)** *Assume that* PA *is consistent. Then* PA *cannot prove* Con(PA).

Here the statement Con(PA) is a $\Pi^0_1$-statement (which informally asserts "for all natural numbers $n$, $n$ does not code a proof of a contradiction from the axioms of PA"). Thus, granting the consistency of PA, the second incompleteness theorem shows that there is a $\Pi^0_1$-truth that cannot be proved in PA.

To show that PA cannot prove ¬Con(PA) (and hence that Con(PA) is independent of PA) one needs to assume slightly more than the consistency of PA, namely, as above, one needs to assume that PA is $\Sigma^0_1$-sound.

It is important to stress that the incompleteness theorems are perfectly general. They apply not just to PA but rather to any sufficiently strong formal system $T$. For example, they apply to very weak systems like Q and to very strong systems like ZFC and ZFC supplemented with large cardinal axioms.

## 7.5 Epistemic Arithmetic

The language $L_{\text{EA}}$ of EA is the language $L_{\text{PA}}$ augmented with a unary sentential operator '$K$'.

The axioms of EA fall into two categories—the axioms of arithmetic and the axioms of absolute provability. The *axioms of arithmetic* are simply those of PA, only now the induction scheme is taken to hold for all formulas in $L_{\text{EA}}$. The *basic axioms of absolute provability* are:

($E_1$) Universal closures of formulas of the form

$$K\varphi,$$

where $\varphi$ is a first-order validity.

($E_2$) Universal closures of formulas of the form

$$(K(\varphi \to \psi) \wedge K\varphi) \to K\psi.$$

($E_3$) Universal closures of formulas of the form

$$K\varphi \to \varphi.$$

($E_4$) Universal closures of formulas of the form

$$K\varphi \rightarrow KK\varphi.$$

For a collection $\Gamma$ of formulas in $L_{\mathrm{EA}}$ we use '$K\Gamma$' to denote the collection of formulas '$K\varphi$' where '$\varphi$' is in $\Gamma$.

The system EA is the theory axiomatized by $\Sigma \cup K\Sigma$, where $\Sigma$ consists of the axioms of PA (in the language $L_{\mathrm{EA}}$) and the basic axioms of absolute provability.

It is important to notice that we have formulated epistemic arithmetic with an *operator* '$K$' and not a *predicate* '$K$'. The motivation for this restriction is the limitative results of Gödel [12], Myill [26], Montague [25], Thomason [46], and others. Here are two representative samples:

**Theorem 7.5.1 Montague, [25]** *Suppose $T$ is a theory and $K(x)$ is a formula such that for all sentences in the language of $T$,*

(1) $T \vdash (K(\ulcorner\varphi\urcorner) \rightarrow \varphi)$,
(2) $T \vdash K(\ulcorner\varphi\urcorner)$, *whenever* $T \vdash \varphi$, *and*
(3) $Q \subseteq T$.

*Then $T$ is inconsistent.*

**Theorem 7.5.2 Montague, [25]** *Suppose $T$ is a theory and $K(x)$ is a formula such that for all sentences $\varphi$ and $\psi$ in the language of $T$ we have*

(1) $T \vdash (K(\ulcorner\varphi\urcorner) \rightarrow \varphi)$,
(2) $T \vdash K(\ulcorner K(\ulcorner\varphi\urcorner) \rightarrow \varphi\urcorner)$,
(3) $T \vdash K(\ulcorner\varphi\urcorner)$, *if $\varphi$ is a logical axiom,*
(4) $T \vdash (K(\ulcorner\varphi \rightarrow \psi\urcorner) \rightarrow (K(\ulcorner\varphi\urcorner) \rightarrow K(\ulcorner\psi\urcorner)))$, *and*
(5) $Q \subseteq T$.

*Then $T$ is inconsistent.*

These results show that if we wish to meet the above principles then we cannot treat '$K$' as a predicate.

In this regard the concept of *absolute* provability behaves quite differently than the concept of *relative* provability. Recall that with each formal system $T$ there is an associated concept of *relative provability*. Let $\Box_T$ express this concept. More precisely, let $\Box_T$ be a provability operator such that $\Box_T\varphi$ holds if and only if the sentence $\varphi$ is provable relative to $T$. This provability operator satisfies the following axiom schemes, known as GL:

(1) Boolean tautologies
(2) $\Box_T(\varphi \rightarrow \psi) \rightarrow (\Box_T\varphi \rightarrow \Box_T\psi)$
(3) $\Box_T(\Box_T\varphi \rightarrow \varphi) \rightarrow \Box_T\varphi$.

The rules of inference are modus ponens and necessitation (if $\varphi$ has been proved then one can infer $\Box_T(\varphi)$). In this system one can prove $\Box_T\varphi \rightarrow \Box_T\Box_T\varphi$. Thus, the system extends K4. However, GL is incompatible with the reflexivity axiom $\Box_T\varphi \rightarrow \varphi$ and is therefore incompatible with the system **S4**.

The key difference between the above system concerning $K$ (absolute provability) and the system concerning $\Box_T$ (relative provability) is that it *is* possible in the latter to treat $\Box_T$ as a predicate, while, as we have seen, this is not possible in the former case. In fact, results of Halbach and Welch [16] show that if one wishes to treat $K$ as a predicate then one must admit only principles that are "close" to those of GL (that is, the concept expressed must have properties "close" to those of the concept of relative provability). Later, when we fold in the concept of truth in a type-free setting we will see that there is a different perspective, one in which it *is* possible to treat '$K$' as a predicate. But for the time being let us start by folding in the typed concept of truth.

## 7.6 Epistemic Arithmetic with Typed Truth

The language $L_{\mathrm{EA_T}}$ of $\mathrm{EA_T}$ is the language $L_{\mathrm{EA}}$ augmented with a unary predicate '$T$'. The predicate '$T$' is intended to be a (Tarskian) truth predicate that applies to sentences of the sublanguage $L_{\mathrm{EA}}$ where '$T$' is omitted. Hence we are dealing here with a typed truth predicate.

We define a valuation function Val on the closed terms in such a way that

$$(V_1)\quad (\forall x)[\mathrm{Val}(\dot{\bar{x}}) = x]$$

$$(V_2)\quad (\forall x_1)\cdots(\forall x_n)[\mathrm{CTerm}(x_1) \wedge \cdots \wedge \mathrm{CTerm}(x_n) \\ \rightarrow (\mathrm{Val}(f(x_1,\ldots,x_n)) = f(\mathrm{Val}(x_1),\ldots,\mathrm{Val}(x_n))].$$

The truth axioms are

$$(T_1)\quad (\forall x)[T(x) \rightarrow \mathrm{Sent}(x)]$$

$(T_2)$ For each atomic formula $R(x_1,\ldots,x_n)$ of $L_{\mathrm{PA}}$:

$$(\forall x_1)\cdots(\forall x_n)[\mathrm{CTerm}(x_1) \wedge \cdots \wedge \mathrm{CTerm}(x_n) \\ \rightarrow (T(\underset{\cdot}{R}(x_1,\ldots,x_n)) \leftrightarrow R(\mathrm{Val}(x_1),\ldots,\mathrm{Val}(x_n)))]$$

$$(T_3)\quad (\forall x)[\mathrm{Sent}(x) \rightarrow (T(\underset{\cdot}{\neg}x) \leftrightarrow \neg T(x))]$$

$$(T_4)\quad (\forall x)(\forall y)[\mathrm{Sent}(x) \wedge \mathrm{Sent}(y) \rightarrow (T(x\underset{\cdot}{\vee}y) \leftrightarrow T(x) \vee T(y))]$$

$$(T_5)\quad (\forall x)(\forall y)[\mathrm{Sent}(x) \wedge \mathrm{Sent}(y) \rightarrow (T(x\underset{\cdot}{\rightarrow}y) \leftrightarrow T(x) \rightarrow T(y))]$$

$$(T_6)\quad (\forall x)(\forall z)[\mathrm{Var}(z) \wedge \mathrm{Sent}((\underset{\cdot}{\forall}z)x) \rightarrow (T((\underset{\cdot}{\forall}z)x) \leftrightarrow \forall y T(x(\dot{\bar{y}}/z))]$$

$$(T_7)\quad (\forall x)[\mathrm{Sent}(x) \rightarrow (T(\underset{\cdot}{K}x) \leftrightarrow KT(x))].$$

Notice that in the case of $T_7$ we cannot follow the pattern of the other axioms by putting '$(\forall x)[\mathrm{Sent}(x) \rightarrow (T(\underset{\cdot}{K}x) \leftrightarrow Kx)]$' since $K$ is a sentential operator and so we cannot quantify over its argument. (To see this, consider an instance where $x$ is $\ulcorner\varphi\urcorner$. For this instance, the right-hand side of the biconditional becomes '$K\ulcorner\varphi\urcorner$' but that is nonsensical (just as '$\neg\ulcorner\varphi\urcorner$' is nonsensical). In contrast, in our formulation, '$KT(x)$' does make sense (just as '$\neg P(x)$' makes sense).) Another option would be to state the axiom as a scheme: For all $\varphi$ in $L_{\mathrm{EA}}$, $T(\dot{K}\ulcorner\varphi\urcorner) \leftrightarrow K\varphi$. Instead we have followed Reinhardt [39], pp. 436–438.

The system $\mathrm{EA_T}$ is the theory axiomatized by $\Sigma \cup K\Sigma$, where $\Sigma$ consists of the axioms of PA (in the language $L_{\mathrm{EA_T}}$), the basic axioms of absolute provability (in the language $L_{\mathrm{EA_T}}$), and the axioms of truth (for the language $L_{\mathrm{EA}}$).

## 7.7 The Disjunction in $\mathrm{EA}_T$

Now that all of the pieces are in place we can consider the disjunction in this precise setting.

To begin with, Reinhardt showed that in this setting one can prove versions of the incompleteness theorem that pertain to the concept of absolute provability. The absolute version of the first incompleteness theorem is the following:

**Theorem 7.7.1 (Reinhardt, [37])** *Assume that S includes* EA. *Suppose $F(x)$ is a formula with one free variable and the feature that for each sentence $\varphi$*

$$S \vdash K(F(\ulcorner\varphi\urcorner) \rightarrow \varphi).$$

*Then there is a sentence $\bar{\varphi}$ such that*

$$S \vdash K\bar{\varphi} \wedge K\neg F(\ulcorner\bar{\varphi}\urcorner).$$

Here the intention is that $F$ is the presentation of a system. There is no assumption to the effect that $F$ is a $\Sigma^0_1$-formula, that is, that it defines a recursively enumerable set of sentences. The theorem says that if it is absolutely provable (pointwise) that $F$ is sound, then there are statements that are absolutely provable but which cannot be proved by the system $F$.

The absolute version of the second incompleteness theorem is the following:

**Theorem 7.7.2 (Reinhardt, [37])** *Assume that S includes* EA. *Suppose $F(x)$ is a formula with one free variable and the feature that for each sentence $\varphi$*

$$S \vdash K(K\varphi \rightarrow F(\ulcorner\varphi\urcorner)).$$

*Then*

$$S \vdash K\neg K(\mathrm{Con}(F)).$$

In other words, if it is absolutely provable (pointwise) that the system $F$ captures everything that is absolutely provable then (it is absolutely provable that) the consistency of $F$ is not absolutely provable.

We now show that Gödel's disjunction can be formulated and proved in $\mathrm{EA_T}$. Recall that the disjunction is: Either (1) the statements of mathematics which are absolutely provable outstrip those that are relatively provable with regard to any given formal system, or (2) there exist true arithmetical statements $\varphi$ such that neither $\varphi$ nor $\neg\varphi$ is absolutely provable.

In the setting of $EA_T$ this can be formalized as follows. Let us use '$K_p(x)$' as shorthand for '$T(\underset{\cdot}{K}x) \wedge \mathrm{Sent}_{L_{EA}}(x)$' and let us use '$x \in W_e$' as shorthand for the statement "$x$ is in the $e^{\text{th}}$ computably enumerable set." In the language of $EA_T$ we can express the first disjunct as

$$(\forall e)\big[(\forall x)((\mathrm{Sent}_{L_{PA}}(x) \wedge x \in W_e) \rightarrow T(x)) \\ \rightarrow (\exists y)(\mathrm{Sent}_{L_{PA}}(y) \wedge K_p(y) \wedge y \notin W_e)\big]$$

and the second disjunct as

$$(\exists x)\big[\mathrm{Sent}_{L_{PA}}(x) \wedge T(x) \wedge \neg T(\underset{\cdot}{K}x) \wedge \neg T(\underset{\cdot}{K}\underset{\cdot}{\neg}x)\big].$$

Let GD be the disjunction of these two statements.

**Lemma 7.7.3** *Assume* $EA_T$. $(\forall x)[K_p(x) \rightarrow T(x)]$.

***Proof*** Suppose $K_p(x)$. Thus, by the definition of '$K_p(x)$', we have $T(\underset{\cdot}{K}x)$. It follows that $KT(x)$ (by $T_7$) and $T(x)$ (by $E_3$). ■

**Theorem 7.7.4 (Reinhardt)** *Assume* $EA_T$. *Then* GD *holds.*

***Proof*** The following proof is sketched in [39]: Let '$K_p = T$' be shorthand for

$$(\forall x)[\mathrm{Sent}_{L_{PA}}(x) \rightarrow (K_p(x) \leftrightarrow T(x))].$$

There are two cases:

**Case 1**: $K_p = T$.

In this case we argue that the first disjunction holds. The point is that because we have allowed '$T$' to figure in the induction scheme, the proof of the first incompleteness theorem gives us

$$(\forall e)\big[(\forall x)((\mathrm{Sent}_{L_{PA}}(x) \wedge x \in W_e) \rightarrow T(x)) \\ \rightarrow (\exists y)(\mathrm{Sent}_{L_{PA}}(y) \wedge T(y) \wedge y \notin W_e)\big].$$

Since our case hypothesis is $K_p = T$, we can replace '$T(y)$' with '$K_p(y)$' to obtain the first disjunct.

**Case 2**: $K_p \neq T$.

We are assuming

$$(\exists x)[\mathrm{Sent}_{L_{PA}}(x) \wedge \neg(K_p(x) \leftrightarrow T(x))].$$

By Lemma 7.7.3 it follows that

$$(\exists x)[\mathrm{Sent}_{L_{PA}}(x) \wedge T(x) \wedge \neg K_p(x)].$$

Fix an instance $x$. The last conjunct implies $\neg T(\underset{\cdot}{K}x)$. But we also have $\neg K_p(\underset{\cdot}{\neg}x)$. (Suppose, for contradiction, that $K_p(\underset{\cdot}{\neg}x)$ holds. Then $T(\underset{\cdot}{\neg}x)$, which contradicts $T(x)$.) Since $\mathrm{Sent}_{L_{PA}}(\underset{\cdot}{\neg}x)$ it follows that $\neg T(\underset{\cdot}{K}\underset{\cdot}{\neg}x)$. Thus, we have shown

$$(\exists x)\left[\mathrm{Sent}_{L_{\mathrm{PA}}}(x) \wedge T(x) \wedge \neg T(\underset{\cdot}{K}x) \wedge \neg T(\underset{\cdot}{K}\underset{\cdot}{\neg}x)\right],$$

which is the second disjunct. ■

In this sense, Gödel was correct in claiming that the disjunction is a "mathematically established fact." This raises the question of which disjunct holds. In particular, is the first disjunct a "mathematically established fact"?

## 7.8 The Classic Argument for the First Disjunct

The first generation of arguments for the first disjunct (due to Lucas, and Penrose in his early writings) have been extensively discussed in the literature.[7] In our present setting, with the above technical apparatus at hand, we can be rather brief in our discussion of these arguments and we can present a critique that is a good deal stronger than the critiques contained in the standard literature.

It turns out that the first generation of arguments for the first disjunct are actually versions of Gödel's argument, to the effect that *if* the correctness of $F$ is absolutely provable then $F \subsetneq K$. In the words of Penrose: "Human mathematicians are not using a knowably sound algorithm in order to ascertain mathematical truth." ([33], p. 76). As we have seen, this conclusion, when suitably formalized, is provable in $\mathrm{EA}_{\mathrm{T}}$. It is simply Theorem 7.7.1.

However, as pointed out by many logicians, the argument does not yield the first disjunct; rather, it provides us with a *conditional* statement and to arrive at the consequent of the conditional one needs to discharge the antecedent, that is, one needs the additional premise that the soundness of the system is absolutely provable.

The question, then, is whether for any given system $F$ one can absolutely prove that $F$ is sound. This is no small task. For example, let $S$ be the system PA + $R$ where '$R$' stands for the famous open problem known as the Riemann hypothesis. A result of Kreisel shows that $R$ can be formulated as a $\Pi^0_1$-sentence. If PA + $R$ is consistent then $R$ must be true.[8] And if PA + $R$ is inconsistent then $R$ must be false. So to know whether or not PA + $R$ is consistent is to know whether or not $R$ is true. But $R$ is a major outstanding problem in mathematics, so outstanding that the Clay Institute has offered one million dollars for its resolution. No one at present knows the answer to $R$ and it is no small task finding the answer. So neither Lucas nor Penrose knows whether PA + $R$ is consistent.

Now, one might question this example, pointing out that although Lucas and Penrose do not know whether PA + $R$ is sound, they might plausibly maintain that the answer is indeed within the reach of what is absolutely provable. But the choice of $R$ was merely representative and the problem is much more general. For example, by climbing the hierarchy of consistency strength one can choose examples—such as the statement that 'ZFC + There is a supercompact cardinal' is consistent or the statement that 'ZF + There is a Reinhardt cardinal' is consistent—which are provably very difficult to resolve. Indeed to know of every system $F$, whether or not $F$ is consistent, is to have an oracle for $\Pi^0_1$-truth and that is not something we can claim to have at the *start* of an argument for the first disjunct, since it trivially contains the conclusion of the argument.

These considerations show that this *particular* argument for the first disjunct fails, but perhaps there is another argument . . .

## 7.9 The First Disjunct in $EA_T$

. . . In fact, one can show that there is *no* argument for the first disjunct that proceeds on the basis of the principles of $EA_T$, and since these principles would seem to encompass everything that the proponents of the first disjunct accept, this shows that the obstacle is much deeper. To describe the limitative results and the subtle issues involved it is useful to distinguish (following Reinhardt) three grades of the mechanistic thesis:

(1) (WMT) $\exists e(K_p = W_e)$
(2) (SMT) $K\exists e(K_p = W_e)$
(3) (SSMT) $\exists eK(K_p = W_e)$.

The first is the *weak mechanistic thesis.* It asserts that there is a Turing machine whose outputs are exactly the arithmetical statements that are absolutely provable. In short, it asserts that the idealized human mind is a Turing machine. The second is the *strong mechanistic thesis.* It asserts that it is absolutely provable that there is a Turing machine whose outputs are exactly the arithmetical statements that are absolutely provable. In short, it asserts that the idealized human mind knows that it is a Turing machine. The third is the *super strong mechanistic thesis.* It asserts that there is a particular Turing machine such that it is absolutely provable that the outputs of *that* Turing machine are exactly the arithmetical statements that are absolutely provable. In short, it asserts of a given Turing machine that the idealized human mind knows that it is *that* Turing machine.

The first result that is of relevance to this discussion is the following:

**Theorem 7.9.1 (Reinhardt, [37])** *'$EA_T$ + SSMT' is inconsistent.*

In other words, in the context of $EA_T$, it is true that one cannot know of a *given* Turing machine that the mind knows that it is *that* machine.

This fact (known already to Gödel) does not vindicate the proponents of the first disjunct. For they are claiming that $\neg$WMT holds, that is, they are claiming that there is *no* Turing machine that produces exactly the same arithmetical outputs as the idealized human mind. But this stronger conclusion cannot be obtained in $EA_T$, as the following result demonstrates:

**Theorem 7.9.2 (Reinhardt, [38])** *'$EA_T$ + WMT' is consistent.*

This shows that there is no argument for the first disjunct in $EA_T$ and since $EA_T$ would seem to embody all of the assumptions held by the proponents of the first disjunct, it shows that there is a fundamental obstacle.

In fact, there is an even stronger conclusion. Reinhardt conjectured that even SMT is consistent with $EA_T$ and Carlson proved this conjecture, via a sophisticated construction:

**Theorem 7.9.3 (Carlson, [2])** '$\text{EA}_\text{T}$ + SMT$'$' *is consistent.*

In other words, from the point of view of $\text{EA}_T$ it is entirely possible that the idealized human mind *knows* that it is a Turing machine. It just can't know which one!

These results show that if one is to have a hope of establishing the first disjunct, one is either going to have to invoke stronger assumptions or shift perspective entirely.

## 7.10 Penrose's New Argument

There are independent reasons for shifting perspective. For in the above we have treated the truth predicate as *typed*. But the truth predicate that we employ in natural language appears to be *type-free*. It turns out that Penrose's new argument—which is the most interesting and promising of the arguments for the first disjunct—actually requires a type-free truth predicate and so it requires a shift to an entirely new framework.

Penrose gave his new argument in [33] and after much discussion he distilled the essence of the argument as follows:

> Though I don't know that I necessarily am $F$, I conclude that if I were, then the system $F$ would have to be sound and, more to the point, $F'$ would have to be sound, where $F'$ is $F$ supplemented by the further assertion "I am $F$". I perceive that it follows from the assumption that I am $F$ that the Gödel statement $G(F')$ would have to be true and, furthermore, that it would not be a consequence of $F'$. But I have just perceived that "if I happened to be $F$, then $G(F')$ would have to be true", and perceptions of this nature would be precisely what $F$ is supposed to achieve. Since I am therefore capable of perceiving something beyond the powers of $F$, I deduce that I cannot be $F$ after all. ([34], §3.2).

This argument has a convincing air about it. I think it is quite subtle and I do not believe that the logical subtleties have been fully appreciated.

The argument has been criticized by many people, most notably, Chalmers [3], Feferman [6], Lindström [22], [23], and Shapiro [41], [42]. I am in agreement with many of the critical points in these papers but I think that they all fail to be fully charitable in at least two respects. First, in reconstructing Penrose's argument they employ systems that formalize absolute provability with a predicate. But, as we have seen, if one formalizes absolute provability as a predicate then the admissible principles of knowledge are severely restricted; indeed one cannot accept the standard principles of knowledge. For this reason it is much more charitable to proceed, as we have done, by formalizing absolute provability with an operator. Second, although at times it is noted—for example by Shapiro—that Penrose's argument seems to require a truth predicate that is self-applicable, it is simply remarked that such predicates lead to difficulties. But there are well-known systems of type-free truth. For this reason it is more charitable to proceed by formalizing truth as type-free.

We shall follow the most charitable course. For the strength of a criticism is proportional to the degree to which it is charitable. We will begin by formalizing absolute provability with an operator and by formalizing truth as type-free. But first let us go through the above argument line by line, formulating it in terms of the concise notation that we have already employed, where, for example, '$K = W_n$' is taken to be shorthand for

$$\forall \ulcorner\varphi\urcorner (K(\ulcorner\varphi\urcorner) \leftrightarrow \ulcorner\varphi\urcorner \in W_n).$$

(In the case where '$K$' is treated as an operator one should think of '$K$' here as '$K_p$'.) The argument will invoke several principles of mathematics, truth, and knowledge. There are three such principles that it will be useful to state at the outset:

$$K(\ulcorner\varphi\urcorner) \rightarrow T(\ulcorner\varphi\urcorner) \qquad (K \rightarrow T)$$

$$\varphi \rightarrow T(\ulcorner\varphi\urcorner) \qquad (T\text{-In})$$

$$\frac{\varphi}{K(\ulcorner\varphi\urcorner)}. \qquad (K\text{-Intro})$$

Here is the argument, step by step:

(1) $K = W_n \rightarrow W_n \subseteq T$.
[This follows by the rule '$K \rightarrow T$'.]

(2) $K = W_n \rightarrow W_{n^+} \subseteq T$.
[Here $n^+$ is the index of the machine obtained by adding the sentence '$K = W_n$' to $W_n$. This line follows from (1) by the rule '$T$-In'.]

(3) $K = W_n \rightarrow G(W_{n^+})$.
[Here '$G(W_{n^+})$' is the Gödel sentence for $W_{n^+}$. This line follows from (2) by reflection and the nature of the Gödel sentence.] theorem.]

(4) $K = W_n \rightarrow W_{n^+} \nvdash G(W_{n^+})$.
[This line follows from (2) by the incompleteness theorem.]

(5) $K(K = W_n \rightarrow G(W_{n^+}))$.
[This line follows from (3) by the rule '$K$-Intro'.]

(6) $K = W_n \rightarrow (W_n \nvdash (K = W_n \rightarrow G(W_{n^+})))$.
[This line follows from (4).]

(7) $K = W_n \rightarrow K \neq W_n$.
[This line follows from (5) and (6) since under the assumption that $K = W_n$ lines (5) and (6) show that $K$ and $W_n$ disagree on "$K = W_n \rightarrow G(W_{n^+})$."]

(8) $K \neq W_n$
[This line follows from (7) by logic.]

Notice that in (2) '$T$' is applied to '$K$' and in the rule '$K$-Intro' '$K$' is applied to '$T$'. So '$T$' is applied to '$T$' and '$K$' is applied to '$K$'. Now, in $EA_T$ we had a certain degree of self-applicability but not full self-applicability since, although '$T$' could be applied to some formulas involving '$K$', it could not be applied to formulas in which '$K$' was in turn applied to '$T$'. Here, however, we have full self-applicability, and so any formal system in which the above argument can be implemented will be one involving a type-free theory of $K$ and a type-free theory of $T$.

The theory $EA_T$ of the last section involved a notion of knowledge which was self-applicable but the theory of truth that was adjoined to this was a *typed* theory of truth. Our goal now is to replace the typed theory of truth with a *type-free* theory of truth.

But now we face an obstacle. For in contrast to typed theories of truth (of which there is essentially only one, namely, Tarski's), there is a wide variety of type-free theories of truth. It would be ideal if we could quantify over "all possible type-free theories of truth" and show that no such theory yields a system that provides a convincing argument for the first disjunct. But given the open-endedness of the motion of "a possible type-free theory" it is hard to see how to do this. What I shall do instead is select one of the best current candidates of a type-free theory of truth—namely, Feferman's theory of determinate truth, DT—and use it to build a type-free analogue of $EA_T$. The results will be proved with regard to this system.

## 7.11 Type-Free Truth

There is a nice discussion in Leitgeb [20] of the choices one must make when constructing a theory of type-free truth. Leitgeb distinguishes eight norms governing the informal concept of truth in natural discourse. It turns out (by an argument due to Tarski) that there is no concept of truth that meets all eight norms. The aim of a type-free theory of truth is to capture as many of these norms as possible.

A theory that does this quite nicely is Feferman's recent theory DT, introduced in [8]. It is based on the view that once we enter the realm of type-free truth and encounter the paradoxical sentences that come with it, we should expect that some sentences are *indeterminate*—that is, neither true nor false—and that once we take this into account we should realize that our endorsement of the above norms was rather naive; for example, once indeterminate sentences enter the picture it is natural to revise some of the norms—in particular, norms (4), (5), and (7) in [20]—so that they are only expected to hold for *determinate* sentences. This is precisely what the theory DT does: It meets all of the norms (1)–(8) with the caveat that it only meets (4), (5), and (7) when these norms are restricted to determinate sentences, a restriction, it is argued, that is justified.

Feferman starts with the base language $L_{PA}$ and the system PA. He then extends the language by adding a (type-free) truth predicate '$T$' and he extends the system PA by axioms governing truth.

It will be useful to introduce two defined concepts: Let '$D(x)$' be short for '$T(x) \vee T(\dot{\neg}x)$' (which asserts that $x$ is true or false). The axiomatization of DT is most perspicuously formulated in terms of these derived concepts. We use 'Sent($x$)' (without a subscript) to indicate that $x$ ranges over (Gödel numbers of) sentences in the full language $L_{EA_T}$.

We define a valuation function Val on the closed terms in such a way that

$$(V_1)\ (\forall x)[\mathrm{Val}(\dot{\dot{x}}) = x].$$

$$(V_2)\ (\forall x_1)\cdots(\forall x_n)[\mathrm{CTerm}(x_1) \wedge \cdots \wedge \mathrm{CTerm}(x_n) \rightarrow (\mathrm{Val}(\dot{f}(x_1,\ldots,x_n)) = f(\mathrm{Val}(x_1),\ldots,\mathrm{Val}(x_n))].$$

The axioms of determinateness are:

$(D_1)$ $(\forall x)[\text{Sent}_{L_{PA}}(x) \rightarrow D(x)]$.
$(D_2)$ $(\forall x)[\text{Sent}(x) \rightarrow (D(\underset{\cdot}{\neg} x) \leftrightarrow D(x))]$.
$(D_3)$ $(\forall x)(\forall y)[\text{Sent}(x) \wedge \text{Sent}(y) \rightarrow (D(x \underset{\cdot}{\vee} y) \leftrightarrow D(x) \wedge D(y))]$.
$(D_4)$ $(\forall x)(\forall y)[\text{Sent}(x) \wedge \text{Sent}(y) \rightarrow (D(x \underset{\cdot}{\rightarrow} y) \leftrightarrow D(x) \wedge (T(x) \rightarrow D(y)))]$.
$(D_5)$ $(\forall x)(\forall z)[\text{Var}(z) \wedge \text{Sent}((\underset{\cdot}{\forall} z)x) \rightarrow (D((\underset{\cdot}{\forall} z)x) \leftrightarrow \forall y D(x(\dot{\bar{y}}/z)))]$.
$(D_6)$ $(\forall x)[D(\underset{\cdot}{T}(\dot{\bar{x}})) \leftrightarrow D(x)]$.

The axioms of truth are:

$(T_1)$ For each atomic formula $R(x_1, \ldots, x_n)$ of $L_{\text{PA}}$:

$$(\forall x_1) \cdots (\forall x_n)[\text{CTerm}(z_1) \wedge \cdots \wedge \text{CTerm}(z_n) \rightarrow (T(\underset{\cdot}{R}(z_1, \ldots, z_n)) \leftrightarrow R(\text{Val}(z_1), \ldots, \text{Val}(z_n)))].$$

$(T_2)$ $(\forall x)[\text{Sent}(x) \wedge D(x) \rightarrow (T(\underset{\cdot}{\neg} x) \leftrightarrow \neg T(x))]$.
$(T_3)$ $(\forall x)(\forall y)[\text{Sent}(x) \wedge \text{Sent}(y) \wedge D(x \underset{\cdot}{\vee} y) \rightarrow (T(x \underset{\cdot}{\vee} y) \leftrightarrow T(x) \vee T(y))]$.
$(T_4)$ $(\forall x)(\forall y)[\text{Sent}(x) \wedge \text{Sent}(y) \wedge D(x \underset{\cdot}{\rightarrow} y) \rightarrow (T(x \underset{\cdot}{\rightarrow} y) \leftrightarrow T(x) \rightarrow T(y))]$.
$(T_5)$ $(\forall x)(\forall z)[\text{Var}(z) \wedge \text{Sent}((\underset{\cdot}{\forall} z)x) \wedge D((\underset{\cdot}{\forall} z)x) \rightarrow (T((\underset{\cdot}{\forall} z)x) \leftrightarrow \forall y T(x(\dot{\bar{y}}/z)))]$.
$(T_6)$ $(\forall x)[D(x) \rightarrow (T(\underset{\cdot}{T}(\dot{\bar{x}})) \leftrightarrow T(x))]$.

The system DT consists of these axioms.

Feferman uses a fixed-point construction to show that DT is consistent and one of his key results is the following:

**Lemma 7.11.1 (Feferman)** *For each $\varphi$ in the language of* DT,

$$\text{DT} \vdash D(\ulcorner \varphi \urcorner) \rightarrow (T(\ulcorner \varphi \urcorner) \leftrightarrow \varphi).$$

That is, in DT one can prove each $T$-biconditional *conditioned on determinateness.*

## 7.12 A Failed Attempt

We thus have two systems extending PA—the system EA governing absolute provability and the system DT governing the notion of type-free truth—each of which is consistent. It is natural to simply amalgamate them.

We add the natural axioms of determinateness and truth for absolute provability:

$(D_7)$ $(\forall x)[\text{Sent}(x) \rightarrow (D(\underset{\cdot}{K} x) \leftrightarrow D(x))]$.
$(T_7)$ $(\forall x)[\text{Sent}(x) \wedge D(x) \rightarrow (T(\underset{\cdot}{K} x) \leftrightarrow K T(x))]$.

(Notice that since $K$ is an operator we have had to state these axioms in the manner we have.) And we also need the obvious axiom-scheme governing the interaction of absolute provability and truth, a scheme that is closely related to $E_3$:

($I$) For each sentence $\varphi$:

$$K\varphi \rightarrow T(\ulcorner\varphi\urcorner).$$

This is a very plausible axiom. If $K\varphi$ then it seems pretty clear that we must have $D(\ulcorner\varphi\urcorner)$ (since it is hard to see how $\varphi$ could be ideally known if it were *in*determinate). But then either $T(\ulcorner\varphi\urcorner)$ or $T(\dot{\neg}\ulcorner\varphi\urcorner)$. We clearly cannot have the latter. So we must have $T(\ulcorner\varphi\urcorner)$.

The system $\mathrm{EA}_{\mathrm{DT}}$ is the theory axiomatized by $\Sigma \cup K\Sigma$, where $\Sigma$ consists of the axioms of PA (in the language $L_{\mathrm{EA_T}}$), the basic axioms of absolute provability (in the language $L_{\mathrm{EA_T}}$), the axioms of determinateness and truth (for the language $L_{\mathrm{EA_T}}$), and the interaction axiom $I$. This is a very natural system. However, there is a problem.

**Theorem 7.12.1** $\mathrm{EA}_{\mathrm{DT}}$ *is inconsistent.*

***Proof*** The main point is that for each $\varphi$ in $L_{\mathrm{EA_{DT}}}$

$$\mathrm{EA}_{\mathrm{DT}} \vdash (K\varphi \leftrightarrow K_p(\ulcorner\varphi\urcorner)).$$

We show this in two claims. Let $\varphi$ be a sentence of $L_{\mathrm{EA_{DT}}}$.
CLAIM 1. $\mathrm{EA}_{\mathrm{DT}} \vdash (K\varphi \rightarrow K_p(\ulcorner\varphi\urcorner))$.

***Proof*** By ($E_4$)

$$K\varphi \rightarrow KK\varphi. \quad (1)$$

By ($I$), for each $\psi$ in $L_{\mathrm{EA_{DT}}}$,

$$K\psi \rightarrow T(\ulcorner\psi\urcorner)$$

and so, in particular (taking '$\psi$' to be '$K\varphi$'),

$$KK\varphi \rightarrow T(\ulcorner K\varphi\urcorner). \quad (2)$$

By (1) and (2),

$$K\varphi \rightarrow T(\ulcorner K\varphi\urcorner)$$

and so

$$K\varphi \rightarrow T(\ulcorner K\varphi\urcorner) \wedge \mathrm{Sent}_{\mathrm{EA_{DT}}}(\ulcorner\varphi\urcorner)$$
$$\rightarrow K_p(\ulcorner\varphi\urcorner).$$ ■

CLAIM 2. $\mathrm{EA}_{\mathrm{DT}} \vdash (K_p(\ulcorner\varphi\urcorner) \rightarrow K\varphi)$.

***Proof*** We want to show that

$$\mathrm{EA}_{\mathrm{DT}} \vdash T(\ulcorner K\varphi\urcorner) \wedge \mathrm{Sent}_{\mathrm{EA_{DT}}}(\ulcorner\varphi\urcorner) \rightarrow K\varphi.$$

By Lemma 7.7.3 we have

$$\mathrm{EA}_{\mathrm{DT}} \vdash T(\ulcorner\varphi\urcorner) \rightarrow \varphi$$

and so

$$\mathrm{EA}_{\mathrm{DT}} \vdash T(\ulcorner K\varphi \urcorner) \rightarrow K\varphi,$$

as desired. ■

We can now use the main claim to turn the operator $K$ into a predicate (namely, $K_p$) and show that the predicate has exactly the properties that $K$ has, which violates the limitative results discussed in §7.5. Thus, $\mathrm{EA}_{\mathrm{DT}}$ is inconsistent, which competes the proof. ■

(There is an even quicker way to see the inconsistency in $\mathrm{EA}_{\mathrm{DT}}$ as follows. Let $\lambda$ be the liar sentence, $\lambda \vee \neg\lambda$ is valid. So by $E_1$ we have $K(\ulcorner \lambda \vee \neg\lambda \urcorner)$ and by $I$ we have $T(\ulcorner \lambda \vee \neg\lambda \urcorner)$. But $T(\ulcorner \lambda \vee \neg\lambda \urcorner)$ is equivalent to $T(\ulcorner \lambda \urcorner) \vee T(\dot{\neg}\ulcorner \lambda \urcorner)$ which, by the definition of '$D$', is equivalent to $D(\ulcorner \lambda \urcorner)$. Hence we have $D(\ulcorner \lambda \urcorner)$, which is a contradiction.)

This result is of interest in itself. $\mathrm{EA}_{\mathrm{DT}}$ has two parts—the axioms governing $K$ and the axioms governing $T$. When one adds each part separately one arrives at a consistent system. But when one adds both parts and the interaction axiom $I$, inconsistency arises.

## 7.13 The System DTK

This is perhaps not so surprising. The trouble is that once we enter the realm where indeterminate sentences arise, it is no longer plausible to maintain that one knows logical validities if the validities in question are themselves indeterminate. This motivates a natural modification of the above system. And once one makes this modification it turns out to be possible to treat '$K$' as a predicate. On this approach the work in circumventing the paradoxes is carried by the theory of truth and transfered over to the theory of absolute provability (where '$K$' is now treated as a predicate) by the manner in which absolute provability is linked to truth.[9]

The language of DTK is $L_{\mathrm{DTK}}$ and the logic is first-order logic. As above, '$D(x)$' is shorthand for '$T(x) \vee T(\dot{\neg}x)$'.

The system DTK has four groups of axioms (in addition to our fixed set of axioms for first-order logic):

I. ARITHMETIC

The axioms of arithmetic are those of PA except where the induction scheme is extended to the entire language $L_{\mathrm{DTK}}$ (so that '$T$' and '$K$' are allowed to figure in induction).

II. DETERMINATENESS

The axioms of determinateness are:

($D_1$) $(\forall x)[\text{At-Sent}_{L_{\mathrm{PA}}}(x) \rightarrow D(x)]$.

($D_2$) $(\forall x)[\mathrm{Sent}(x) \rightarrow (D(\dot{\neg}x) \leftrightarrow D(x))]$.

($D_3$) $(\forall x)(\forall y)[\mathrm{Sent}(x) \wedge \mathrm{Sent}(y) \rightarrow (D(x \dot{\vee} y) \leftrightarrow D(x) \wedge D(y))]$.

$(D_4)$ $(\forall x)(\forall y)[\mathrm{Sent}(x) \wedge \mathrm{Sent}(y) \to (D(x \dot{\to} y) \leftrightarrow D(x) \wedge (T(x) \to D(y)))]$.

$(D_5)$ $(\forall x)(\forall z)[\mathrm{Var}(z) \wedge \mathrm{Sent}((\dot{\forall} z)x) \to (D((\dot{\forall} z)x) \leftrightarrow \forall y D(x(\dot{\bar{y}}/z)))]$.

$(D_6)$ $(\forall x)[D(\dot{T}(\dot{\bar{x}})) \leftrightarrow D(x)]$.

$(D_7)$ $(\forall x)[D(\dot{K}(\dot{\bar{x}})) \leftrightarrow D(x)]$.

III. TRUTH

The axioms of truth are:

$(T_1)$ For each atomic formula $R(x_1, \ldots, x_n)$ of $L_{\mathrm{PA}}$:

$$(\forall x_1) \cdots (\forall x_n)[T(\dot{R}(\dot{\bar{z}}_1, \ldots, \dot{\bar{z}}_n)) \leftrightarrow R(z_1, \ldots, z_n)].$$

$(T_2)$ $(\forall x)[\mathrm{Sent}(x) \wedge D(x) \to (T(\dot{\neg} x) \leftrightarrow \neg T(x))]$.

$(T_3)$ $(\forall x)(\forall y)[\mathrm{Sent}(x) \wedge \mathrm{Sent}(y) \wedge D(x \dot{\vee} y) \to (T(x \dot{\vee} y) \leftrightarrow T(x) \vee T(y))]$.

$(T_4)$ $(\forall x)(\forall y)[\mathrm{Sent}(x) \wedge \mathrm{Sent}(y) \wedge D(x \dot{\to} y) \to (T(x \dot{\to} y) \leftrightarrow T(x) \to T(y))]$.

$(T_5)$ $(\forall x)(\forall z)[\mathrm{Var}(z) \wedge \mathrm{Sent}((\dot{\forall} z)x) \wedge D((\dot{\forall} z)x) \to (T((\dot{\forall} z)x) \leftrightarrow \forall y T(x(\dot{\bar{y}}/z)))]$.

$(T_6)$ $(\forall x)[D(x) \to (T(\dot{T}(\dot{\bar{x}})) \leftrightarrow T(x))]$.

$(T_7)$ $(\forall x)[D(x) \to (T(\dot{K}(\dot{\bar{x}})) \leftrightarrow K(x))]$.

IV. KNOWLEDGE

The axioms of knowledge are:

$(K_1)$ $(\forall x)[\mathrm{Sent}(x) \to (K(x) \to T(x))]$.

$(K_2)$ $(\forall x)(\forall y)[\mathrm{Sent}(x) \wedge \mathrm{Sent}(y) \to (K(x \dot{\to} y) \wedge K(x) \to K(y))]$.

$(K_3)$ $(\forall x)[\mathrm{Sent}(x) \to (K(x) \to K(\dot{K}(\dot{\bar{x}})))]$.

RULES

The rules are

$$\frac{\varphi \wedge D(\ulcorner \varphi \urcorner)}{K(\ulcorner \varphi \urcorner)} \qquad (DK\text{-Intro})$$

and

$$\frac{\varphi \wedge D(\ulcorner \varphi \urcorner)}{T(\ulcorner \varphi \urcorner)}. \qquad (DT\text{-Intro})$$

**Remark 7.13.1** *The rule DK-Intro enables us to conclude that the determinate statements of logic and arithmetic are absolutely provable.*

## 7.14 Basic Results in DTK

**Theorem 7.14.1** DTK *is consistent.*

***Proof*** Feferman used a fixed-point construction to show that DT is consistent. We show that DTK holds in this model by interpreting DTK in DT. The interpretation is simply the one that interprets '$K$' as '$T$'.

Under this interpretation groups I–III are immediately satisfied since $\text{PA}_{\text{DTK}}$ becomes $\text{PA}_{\text{DT}}$, $D_7$ becomes $D_6$, and $T_7$ becomes $T_6$, which are all principles of DT.

The axioms of group IV are also immediate: $K_1$ is trivial. $K_2$ follows from $T_4$. $K_3$ follows from $T_6$.

It remains to show that the rules hold under this interpretation. Under the interpretation there is only one rule, namely,

$$\frac{\varphi \wedge D(\ulcorner\varphi\urcorner)}{T(\ulcorner\varphi\urcorner)}.$$

So it suffices to show that this rule holds in DT for each $\varphi \in L_{\text{DT}}$. The key to this is the fact that DT satisfies the principle "$T$-Out," as Feferman observed:

**Lemma** For each $\varphi \in L_{\text{DT}}$

$$\text{DT} \vdash T(\ulcorner\varphi\urcorner) \rightarrow \varphi. \qquad (T\text{-Out})$$

Now assume

(1) $\text{DT} \vdash \varphi \wedge D(\ulcorner\varphi\urcorner)$,
that is, assume

(2) $\text{DT} \vdash \varphi \wedge (T(\ulcorner\varphi\urcorner) \vee T(\ulcorner\neg\varphi\urcorner))$.
We have

(3) $\text{DT} \vdash T(\ulcorner\neg\varphi\urcorner) \rightarrow \neg\varphi$ ($T$-Out)

(4) $\text{DT} \vdash \varphi \rightarrow \neg T(\ulcorner\neg\varphi\urcorner)$ (Logic)

(5) $\text{DT} \vdash \varphi \wedge \neg T(\ulcorner\neg\varphi\urcorner)$ (2,4)

(6) $\text{DT} \vdash T(\ulcorner\varphi\urcorner)$ (2,5)

and so the rule holds in DT. ■

**Remark 7.14.2** *The above proof shows that* DTK *is proof-theoretically equivalent to* DT. *It follows that the proof-theoretic ordinal of* DTK *is the same as that of* DT, *namely,* $\varphi_{\varepsilon}0$ *(something that was conjectured for* DT *by Feferman [8] and proved by Fujimoto [9].)*

**Remark 7.14.3** *There is another consistency proof of* DTK *(given in Theorem 7.16.2) which does not involve interpreting '$K$' as '$T$', but instead has a more restricted interpretation of '$K$', one that allows there to be absolutely undecidable sentences.*

**Lemma 7.14.4** *For each* $\varphi \in L_{\text{DTK}}$

$$\text{DTK} \vdash D(\ulcorner\varphi\urcorner) \rightarrow (T(\ulcorner\varphi\urcorner) \leftrightarrow \varphi),$$

*and hence*

$$\text{DTK} \vdash T(\ulcorner\varphi\urcorner) \rightarrow \varphi. \qquad (T\text{-Out})$$

***Proof*** More generally, we show that for each formula $\varphi(x_1, \ldots, x_n)$

$$\text{DTK} \vdash D(\ulcorner \varphi(\dot{\dot{x}}_1, \ldots, \dot{\dot{x}}_n) \urcorner) \\ \rightarrow (T(\ulcorner \varphi(\dot{\dot{x}}_1, \ldots, \dot{\dot{x}}_n) \urcorner) \leftrightarrow \varphi(x_1, \ldots, x_n)).$$

The proof is by induction on formula complexity.

The induction hypothesis is only used in the clauses of the $T$-axioms that have the same number of '$T$'s on the left-hand side and the right-hand side of the biconditional—it is used inductively to erase the '$T$'s on the right-hand side. But with $T_1$, $T_6$, and $T_7$ the extra '$T$' on the right-hand side is already erased; so these are our "base" cases and there is no appeal to the induction hypothesis. For $T_5$—the clause for the universal quantifier—we use universal introduction to get the universal closure of the induction hypothesis.

Here are the details for the step involving '$K$': Suppose the induction hypothesis holds for $\varphi(x_1, \ldots, x_n)$. We have to show that

$$\text{DTK} \vdash D(\underset{\cdot}{K}(\ulcorner \varphi(\dot{\dot{x}}_1, \ldots, \dot{\dot{x}}_n) \urcorner)) \\ \rightarrow (T(\underset{\cdot}{K}(\ulcorner \varphi(\dot{\dot{x}}_1, \ldots, \dot{\dot{x}}_n) \urcorner)) \leftrightarrow K(\ulcorner \varphi(\dot{\dot{x}}_1, \ldots, \dot{\dot{x}}_n) \urcorner)).$$

By $D_7$ the antecedent of this statement gives

$$D(\ulcorner \varphi(\dot{\dot{x}}_1, \ldots, \dot{\dot{x}}_n) \urcorner)$$

and $T_7$ gives

$$T(\underset{\cdot}{K}(\ulcorner \varphi(\dot{\dot{x}}_1, \ldots, \dot{\dot{x}}_n) \urcorner)) \leftrightarrow K(\ulcorner \varphi(\dot{\dot{x}}_1, \ldots, \dot{\dot{x}}_n) \urcorner)),$$

which completes the proof. ■

**Theorem 7.14.5** *For each* $\varphi \in L_{\text{DTK}}$

$$\text{DTK} \vdash \text{Sent}(\ulcorner \varphi \urcorner) \rightarrow (K(\ulcorner \varphi \urcorner) \rightarrow \varphi).$$

***Proof*** This is immediate, by Lemma 7.14.4. ■

**Lemma 7.14.6** *For each* $\varphi \in L_{\text{PA}}$ *we have the following:*

(1) $\text{DTK} \vdash D(\ulcorner \varphi \urcorner)$.

(2) *If PA* $\vdash \varphi$ *then* $\text{DTK} \vdash K(\ulcorner \varphi \urcorner)$.

***Proof***

(1). It is straightforward to show by induction that for each formula $\varphi(x_1, \ldots, x_n) \in L_{\text{PA}}$,

$$\text{DTK} \vdash D(\ulcorner \varphi(\dot{\dot{x}}_1, \ldots, \dot{\dot{x}}_n) \urcorner).$$

(2). For this one simply applies the rule *DK*-Intro to (1).

■

**Theorem 7.14.7** *For each $\psi(x) \in L_{PA}$ there exists a sentence $\varphi \in L_{PA}$ such that*

$$\mathrm{DTK} \vdash K(\ulcorner \varphi \leftrightarrow \psi(\ulcorner \varphi \urcorner) \urcorner).$$

***Proof*** This is immediate, by Lemma 7.14.6 and the fixed point lemma. ■

**Remark 7.14.8** *For $\varphi(x) \in L_{PA}$ we have $D(\ulcorner \varphi(\dot{x}) \urcorner)$ and hence $D(\underset{\cdot}{T}(\ulcorner \varphi(\dot{x}) \urcorner))$ by $D_6$. This can be iterated and so we have the corresponding Tarski biconditionals (since we remain in the realm of determinateness). In this way we can simulate the Tarski hierarchy, avoiding pathological self-referential sentences involving 'T'. Of course, we have sentences like '$T(\underset{\cdot}{T}(\ulcorner \varphi \urcorner))$' in which 'T' is applied to 'T', but we can consistently assign subscripts that capture the "level" of the application of 'T'. We don't have genuine self-referential applications of T like*

$$\lambda \leftrightarrow \neg T(\ulcorner \lambda \urcorner)$$

*when we climb the hierarchy in this way, that is, $\lambda$ and $\neg T(\ulcorner \lambda \urcorner)$ never appear on the same level.*

**Theorem 7.14.9** DTK $\vdash (\forall x)[\mathrm{Sent}(x) \rightarrow \neg(T(\underset{\cdot}{\neg} x) \wedge T(x))]$.

***Proof*** Work in DTK and let $x$ be a sentence. Assume, for contradiction $T(\underset{\cdot}{\neg} x) \wedge T(x)$. Thus, $D(x)$ and so, by $T_2$, $T(\underset{\cdot}{\neg} x) \leftrightarrow \neg T(x)$. So we have $T(x)$ and $\neg T(x)$. ■

In the system DTK we can recover absolute versions of the incompleteness theorems.

**Theorem 7.14.10** *Suppose that S includes* DTK*. Suppose $F(x)$ is a formula in $L_{PA}$ such that*

$$S \vdash K(\ulcorner F(\ulcorner \varphi \urcorner) \rightarrow \varphi \urcorner),$$

*for all $\varphi \in L_{PA}$. Then there is a sentence $\psi \in L_{PA}$ such that*

$$S \vdash K(\ulcorner \psi \urcorner) \wedge K(\ulcorner \neg F(\ulcorner \psi \urcorner) \urcorner).$$

***Proof*** By Theorem 7.14.7 there exists $\varphi \in L_{PA}$ such that

(1) $S \vdash K(\ulcorner \psi \leftrightarrow \neg F(\ulcorner \psi \urcorner) \urcorner)$.
By hypothesis

(2) $S \vdash K(\ulcorner F(\ulcorner \psi \urcorner) \rightarrow \psi \urcorner)$.
Furthermore, we have

(3) $S \vdash K(\ulcorner (F(\ulcorner \psi \urcorner) \rightarrow \psi) \rightarrow (\neg F(\ulcorner \psi \urcorner) \rightarrow \psi) \rightarrow \psi \urcorner)$,
since the quoted sentence is a logical validity and provably determinate (as it is in $L_{PA}$), so we can apply *DK*-Intro. Distributing '*K*' in (3) and discharging (1) and (2) gives

(4) $S \vdash K(\ulcorner \psi \urcorner)$.
Distributing *K* in (1) gives

(5) $S \vdash K(\ulcorner \neg F(\ulcorner \psi \urcorner) \urcorner)$,
which completes the proof. ■

**Theorem 7.14.11** *Suppose that S includes* DTK. *Suppose $F(x)$ is a formula in $L_{PA}$ such that*

$$S \vdash K(\ulcorner K(\ulcorner \varphi \urcorner) \rightarrow F(\ulcorner \varphi \urcorner) \urcorner),$$

*for all $\varphi \in L_{PA}$. Then*

$$S \vdash K(\ulcorner \neg K(\ulcorner \mathrm{Con}(F) \urcorner) \urcorner).$$

***Proof*** The proof follows Reinhardt's proof of Theorem 7.7.2, noting along the way that all sentences in all steps are provably determinate and so *DK*-Intro applies and we can prefix and distribute '*K*' as we did in the previous proof. ■

**Remark 7.14.12** *In each of the above theorems we restricted F to be a formula of $L_{PA}$ which holds of (the Gödel numbers of) formulas of $L_{PA}$ since such statements are* provably determinate. *But the theorems are more general. For example, the first holds for any F for which the diagonal lemma (without 'K') yields a provably determinate sentence $\psi$ for $\neg F(x)$ and for which $F(\ulcorner \psi \urcorner)$ and $\neg F(\ulcorner \psi \urcorner)$ are provably determinate.*

*Notice, however, that the theorems do not apply* without restriction *on F. For example, if $\varphi = \lambda$ and $S = T$ (truth) then the hypothesis of Theorem 7.14.10 cannot hold since $T(\ulcorner \lambda \urcorner) \rightarrow \lambda$ is not determinate and the hypothesis of Theorem 7.14.11 cannot hold since $K(\ulcorner \lambda \urcorner) \rightarrow T(\ulcorner \lambda \urcorner)$ is not determinate.*

## 7.15 The Disjunction in DTK

Let us begin by recalling the three versions of the mechanistic thesis: WMT, SMT, and SSMT. In our present context the results will be sensitive (as we have seen above) to the language in the scope of the claims (in particular, it will be sensitive to whether or not the sentences in the language are determinate). We will specify the language with a subscript and when there is no subscript we intend the language to be the full language. We use '$K =_L W_e$' in the context of $L$ to abbreviate

$$(\forall x)(\mathrm{Sent}_L(x) \rightarrow (K(x) \leftrightarrow W_e(x)))$$

and likewise for other predicates—most notably '$T$'—and other set-theoretic operations. The three theses are thus (relative to $L$):

($\mathrm{WMT}_L$) $\exists e\,(K =_L W_e)$.
($\mathrm{SMT}_L$) $K(\ulcorner \exists e\,(K =_L W_e) \urcorner)$.
($\mathrm{SSMT}_L$) $\exists e\, K(\ulcorner K =_L W_e \urcorner)$.

Whether these theses turn out to be determinate in DTK depends on the parameter $L$. Let us first take $L = L_{PA}$. There are two natural readings of each thesis. We will illustrate them by focusing on the case of $\mathrm{WMT}_{PA}$. The first reading of $\mathrm{WMT}_{PA}$ is

$$(\exists e)\big[(\forall x)\,(\mathrm{Sent}_{\mathrm{PA}}(x) \to (K(x) \to W_e(x)))$$
$$\wedge\,(\forall x)\,(\mathrm{Sent}_{\mathrm{PA}}(x) \to (W_e(x) \to K(x)))\big].$$

We need to see that each instance is determinate. For

$$\mathrm{Sent}_{\mathrm{PA}}(\ulcorner\varphi\urcorner) \to (K(\ulcorner\varphi\urcorner) \to W_e(\ulcorner\varphi\urcorner))$$

we have that the antecedent is always determinate and we need to see that the consequent is determinate if the antecedent is true, i.e. in those cases where $\varphi \in L_{\mathrm{PA}}$. For these cases it is indeed determinate since $K(\ulcorner\varphi\urcorner)$ is always determinate when $\varphi$ is determinate and $W_e(\ulcorner\varphi\urcorner)$ is determinate for any $\varphi$. The same applies to the second conjunct. Thus, $\mathrm{WMT}_{\mathrm{PA}}$ is determinate. In a similar fashion one can also see that $\mathrm{SMT}_{\mathrm{PA}}$ and $\mathrm{SSMT}_{\mathrm{PA}}$ are determinate on the analogous reading.

A second reading of $\mathrm{WMT}_{\mathrm{PA}}$ is

$$(\exists e)\,\neg(\exists x)\,\big[\mathrm{Sent}_{\mathrm{PA}}(x) \wedge (K(x) \wedge \neg W_e(x)) \wedge (W_e(x) \wedge \neg K(x))\big].$$

But now the instance

$$\mathrm{Sent}_{\mathrm{PA}}(\ulcorner\lambda\urcorner) \wedge (K(\ulcorner\lambda\urcorner) \wedge \neg W_e(\ulcorner\lambda\urcorner)) \wedge (W_e(\ulcorner\lambda\urcorner) \wedge \neg K(\ulcorner\lambda\urcorner))$$

is *not* determinate since $K(\ulcorner\lambda\urcorner)$ is not determinate and this infects the determinateness of the entire statement. In effect, we need, for determinateness, $\mathrm{Sent}_{\mathrm{PA}}(\cdot)$ to play a *restrictive role*, restricting the determinate instances. Similarly, the second readings of $\mathrm{SMT}_{\mathrm{PA}}$ and $\mathrm{SSMT}_{\mathrm{PA}}$ are not determinate.

The full versions WMT, SMT, and SSMT are not determinate (on either reading) since, for example, in the case of the first reading, the component

$$\mathrm{Sent}(\ulcorner\lambda\urcorner) \to (K(\ulcorner\lambda\urcorner) \to W_e(\ulcorner\lambda\urcorner))$$

is not determinate. Moreover, there is no way of repairing this defect and restoring determinateness by reformulating the statements, for the only way to improve the chances of determinateness is by *restricting* to determinate sentences. But in this case $L = L_{\mathrm{DTK}}$ and so we must restrict to $\varphi$ such that $D(\ulcorner\varphi\urcorner)$. However, the *only* way of doing this is by relativizing with '$D(\ulcorner\varphi\urcorner) \to$' but even this will not work since, for example, '$D(\ulcorner\lambda\urcorner)$' is *false*.

Henceforth our official construal of WMT, SMT, and SSMT (and their relativizations) will follow the first reading (since this reading has the greatest chance of determinateness). In summary, we have shown:

**Theorem 7.15.1** *In* DTK *the statements* $\mathrm{WMT}_{\mathrm{PA}}$, $\mathrm{SMT}_{\mathrm{PA}}$, *and* $\mathrm{SSMT}_{\mathrm{PA}}$ *are provably determinate, while the statements* WMT, SMT, *and* SSMT *are provably indeterminate.*

Gödel's disjunction evidently concerns ¬WMT but it is unclear when he speaks of proving ¬WMT using an adequate type-free theory of the "intensional notions"—in this context, the concept of truth and the concept of absolute provability—he means that the concepts are used to state and prove $\neg\text{WMT}_{\text{PA}}$ or whether he means that these concepts are used to state and prove $\neg$WMT. We shall cover both cases and accordingly distinguish $\text{GD}_L$ for various $L$, again with $L = L_{\text{PA}}$ and $L = L_{\text{DTK}}$ being the cases of interest. Let

$$\text{GD}_L = \neg\text{WMT}_L \vee \text{AU}_L,$$

where

$$\text{AU}_L = (\exists x)\left(\text{Sent}_L(x) \wedge T(x) \wedge \neg K(x) \wedge \neg K(\dot\neg x)\right).$$

**Remark 7.15.2** *In our present context—where we have indeterminate sentences like λ—we cannot simply let*

$$\text{AU}_L = (\exists x)\left(\text{Sent}_L(x) \wedge \neg K(x) \wedge \neg K(\dot\neg x)\right),$$

*since then for $L = L_{\text{DTK}}$ the statement $\text{AU}_L$ would be trivially satisfied by λ.*

**Theorem 7.15.3** DTK $\vdash$ GD.

***Proof*** The proof follows that of Gödel and Reinhardt.

**Case 1:** $K = T$.

In this case we argue that $\neg$WMT holds. By the second incompleteness theorem we have

$$(\forall w)\,(W_e \subseteq T \rightarrow W_e \subsetneq T)$$

and so

$$(\forall w)\,(W_e \subseteq K \rightarrow W_e \subsetneq K)$$

follows by our case assumption.

**Case 2:** $K \neq T$.

In this case we argue that AU holds. Since $K \subseteq T$ we have $K \subsetneq T$. So there exists $\varphi$ such that $T(\ulcorner\varphi\urcorner) \wedge \neg K(\ulcorner\varphi\urcorner)$. But we can't have $K(\ulcorner\neg\varphi\urcorner)$ since then $T(\ulcorner\neg\varphi\urcorner)$ and yet

$$\text{DTK} \vdash (\forall x)\left(\text{Sent}(x) \rightarrow \neg(T(x) \wedge T(\dot\neg x))\right).$$

So $\neg K(\dot\neg\ulcorner\varphi\urcorner)$ must hold and hence AU holds. ■

**Corollary 7.15.4** DTK $\vdash$ GD $\wedge \neg D(\ulcorner\text{GD}\urcorner)$.

**Corollary 7.15.5** DTK $\vdash \text{GD}_{\text{PA}} \wedge D(\ulcorner\text{GD}_{\text{PA}}\urcorner) \wedge K(\ulcorner\text{GD}_{\text{PA}}\urcorner)$.

## 7.16 The Disjuncts in DTK

We now wish to show that although DTK can prove $\text{GD}_{\text{PA}}$ it can neither prove nor refute either disjunct.

**Theorem 7.16.1** *Assume that* DTK *is consistent. Then* DTK *can neither refute the first disjunct nor prove the second disjunct of* $\mathrm{GD_{PA}}$, *that is,*

*(1)* $\mathrm{DTK} \nvdash \mathrm{WMT_{PA}}$ *and*
*(2)* $\mathrm{DTK} \nvdash \mathrm{AU_{PA}}$.

***Proof*** This follows since $\neg\mathrm{AU_{PA}}$ and $\neg\mathrm{WMT_{PA}}$ hold in the "$K = T$"-model of Theorem 7.14.1. ■

**Theorem 7.16.2** *Assume that* DTK *is* $\Sigma^0_n$*-sound, for each* $n < \omega$. *Then* DTK *can neither prove the first disjunct nor refute the second disjunct of* $\mathrm{GD_{PA}}$, *that is,*

*(1)* $\mathrm{DTK} \nvdash \neg\mathrm{WMT_{PA}}$
*(2)* $\mathrm{DTK} \nvdash \neg\mathrm{AU_{PA}}$.

***Proof*** Let $M$ be the fixed point model from Theorem 7.14.1 and work in $M$. There we have used an interpretation of '$K$' as '$T$' and showed that the $K$-axioms hold. Now we require a much more restrictive interpretation of '$K$'. We shall interpret '$K(x)$' as

$$\mathrm{Prov_{DT}}(x \,\dot{\wedge}\, \dot{D}(x)) \wedge T(x).$$

The translation

$$`K(\ulcorner\varphi\urcorner)\text{'} \mapsto `\mathrm{Prov_{DT}}(\ulcorner\varphi \wedge D(\ulcorner\varphi\urcorner)\urcorner) \wedge T(\ulcorner\varphi\urcorner)\text{'}$$

of $L_{\mathrm{DTK}}$ to $L_{\mathrm{DT}}$ proceeds inductively on the complexity of the formulas.

We can, in our proof, always appeal to $D$- and $T$-principles not involving '$K$' without assuming the induction hypothesis. In the proof we actually verify each $K$-axiom completely for each formula at each step.

We now verify that under this interpretation the axioms of DTK which are not in DT hold in $M$.

$(K_1)$ $(\forall x)[\mathrm{Sent}(x) \rightarrow (K(x) \rightarrow T(x))]$.

***Proof*** Suppose $\varphi$ is such that $\mathrm{Sent}(\ulcorner\varphi\urcorner) \wedge K(\ulcorner\varphi\urcorner)$. By the interpretation of '$K$' we have

$$\mathrm{Prov_{DT}}(\ulcorner\varphi \wedge D(\ulcorner\varphi\urcorner)\urcorner) \wedge T(\ulcorner\varphi\urcorner),$$

and so, trivially, $T(\ulcorner\varphi\urcorner)$. ■

$(K_2)$ $(\forall x)(\forall y)[\mathrm{Sent}(x) \wedge \mathrm{Sent}(y) \rightarrow (K(x \dot{\rightarrow} y) \wedge K(x) \rightarrow K(y))]$.

***Proof*** Suppose $\varphi$ and $\psi$ are such that $\mathrm{Sent}(\ulcorner\varphi\urcorner)$, $\mathrm{Sent}(\ulcorner\psi\urcorner)$, $K(\ulcorner\varphi \rightarrow \psi\urcorner)$, and $K(\ulcorner\varphi\urcorner)$. By the interpretation of '$K$' we have

$$\mathrm{Prov_{DT}}(\ulcorner\varphi \rightarrow \psi\urcorner) \wedge \mathrm{Prov_{DT}}(\ulcorner D(\ulcorner\varphi \rightarrow \psi\urcorner)\urcorner) \wedge T(\ulcorner\varphi \rightarrow \psi\urcorner)$$

and

$$\mathrm{Prov_{DT}}(\ulcorner\varphi\urcorner) \wedge \mathrm{Prov_{DT}}(\ulcorner D(\ulcorner\varphi\urcorner)\urcorner) \wedge T(\ulcorner\varphi\urcorner).$$

We need to show

$$\mathrm{Prov_{DT}}(\ulcorner\psi\urcorner) \wedge \mathrm{Prov_{DT}}(\ulcorner D(\ulcorner\psi\urcorner)\urcorner) \wedge T(\ulcorner\psi\urcorner).$$

We get $\text{Prov}_{\text{DT}}(\ulcorner\psi\urcorner)$ from $\text{Prov}_{\text{DT}}(\ulcorner\varphi \to \psi\urcorner)$ and $\text{Prov}_{\text{DT}}(\ulcorner\varphi\urcorner)$ using the properties of the provability predicate. And we get $T(\ulcorner\psi\urcorner)$ from $T(\ulcorner\varphi \to \psi\urcorner)$ and $T(\ulcorner\varphi\urcorner)$ using $T_5$.

It remains to show that $\text{Prov}_{\text{DT}}(\ulcorner D(\ulcorner\psi\urcorner)\urcorner)$. The point is that we have $\text{Prov}_{\text{DT}}(\ulcorner T(\ulcorner\varphi\urcorner)\urcorner)$ since $\text{Prov}_{\text{DT}}(\ulcorner\varphi \wedge D(\ulcorner\varphi\urcorner)\urcorner)$ and Theorem 7.14.4 is provable in DT. So we have $\text{Prov}_{\text{DT}}(\ulcorner D(\ulcorner\varphi\urcorner)\urcorner)$ and $\text{Prov}_{\text{DT}}(\ulcorner T(\ulcorner\varphi\urcorner)\urcorner)$ and also $\text{Prov}_{\text{DT}}(\ulcorner D(\ulcorner\varphi \to \psi\urcorner)\urcorner)$. Thus, by $D_4$, we have $\text{Prov}_{\text{DT}}(\ulcorner D(\ulcorner\psi\urcorner)\urcorner)$. ■

($K_3$) $(\forall x)[\text{Sent}(x) \to (K(x) \to K(\underset{\cdot}{K}(\dot{x})))]$.

***Proof*** Suppose $\varphi$ is such that $\text{Sent}(\ulcorner\varphi\urcorner)$ and $K(\ulcorner\varphi\urcorner)$. By the interpretation of '$K$' we have

$$\text{Prov}_{\text{DT}}(\ulcorner\varphi \wedge D(\ulcorner\varphi\urcorner)\urcorner) \wedge T(\ulcorner\varphi\urcorner).$$

We have to show

(1) $\text{Prov}_{\text{DT}}(\ulcorner\text{Prov}_{\text{DT}}(\ulcorner\varphi\urcorner)\urcorner)$, $\text{Prov}_{\text{DT}}(\ulcorner\text{Prov}_{\text{DT}}(\ulcorner D(\ulcorner\varphi\urcorner)\urcorner)\urcorner)$, and $\text{Prov}_{\text{DT}}(\ulcorner T(\ulcorner\varphi\urcorner)\urcorner)$,

(2) $\text{Prov}_{\text{DT}}(\ulcorner D(\ulcorner\text{Prov}_{\text{DT}}(\ulcorner\varphi\urcorner)\urcorner)\urcorner)$, $\text{Prov}_{\text{DT}}(\ulcorner D(\ulcorner\text{Prov}_{\text{DT}}(\ulcorner D(\ulcorner\varphi\urcorner)\urcorner)\urcorner)\urcorner)$, and $\text{Prov}_{\text{DT}}(\ulcorner D(\ulcorner T(\ulcorner\varphi\urcorner)\urcorner)\urcorner)$, and

(3) $T(\ulcorner\text{Prov}_{\text{DT}}(\ulcorner\varphi\urcorner)\urcorner)$, $T(\ulcorner\text{Prov}_{\text{DT}}(\ulcorner D(\ulcorner\varphi\urcorner)\urcorner)\urcorner)$, and $T(\ulcorner T(\ulcorner\varphi\urcorner)\urcorner)$.

Regarding (1): The first two follow by the reflection property of '$\text{Prov}_{\text{DT}}$'. The third follows since DT satisfies $DT$-Intro (as shown in the proof of Theorem 7.14.1).

Regarding (2): The first two follow since the statements in question are $\Sigma^0_1$, hence provably determinate. The third follows since we have $\text{Prov}_{\text{DT}}(\ulcorner D(\ulcorner\varphi\urcorner)\urcorner)$ and hence $\text{Prov}_{\text{DT}}(\ulcorner D(\ulcorner T(\ulcorner\varphi\urcorner)\urcorner)\urcorner)$, by $D_6$ in DT.

Regarding (3): The first two follow since the statements in question are $\Sigma^0_1$, hence provably determinate, and, since we are assuming $\text{Prov}_{\text{DT}}(\ulcorner\varphi\urcorner)$ and $\text{Prov}_{\text{DT}}(\ulcorner D(\ulcorner\varphi\urcorner)\urcorner)$ we can apply Lemma 7.11.1 to get $T(\ulcorner\text{Prov}_{\text{DT}}(\ulcorner\varphi\urcorner)\urcorner)$, $T(\ulcorner\text{Prov}_{\text{DT}}(\ulcorner D(\ulcorner\varphi\urcorner)\urcorner)\urcorner)$. The third follows since we have $D(\ulcorner\varphi\urcorner)$ and $T(\ulcorner\varphi\urcorner)$ and so, again, can conclude $T(\ulcorner T(\ulcorner\varphi\urcorner)\urcorner)$ by Lemma 7.11.1. ■

We now show that the inference rule for $K$ holds:

$$\frac{\varphi \wedge D(\ulcorner\varphi\urcorner)}{K(\ulcorner\varphi\urcorner)}. \qquad (DK\text{-Intro})$$

***Proof*** By the interpretation of '$K$' the conclusion is: $\text{Prov}_{\text{DT}}(\ulcorner\varphi \wedge D(\ulcorner\varphi\urcorner)\urcorner) \wedge T(\ulcorner\varphi\urcorner)$. The first conjunct follows from the properties of the provability predicate, while the second follows from $DT$-Intro. ■

($D_7$) $(\forall x)[D(\underset{\cdot}{K}(\dot{x})) \leftrightarrow D(x)]$.

***Proof*** Fix $x$. We need to show that

$$D(\ulcorner\text{Prov}_{\text{PA}}(\dot{x}) \wedge \text{Prov}_{\text{DT}}(\underset{\cdot}{D}(\dot{x})) \wedge T(\dot{x})\urcorner) \leftrightarrow D(x).$$

'$\text{Prov}_{\text{PA}}(\dot{x})$' and '$\text{Prov}_{\text{DT}}(\underset{\cdot}{D}(\dot{x}))$' are arithmetical, hence provably determinate. So showing the above reduces to showing that

$$D(\underset{\cdot}{T}(\dot{\dot{x}})) \leftrightarrow D(x),$$

which is just $D_6$. ■

$(T_7)$ $(\forall x)[D(x) \rightarrow (T(\underset{\cdot}{K}(\dot{\dot{x}})) \leftrightarrow K(x))]$.

***Proof*** Fix $x$ such that $D(x)$. We need to show

$$T(\ulcorner \text{Prov}_{\text{PA}}(\dot{\dot{x}}) \wedge \text{Prov}_{\text{DT}}(\underset{\cdot}{D}(\dot{\dot{x}})) \wedge T(\dot{\dot{x}})\urcorner) \leftrightarrow \text{Prov}_{\text{PA}}(x) \wedge \text{Prov}_{\text{DT}}(\underset{\cdot}{D}(\dot{\dot{x}})) \wedge T(x).$$

The first two conjuncts of the left-hand side are equivalent to the first two disjuncts of the right-hand side since the statements are arithmetical, hence determinate, and so we can apply Lemma 7.11.1. So it reduces to showing

$$T(\underset{\cdot}{T}(\dot{\dot{x}})) \leftrightarrow T(x),$$

which is just $T_6$. ■

We are now in a position to prove the main claims of the theorem:

(1) DTK $\nvdash \neg\text{WMT}_{\text{PA}}$.

***Proof*** We are assuming that DT is $\Sigma^0_n$-sound, for all $n < \omega$. So, for $\varphi \in L_{\text{PA}}$,

$$\text{if } \text{Prov}_{\text{DT}}(\ulcorner\varphi\urcorner), \text{ then } \mathbb{N} \models \varphi \text{ and hence } M \models T(\ulcorner\varphi\urcorner).$$

So, if $\text{Prov}_{\text{DT}}(\ulcorner\varphi\urcorner)$ then

$$M \models (\text{Prov}_{\text{DT}}(\ulcorner\varphi \wedge D(\ulcorner\varphi\urcorner)\urcorner) \wedge T(\ulcorner\varphi\urcorner)),$$

that is, $M \models K(\ulcorner\varphi\urcorner)$. And, if $M \models K(\ulcorner\varphi\urcorner)$ then trivially $M \models \text{Prov}_{\text{DT}}(\ulcorner\varphi\urcorner)$. So, for $\varphi \in L_{\text{PA}}$,

$$M \models K(\ulcorner\varphi\urcorner) \text{ iff } M \models \text{Prov}_{\text{PA}}(\ulcorner\varphi\urcorner).$$

Thus,

$$M \models (\exists e)\,(K =_{\text{PA}} W_e).$$

■

(2) DTK $\nvdash \neg\text{AU}_{\text{PA}}$.

***Proof*** Let $\varphi_G$ be the Gödel sentence for DT. Since $\varphi_G$ is arithmetical we have

$$\text{Prov}_{\text{DT}}(\ulcorner D(\ulcorner\varphi_G\urcorner)\urcorner).$$

However, $\neg\text{Prov}_{\text{DT}}(\ulcorner\varphi_G\urcorner)$ and $\neg\text{Prov}_{\text{DT}}(\ulcorner\neg\varphi_G\urcorner)$ (by the $\Sigma^0_1$-soundness of DT). But for $\varphi \in L_{\text{PA}}$

$$M \models T(\ulcorner\varphi\urcorner) \text{ iff } \mathbb{N} \models \varphi.$$

So $M \models T(\ulcorner\varphi_G\urcorner)$. Thus, we have

$$M \models \neg K(\ulcorner\varphi_G\urcorner),$$

that is,

$$M \models \neg(\mathrm{Prov}_{\mathrm{DT}}(\ulcorner \varphi_G \wedge D(\ulcorner \varphi_G \urcorner)\urcorner) \wedge T(\ulcorner \varphi_G \urcorner))$$

(because $M \models \neg\mathrm{Prov}_{\mathrm{DT}}(\ulcorner \varphi_G \urcorner)$) and

$$M \models \neg K(\ulcorner \neg\varphi_G \urcorner),$$

(because $M \models \neg T(\ulcorner \neg\varphi_G \urcorner)$), that is,

$$M \models \neg(\mathrm{Prov}_{\mathrm{DT}}(\ulcorner \neg\varphi_G \wedge D(\ulcorner \neg\varphi_G \urcorner)\urcorner) \wedge T(\ulcorner \neg\varphi_G \urcorner))$$

and yet

$$M \models T(\ulcorner \varphi_G \urcorner). \qquad \blacksquare$$

This completes the proof. ■

**Remark 7.16.3** *We have actually showed that*

$$M \models \mathrm{DTK} \wedge T(\overline{\ulcorner \varphi_G \urcorner}) \wedge \neg K(\overline{\ulcorner \varphi_G \urcorner}) \wedge \neg K(\overline{\ulcorner \neg\varphi_G \urcorner})$$

*and*

$$M \models K = W_{\bar{n}},$$

*where* '$\overline{\ulcorner \varphi_G \urcorner}$' *and* '$\bar{n}$' *are specific standard numerals. So we have shown that*

$$\mathrm{DTK} + T(\overline{\ulcorner \varphi_G \urcorner}) + \neg K(\overline{\ulcorner \varphi_G \urcorner}) + \neg K(\overline{\ulcorner \neg\varphi_G \urcorner}) + K = W_{\bar{n}}$$

*is consistent. Let us abbreviate this theory as* '$\mathrm{DTK}^+$'.

*This might at first glance seem to be problematic. For one might attempt to reason as follows: "Work in* $\mathrm{DTK}^+$. *By the rule DK-Intro*

$$\mathrm{DTK}^+ \vdash K(\ulcorner K = W_{\bar{n}} \urcorner) \qquad \text{(by } DK\text{-Intro)}$$

*and so we have actually established a stronger result—namely, the consistency of* DTK *with a version of the strong mechanistic thesis. This seems a bit too easy. In fact, it indicates a problem. For we have*

$$\mathrm{DTK}^+ \vdash \varphi_G \wedge D(\overline{\ulcorner \varphi_G \urcorner}) \qquad \text{(by Theorem 7.14.4)}$$
$$\mathrm{DTK}^+ \vdash K(\overline{\ulcorner \varphi_G \urcorner}) \qquad \text{(by } DK\text{-Intro)}$$

*and so* $\mathrm{DTK}^+$ *is inconsistent, which contradicts the above result. You claimed to have shown the consistency of the weak mechanistic thesis but actually we have refuted the weak mechanistic thesis."*

*The problem with this reasoning is that we have not shown that the DK-Intro rule holds in the form*

$$\frac{\mathrm{DTK}^+ \vdash \varphi \wedge D(\ulcorner \varphi \urcorner)}{\mathrm{DTK}^+ \vdash K(\ulcorner \varphi \urcorner)}. \qquad (DK\text{-Intro})$$

*This observation simply shows that this rule does not hold.*

*In response the opponent might maintain that this extended rule should hold. And once we grant that this rule holds we have all of the necessary tools to refute the weak mechanistic thesis.*

*But this is based on a confusion. The rules DT-Intro and DK-Intro are uncontentious rules* relative to any set of uncontentious principles. *The base of uncontentious principles that we are accepting is* DTK *(or* DTK *with the principles* $K_4$ *and* $K_5$ *adjoined). And so it is legitimate to accept these rules as applied to* DTK *(or* DTK *with the principles* $K_4$ *and* $K_5$ *adjoined). What we have shown is that there is no way to prove or refute of the weak mechanistic thesis using these uncontentious principles. But it is not reasonable to suppose that these rules apply to* $\text{DTK}^+$*, for the additional principles in this system are far from being uncontentious—indeed they are the very principles that we are trying to prove or refute. Our goal was to show that if one accepts uncontentious principles of knowledge and truth then one cannot prove or refute either disjunct. And we have done that for* DTK*. (It is true that although* $\text{DTK}^+$ *is consistent one cannot consistently maintain that one knows the additional principles. But the consistency result shows (among other things) that one can nevertheless* believe *them.)*

The above consistency results show that the principles of DTK are insufficient to prove or refute either disjunct; indeed they are insufficient to distinguish between the "$K$ equals truth" interpretation (rational optimism) and the "$K$ equals relative provability" interpretation (pessimistic mechanism). One natural response to this—on behalf of the proponent of the first disjunct—is that the principles of knowledge embodied in DTK are merely a fragment of the acceptable principles of knowledge, and that when we supplement DTK with the missing principles of knowledge we will be in a position to prove or refute the disjuncts.

Let us, then, consider what happens when we enhance DTK by supplementing new principles of knowledge. Here are two (arguably plausible) principles that one might add:

($K_4$) $(\forall x)[\mathrm{Sent}(x) \wedge K(\ulcorner D(x) \urcorner) \rightarrow K(\ulcorner K(\dot{x}) \rightarrow T(\dot{x}) \urcorner)]$.

($K_5$) The scheme: For each formula $\varphi$

$$K(\ulcorner (\forall x)\varphi(x) \urcorner) \rightarrow (\forall x)\, K(\ulcorner \varphi(\dot{x}) \urcorner).$$

Let DTK′ be the result of supplementing DTK with $K_4$ and $K_5$.

**Theorem 7.16.4** DTK′ *is consistent.*

***Proof*** The proof is continuation of the proof of Theorem 7.14.1.

We show that under the $K = T$ interpretation the principles $K_4$ and $K_5$ hold in DT. Work in DT. $K_5$ follows immediately from $T_5$. For $K_4$ note that if $D(x)$ then $D(\underset{\cdot}{T}(\dot{x}))$ and $D(\ulcorner T(\dot{x}) \rightarrow T(\dot{x}) \urcorner)$, for each $x$. But $T(x) \rightarrow T(x)$ is a logical validity. But Feferman showed that for all $\varphi \in L_{\mathrm{DT}}$,

$$\mathrm{DT} \vdash D(\ulcorner \varphi \urcorner) \rightarrow (T(\ulcorner \varphi \urcorner) \leftrightarrow \varphi).$$

So we have

$$\mathrm{DT} \vdash (\forall x)[\mathrm{Sent}(x) \wedge D(x) \rightarrow T(\ulcorner T(\dot{x}) \rightarrow T(\dot{x})\urcorner)],$$

which completes the proof. ■

**Theorem 7.16.5** *Assume that* DTK′ *is consistent. Then* DTK′ *can neither refute the first disjunct nor prove the second disjunct of* $\mathrm{GD_{PA}}$, *that is,*

(1) $\mathrm{DTK}' \nvdash \mathrm{WMT_{PA}}$ *and*

(2) $\mathrm{DTK}' \nvdash \mathrm{AU_{PA}}$.

***Proof*** The proof is the same as the proof of Theorem 7.16.1. ■

**Theorem 7.16.6** *Assume that* DTK′ *is* $\Sigma^0_n$*-sound, for each* $n < \omega$. *Then* DTK′ *can neither prove the first disjunct nor refute the second disjunct of* $\mathrm{GD_{PA}}$, *that is,*

(1) $\mathrm{DTK}' \nvdash \neg\mathrm{WMT_{PA}}$

(2) $\mathrm{DTK}' \nvdash \neg\mathrm{AU_{PA}}$.

***Proof*** The proof is continuation of the proof of Theorem 7.16.2. It remains to prove the inductive clauses for $K_4$ and $K_5$.

$$(K_4)\ \ (\forall x)[\mathrm{Sent}(x) \wedge K(\ulcorner D(x)\urcorner) \rightarrow K(\ulcorner K(\dot{x}) \rightarrow T(\dot{x})\urcorner)].$$

***Proof*** Assume $\ulcorner\varphi\urcorner$ is such that $\mathrm{Sent}(\ulcorner\varphi\urcorner)$ and $K(\ulcorner D(\ulcorner\varphi\urcorner)\urcorner)$, that is,

$$\mathrm{Prov_{DT}}(\ulcorner D(\ulcorner\varphi\urcorner)\urcorner) \wedge \mathrm{Prov_{DT}}(\ulcorner D(\ulcorner D(\ulcorner\varphi\urcorner)\urcorner)\urcorner) \wedge T(\ulcorner D(\ulcorner\varphi\urcorner)\urcorner).$$

Let '$\psi$' abbreviate

$$\big(\mathrm{Prov_{DT}}(\ulcorner\varphi\urcorner) \wedge \mathrm{Prov_{DT}}(\ulcorner D(\ulcorner\varphi\urcorner)\urcorner) \wedge T(\ulcorner\varphi\urcorner)\big) \rightarrow T(\ulcorner\varphi\urcorner).$$

We have to show

(1) $\mathrm{Prov_{DT}}(\ulcorner\psi\urcorner)$,

(2) $\mathrm{Prov_{DT}}(\ulcorner D(\ulcorner\psi\urcorner)\urcorner)$, and

(3) $T(\ulcorner\psi\urcorner)$.

Regarding (1): This follows since '$\psi$' is a logical validity.

Regarding (2): We have $\mathrm{Prov_{DT}}(\ulcorner D(\ulcorner\varphi\urcorner)\urcorner)$ and $\mathrm{Prov_{DT}}(\ulcorner D(\ulcorner D(\ulcorner\varphi\urcorner)\urcorner)\urcorner)$ since $\mathrm{Prov_{DT}}(\ulcorner\varphi\urcorner)$ and $\mathrm{Prov_{DT}}(\ulcorner D(\ulcorner\varphi\urcorner)\urcorner)$ are arithmetical. Furthermore, we have $\mathrm{Prov_{DT}}(\ulcorner D(\ulcorner T(\ulcorner\varphi\urcorner)\urcorner)\urcorner)$ since we are assuming $K(\ulcorner D(\ulcorner\varphi\urcorner)\urcorner)$, which gives $\mathrm{Prov_{DT}}(\ulcorner D(\ulcorner\varphi\urcorner)\urcorner)$ and hence $\mathrm{Prov_{DT}}(\ulcorner D(\ulcorner T(\ulcorner\varphi\urcorner)\urcorner)\urcorner)$ by $D_6$ in DT.

Regarding (3): Since $\psi$ is logical it suffices to see that we have $D(\ulcorner\varphi\urcorner)$ since then we can apply Lemma 7.11.1 to get $T(\ulcorner\psi\urcorner)$. To see that $D(\ulcorner\varphi\urcorner)$ holds it suffices to note that we have $D(\mathrm{Prov_{DT}}(\ulcorner\varphi\urcorner))$ and $D(\mathrm{Prov_{DT}}(\ulcorner D(\ulcorner\varphi\urcorner)\urcorner)$ since $\mathrm{Prov_{DT}}(\ulcorner\varphi\urcorner)$ and $\mathrm{Prov_{DT}}(\ulcorner D(\ulcorner\varphi\urcorner)\urcorner$ are arithmetical and, finally, we have $D(\ulcorner T(\ulcorner\varphi\urcorner)\urcorner)$ by $D_6$ in DT since we are assuming $D(\ulcorner\varphi\urcorner)$. ■

($K_5$) The scheme: For each formula $\varphi$,

$$K(\ulcorner(\forall x)\varphi(x)\urcorner) \rightarrow (\forall x)\, K(\ulcorner\varphi(\dot{x})\urcorner).$$

***Proof*** Assume $\varphi$ is such that $K(\ulcorner(\forall x)\varphi(x)\urcorner)$, that is,

$$\mathrm{Prov}_{\mathrm{DT}}(\ulcorner(\forall x)\varphi(x)\urcorner) \wedge \mathrm{Prov}_{\mathrm{DT}}(\ulcorner D(\ulcorner(\forall x)\varphi(x)\urcorner)\urcorner) \wedge T(\ulcorner(\forall x)\varphi(x)\urcorner).$$

Fix $x$. We have to show $K(\ulcorner\varphi(\dot{x})\urcorner)$, that is,

$$\mathrm{Prov}_{\mathrm{DT}}(\ulcorner\varphi(\dot{x})\urcorner) \wedge \mathrm{Prov}_{\mathrm{DT}}(\ulcorner D(\ulcorner(\varphi(\dot{x})\urcorner)\urcorner) \wedge T(\ulcorner\varphi(\dot{x})\urcorner).$$

Each conjunct follows from the corresponding conjunct of the previous displayed formula. The first follows by the properties of '$\mathrm{Prov}_{\mathrm{DT}}$', the second follows by $D_5$ in DT, and the third follows by $T_5$ in DT. ■

This completes the proof. ■

Thus, even if one supplements DTK with the additional principles of knowledge $K_4$ and $K_5$, the resulting system, DTK′, is still unable to prove or refute either disjunct; indeed it is still unable to distinguish between the "$K$ equals truth" interpretation (rational optimism) and the "$K$ equals relative provability" interpretation (pessimistic mechanism).

One could try to supplement the principles of knowledge even further, but the most natural candidates are consistent with both the "$K$ equals truth" interpretation and the "$K$ equals relative provability" interpretation and are, consequently, unable to settle either disjunct. But perhaps there are plausible missing principles of knowledge that (in a surprising way) "collapse knowledge to truth" and thereby vindicate rational optimism and enable us to prove the first disjunct. Or perhaps there are plausible missing principles of knowledge that force us into the "border region" between the two extremes—of the "$K$ equals truth" interpretation and the "$K$ equals relative provability" interpretation—and which, in a non-question-begging way, put us in a position to prove the first disjunct. The most that can be said at this stage, with regard to these speculative scenarios, is that it is hard to see how this could be done, and the consistency results proved here are not encouraging.

## 7.17 Conclusion

In the foregoing discussion I have been engaged with my opponent (the proponent of the first disjunct) in his own terms, by granting the concept of *knowability by an idealized human mind* and the concept of *absolute provability*, along with the strong, idealizing assumptions surrounding them. But I should reveal that I think—for the reasons given in [19]—that there are already problems with the element of idealization involved in these concepts, along with the presumption—made at the very start of the attempt to prove the first disjunct—that we can learn something about *our* minds by investigating these concepts in an ideal setting. In particular, in the case of the concept of "an idealized human mind," the dimension along which the idealization is supposed to take place is unclear and, consequently, it is doubtful that any conclusions we arrive at concerning this concept (when treated purely formally in the ideal setting of epistemic arithmetic) can have bearing on the question of real interest, namely, whether *our* minds can be mechanized. If this philosophical critique is correct then the arguments to the effect that the "mind cannot be mechanized" (as understood as a statement about *our* minds) do not even get off the ground. But, not wanting to place the entire weight of a critical case on philosophical misgivings concerning the underlying concepts, I have, for the sake of argument, set these skeptical considerations aside and met my opponent on his own ground, granting him the fundamental concepts (as he understands them) along with the principles he takes to govern them.

Once this is done, and we are explicit about the underlying background assumptions, we can pull the entire debate into a setting where we can prove definitive results. We have seen that there is a natural framework in which the disjunction is indeed provable. But that in this framework the early arguments of Lucas and Penrose fail (for reasons that Gödel appreciated) and, moreover, that there can be *no* argument for the first disjunct. The drawback of this framework, however, was that it treated truth as typed and there are good reasons to shift to a type-free theory of truth. And, interestingly, Penrose's new argument actually requires a type-free theory of truth. So I have shifted to a new framework, one that involves a type-free theory of truth and is suitable for the articulation of Penrose's new argument. We have seen that in this framework Penrose's new argument does not go through and, moreover, that there can be *no* argument for the first disjunct in this framework. In short, even if we grant the opponent the idealizing assumptions that he takes to be unproblematic, it can be shown that in the natural frameworks there can be *no* argument that settles either the question of whether "the mind cannot be mechanized" or the question of whether "there are absolutely undecidable statements."

What are we to make of this? One might argue that the independence results proved here are limited in scope in that they pertain to just one of many possible theories of knowledge and type-free truth. There are two dimensions of variation that one can consider—one can vary the theory of knowledge and one can vary the theory of type-free truth. Here I have considered variations in the theory of knowledge. I have shown that a specific variation—DTK′—is of no help and given reasons for thinking that variations along this dimension will not yield a non-question begging proof of the first disjunct. In the companion paper, I consider variations along the second dimension—the theory of type-free truth. I show that specific variations are of no help and I give reasons for thinking that no variation along this dimension will yield a non-question begging proof of the first disjunct. But perhaps there is a variation along one, or both, of the dimensions that will seal the deal.

Whether this can be done remains to be seen. It is certainly of interest to try since it is what would be required to make a convincing case for the first disjunct. I am, however, doubtful that it can be done. For it seems likely that the above independence results will carry over to new frameworks when one incorporates new theories of knowledge and type-free truth. And for this reason it seems likely that both the question of whether "the mind can be mechanized" and the question of whether "there are absolutely undecidable statements" (in the idealized sense we have been considering) are themselves examples of "absolutely undecidable statements" and, as such, will remain forever undecided and continue to lie outside the scope of human reason.

## Acknowledgement

I began thinking about this topic over 30 years ago when a friend of my father recommended that I read Hofstadter's book *Gödel, Escher, Bach* [17]. He was a scientist and I was an aspiring artist. In addition to finding common ground I think he was probably trying (under the influence of my father) to get this kid to do something other than draw all the time. Anyway, I started with Escher, moved on to Bach, and when I got to Gödel I was hooked.

I am grateful to Leon Horsten and Philip Welch for inviting me to speak on the subject at Bristol, first on March 19, 2010 and then as part of their *Workshop on the Limits and Scope of Mathematical Knowledge*, March 30–31, 2013. I benefited from the comments I received at those talks, as well as the comments received when I spoke on the subject at Barcelona, Duke, MIT, Pittsburgh, and Stanford. Finally, I would like to thank Samuel Alexander, Sol Feferman, Kentaro Fujimoto, Gabriel Goldberg, Wesley Holliday, Leon Horsten, Hannes Leitgeb, and Johannes Stern for helpful comments on the paper.

## Notes

1. It should be noted, however, that although he was not a "reason-transcendent" realist he was nevertheless a realist. He was both a staunch realist and a rational optimist. It is just that his rational optimism was so optimistic that he thought that human reason was sufficiently powerful to discover all mathematical truths.
2. The discussion in the literature appeared in two major waves. The first generation of arguments for the first disjunct began with Nagel and Newman in 1956 ([27], [28]) and continued with Lucas in a talk given in 1959 and a subsequent publication of 1961 ([24]). Nagel and Newman's argument was criticized in correspondence by Gödel (see Feferman [7] for an account) and in print by Putnam [36]. Lucas' argument was widely criticized in the literature (see, for example, Benacerraf [1]). The topic was revisited in 1989 by Penrose in his famous book *The Emperor's New Mind* [29]. This triggered an avalanche in the literature. The following year there was an open peer commentary of the book in *Behavioral and Brain Sciences* (see, in particular, [31], [4], [30], [5], and [32]).

   The second generation of arguments—one argument really—appeared in another book-length account by Penrose in 1994—*Shadows of the Mind: A Search for the Missing Science of Consciousness* [33]. This account has also received a great deal of attention; see, for example, Chalmers [3], Feferman [6], Lindström [22], [23], and Shapiro [41], [42]. Penrose has continued to defend his argument—for example, in his address at the Gödel Centenary in 2006 and in the subsequent publication in 2011 [35].
3. I should stress that in addressing the question of whether the "the mind can be mechanized" I am dealing only with a very specific sense in which this statement can be understood. Like Gödel, Lucas, and Penrose, I am only dealing with the question of whether "the outputs of the idealized human mind outstrip the outputs of any idealized finite machine." I am not dealing with other aspects of the question. For example, I am not taking into account matters such as how a creative

subject arrives at proofs, or our acquaintance with abstract objects, or our capacity for normative judgements, etc. I am simply addressing the question of whether the incompleteness theorems show that in terms of its mathematical outputs the idealized human mind outstrips the outputs of any idealized finite machine. Moreover, in criticizing the arguments that have been given for the first disjunct I should not be taken to be arguing that as far as our mathematical outputs are concerned we are indeed Turing machines. I am, rather, showing that the arguments for the first disjunct do not support such a conclusion and, moreover, I am arguing that there are fundamental limitations in giving any such argument for such a conclusion.

4. Here it is understood that we are working in the language of arithmetic and that the system $F$ contains a fragment of arithmetic sufficiently strong to ensure that the incompleteness theorems apply.
5. Systems of this kind were first introduced by Myhill [26], Reinhardt [38], [37], [39] and Shapiro [40], and have been investigated by many others (see, for example, Horsten [18] and Leitgeb [21] and references therein).
6. See [10], [43], and [44] for further discussion.
7. See, for example, the open peer commentary of Penrose's book [29] in *Behavioral and Brain Sciences*. In particular, see [31], [4], [30], [5], and [32].
8. This is because $R$ is $\Pi^0_1$ and PA proves all true $\Sigma^0_1$-statements.
9. Leon Horsten drew my attention to the work of Johannes Stern [45]. Stern explores systems that combine theories of truth with theories of modality and, as we do here, he arranges things in such a way that the resolution of the paradoxes in the theory of modality is induced by the resolutions of the paradoxes in the theory of truth, via the manner in which modality is linked to truth.

## References

[1] Paul Benaceraff. God, the devil, and Gödel. *The Monist*, **51**(1):9–32, January 1967.

[2] Timothy J. Carlson. Knowledge, machines, and the consistency of Reinhardt's strong mechanistic thesis. *Annals of Pure and Applied Logic*, **105**(1–3):51–81, 2005.

[3] David Chalmers. Minds, machines, and mathematics: A review of *Shadows of the Mind*, by Roger Penrose. *Psyche*, **2**:11–20, 1995.

[4] George Boolos et al. An open peer commentary on *The Emperor's New Mind*. *Behavioral and Brain Sciences*, **13**(4):655–691, 1990.

[5] Martin Davis et al. Continuing commentary on *The Emperor's New Mind*. *Behavioral and Brain Sciences*, **16**(3):611–616, 1993.

[6] Solomon Feferman. Penrose's Gödelian argument. *Psyche*, **2**:21–32, May 1995.

[7] Solomon Feferman. Gödel, Nagel, minds, and machines. *Journal of Philosophy*, **CVI**(4):201–219, August 1999.

[8] Solomon Feferman. Axioms for determinateness and truth. *Review of Symbolic Logic*, **1**(2):204–217, 2008.

[9] Kentaro Fujimoto. Relative truth definability of axiomatic truth theories. *Bulletin of Symbolic Logic*, **16**(3):305–344, 2010.

[10] Robin Gandy. The confluence of ideas in 1936. In Rolf Herken, ed., *The Universal Turing Machine: A Half-Century Survey*, pp. 55–111. Oxford University Press, 1988.

[11] Kurt Gödel. On undecidable sentences. In [15], pp. 31–35. Oxford University Press, *1931?

[12] Kurt Gödel. An interpretation of the intuitionistic propositional calculus. In [14], pp. 301–303. Oxford University Press, 1933.

[13] Kurt Gödel. Some basic theorems on the foundations of mathematics and their implications. In [15], pp. 304–323. Oxford University Press, *1951.

[14] Kurt Gödel. *Collected Works, Volume I: Publications 1929–1936*. Edited by Solomon Feferman, John W. Dawson, Jr., Stephen C. Kleene, Gregory H. Moore, Robert M. Solovay, and Jean van Heijenoort. Oxford University Press, New York and Oxford, 1986.

[15] Kurt Gödel. *Collected Works, Volume III: Unpublished Essays and Lectures*. Oxford University Press, New York and Oxford, 1995.

[16] Volker Halbach and Philip Welch. Necessities and necessary truths: A prolegomenon to the use of modal logic in the analysis of intensional notions. *Mind*, **118**(469):71–100, 2009.

[17] Douglas R. Hofstadter. *Gödel, Escher, Bach: An Eternal Golden Braid*. Basic Books, New York, 1979.

[18] Leon Horsten. In defense of epistemic arithmetic. *Synthese*, **116**(1):1–25, 1998.

[19] Peter Koellner. On the question of whether the mind can be mechanized. Forthcoming, 2014.

[20] Hannes Leitgeb. What theories of truth should be like (but cannot be). *Philosophy Compass*, **2**(2):276–290, 2007.

[21] Hannes Leitgeb. On formal and informal provability. In Otávio Bueno and Øystein Linnebo, eds., *New Waves in the Philosophy of Mathematics*, New Waves in Philosophy, pp. 263–299. Palgrave Macmillan, 2009.

[22] Per Lindström. Penrose's new argument. *Journal of Philosophical Logic*, **30**:241–250, 2001.

[23] Per Lindström. Remarks on Penrose's "new argument". *Journal of Philosophical Logic*, **35**:231–237, 2005.

[24] John Randolph Lucas. Minds, machines, and Gödel. *Philosophy*, **36**:112–137, 1961.

[25] Richard Montague. Syntactical treatments of modality, with corollaries on reflexion principles and finite axiomatizability. *Acta Philosophica Fennica*, **16**:153–167, 1963.

[26] John Myhill. Some remarks on the notion of proof. *Journal of Philosophy*, **LVII**(14):461–471, July 1960.

[27] Ernest R. Nagel and James R. Newman. Gödel's proof. *Scientific American*, New York and London, **3**:1668–1695, 1956.

[28] Ernest R. Nagel and James R. Newman. *Gödel's Proof*. New York University Press, 1958. Revised edition, 2001, edited with a new forward by Douglas R. Hofstadter.

[29] Roger Penrose. *The Emperor's New Mind: Concerning Computers, Minds, and the Laws of Physics*. Oxford University Press, New York, 1989.

[30] Roger Penrose. The nonalgorithmic mind. *Behavioral and Brain Sciences*, **13**(4):692–706, 1990.

[31] Roger Penrose. Précis of *The Emperors New Mind*—concerning computers, minds and the laws of physics. *Behavioral and Brain Sciences*, **13**(4):643–654, 1990.

[32] Roger Penrose. An Emperor still without a mind. *Behavioral and Brain Sciences*, **16**(3):616–622, 1993.

[33] Roger Penrose. *Shadows of the Mind: A Search for the Missing Science of Consciousness*. Oxford University Press, London, 1994.

[34] Roger Penrose. Beyond the doubting of a shadow. *Psyche*, **2**:89–129, 1995.

[35] Roger Penrose. Gödel, the mind, and the laws of physics. In Matthias Baaz, Christos H. Papadimitriou, Hilary W. Putnam, Dana S. Scott, and Charles L. Harper, eds., *Kurt Gödel and the Foundations of Mathematics: Horizons of Truth*, ch. 16, pp. 339–358. Cambridge University Press, 2011.

[36] Hilary Putnam. Minds and machines. In Sydney Hood, ed., *Dimensions of Mind: A Symposium*, pp. 138–164. New York University Press, Washington Square, New York, 1960.

[37] William N. Reinhardt. Absolute versions of incompleteness theorems. *Noûs*, **19**(3):317–346, September 1985.

[38] William N. Reinhardt. The consistency of a variant of Church's thesis with an axiomatic theory of an epistemic notion. In *Special Volume for the Proceedings of the 5th Latin American Symposium on Mathematical Logic, 1981*, vol. 19 of *Revista Colombiana de Matem'aticas*, pp. 177–200, 1985.

[39] William N. Reinhardt. Epistemic theories and the interpretation of Gödel's incompleteness theorems. *Journal of Philosophical Logic*, **15**:427–474, 1986.

[40] Stewart Shapiro. Epistemic and intuitionistic arithmetic. In Stewart Shapiro, ed., *Intensional Mathematics*, vol. 113 of *Studies in Logic and the Foundations of Mathematics*, pp. 11–46. North-Holland, Amsterdam, 1985.

[41] Stewart Shapiro. Incompleteness, mechanism, and optimism. *Bulletin of Symbolic Logic*, **4**(3):273–302, September 1998.

[42] Stewart Shapiro. Mechanism, truth, and Penrose's new argument. *Journal of Philosophical Logic*, **4**(3):273–302, September 2003.

[43] Wilfried Sieg. Mechanical procedures and mathematical experience. In Alexander George, ed., *Mathematics and the Mind*, pp. 71–117. Oxford University Press, New York and Oxford, 1994.

[44] Wilfried Sieg. On computability. In Andrew Irvine, Dov M. Gabbay, Paul Thagard, and John Woods, eds., *Philosophy of Mathematics*, Handbook of Philosophy of Science, pp. 525–621. Elsevier, 2008.

[45] Johannes Stern. *Toward Predicate Approaches to Modality*. PhD thesis, University of Geneva, 2012.

[46] Richmond Thomason. A note on syntactical treatments of modality. *Synthese*, **44**:391–395, 1980.

[47] Hao Wang. *A Logical Journey: From Gödel to Philosophy*. MIT Press, Cambridge, Massachusetts, and London, England, 1996.

# 8

# Idealization, Mechanism, and Knowability

STEWART SHAPIRO

Philosophers, mathematicians, and others have drawn lots of conclusions from Kurt Gödel's incompleteness theorems, and from related results in mathematical logic. Languages, machines, minds, and knowledge are all supposedly illuminated by these theorems. One focus of this volume is a batch of much-discussed—and much maligned—arguments that the incompleteness theorems somehow refute the mechanistic thesis that the human mind is, or can be accurately modeled as, a digital computer or a Turing machine. Prominent advocates of the thesis include J.R. Lucas, the eminent physicist and mathematician Roger Penrose, and, in a carefully attenuated sense, Gödel himself.

The purpose of this chapter is to revisit, develop, and extend some claims and arguments I made in Shapiro [1998], "Incompleteness, mechanism, and optimism." It is a sort of ill-humored, rear-guard action. I claimed there that no one has proposed a mechanist thesis that is both even remotely plausible and sufficiently precise to be undermined (or supported) by the incompleteness theorems. I'll extend that here, and argue that the same goes for a thesis concerning the extent of human (mathematical) knowledge—about what we can and cannot know.

It is my own, admittedly limited experience, that many philosophers dismiss the whole Lucas–Penrose controversy, out of hand. Often this comes with a rolling of the eyes. Penrose [1996] once suggested that some of the nay-sayers just accept, as a matter of faith, that any cognitive process must be computable—something along the lines of Church's thesis, or some extension or misunderstanding thereof. This begs the question. I don't wish to delve into the psychology of the nay-sayers, but in my opinion, the issues do not lie in the neighborhood of Church's thesis or in the nature of cognitive processes, but rather in the *idealizations* we need in order to make sense of the issues and then apply the incompleteness theorems. There are important matters in the vicinity that bear on the nature of mathematical knowledge—real mathematical knowledge in the real world. This chapter is a defense of (some of) the eye-rolling.

## 8.1 Lucas and Penrose

The most famous instance of the anti-mechanistic argument is found in the much reprinted J.R. Lucas [1961]. To summarize, suppose that a mechanist provides us with plans for a

*Gödel's Disjunction*. First Edition. Leon Horsten & Philip Welch (eds).

machine, *M*, and claims that the output of *M* consists of all and only the arithmetic truths that a human (like Lucas), or the totality of all humans (including Lucas), will ever or can ever know. Given that knowledge implies truth, the output of *M* is consistent (assuming that the set of arithmetic truths is itself consistent, contra Priest [2006]). Assume something like Church's thesis, and thus that the output of *M* corresponds, exactly, with the theorems of a particular formal deductive system. Since Lucas understands the proof of the incompleteness theorem, he can study *M* (or the indicated formal deductive system) and construct a Gödel sentence *G*. This is an arithmetic sentence that Lucas knows will never be produced or "asserted" by *M*. He also knows that *G* "says" that *G* will never be produced by *M* (or the deductive system).[1] Thus, Lucas knows that *G* is true. So the mechanist was mistaken in the claim that the output of *M* contains all the truths that (any group containing) Lucas can know.

As George Boolos [1995, p. 295] notes, the Lucas–Penrose arguments "have as yet obtained little credence," and, indeed, there is an extensive literature attacking the position (not to mention the aforementioned eye-rolling). About 200 pages of Penrose [1994] are devoted to responses to various criticisms of the argument and an intriguing new version of it.

In an attempt to delineate the dispute, let us try to put things more succinctly. Let **K** be the collection of sentences in the language of first-order arithmetic that can be proved, either by a human like Lucas or by humanity at large. By "proved" here, I do not mean deduced in a particular, fixed, formal system. As Penrose pointed out, that would beg the question, assuming Church's thesis, since his (and Lucas's) main claim is that what is proved, or at least what is provable, goes beyond any given formal system. "Proved" here means something like "known with unassailable, mathematical certainty, via full mathematical rigor." Let us call **K** the collection of *knowable* arithmetic sentences.

For convenience, we identify sentences with their Gödel numbers, and so we think of **K** ambiguously as a collection of sentences and a collection of natural numbers. The principals to the debate—Lucas, Penrose, and most of their opponents—assume that **K** has sharp boundaries, and so is, in this respect, like any other set of natural numbers. That will be a matter of present contention. For now, note that if **K** does have sharp boundaries, then we can inquire about its arithmetic and computational properties. The mechanist holds that there is a Turing machine that enumerates **K**. In other words, they claim that **K** is recursively enumerable. Lucas and Penrose claim they can refute this contention, by citing the incompleteness theorems at the crucial places.

## 8.2 Gödel

Let **T** be the set of truths of first-order arithmetic. By assumption (or definition, or philosophical analysis), $\mathbf{K} \subseteq \mathbf{T}$: all knowables are true. In his [1951] Gibbs lecture, Gödel refers to **T** as "objective" mathematics and **K** as "subjective" mathematics. We also assume that the first-order arithmetic is bivalent—another matter that is not in dispute among the main protagonists. Suppose that $\mathbf{K} = \mathbf{T}$, that all arithmetic truths are knowable. It then follows that for any sentence $\Phi$ in the language of first-order arithmetic, either $\Phi \in \mathbf{K}$ or $(\neg\Phi) \in \mathbf{K}$. So if the language is bivalent and $\mathbf{K} = \mathbf{T}$, then every arithmetic sentence is what we may call *humanly decidable*, in the sense that, in some relevant sense, humans are capable of determining, for every sentence, whether or not it is true, via a rationally compelling deduction, with full mathematical rigor.

Tarski's theorem is that $\mathbf{T}$ is not definable in the language of arithmetic. *A fortiori*, $\mathbf{T}$ is not recursively enumerable. There is no effective, formal deductive system that has, as theorems, all and only the arithmetic truths. Thus, if $\mathbf{K} = \mathbf{T}$ then $\mathbf{K}$ is not recursively enumerable and the mechanist is wrong.

Taking the contrapositive, mechanism entails that $\mathbf{K} \neq \mathbf{T}$, and so there is an unknowable truth, say $\Phi$. And, of course, $\neg\Phi$ is unknowable, since it is false. In Gödel's terms, the sentence $\Phi$ is *absolutely undecidable*. So, if the mechanism is correct, then there are absolutely undecidable sentences.

As always, Gödel's conclusion is careful:

> ... the following disjunctive conclusion is inevitable: *Either mathematics is incompletable in* [the] *sense that its evident axioms can never be comprised in a finite rule, that is to say, the human mind (even within the realm of pure mathematics) infinitely surpasses the powers of any finite machine, or else there exist absolutely unsolvable diophantine problems of the type specified* ... It is this mathematically established fact which seems to me of great philosophical interest.
>
> (p. 310).

Our present agenda is to explore this item of "great philosophical interest."

In his correspondence with Hao Wang, Gödel at least tentatively endorsed, or at least suggested, the negation of the second of these disjuncts. That is, Gödel proposed that there may be no absolutely undecidable problems of the indicated type: humans may indeed have the ability to decide every arithmetic truth:

> ... human reason is [not] utterly irrational by asking questions it cannot answer, while asserting emphatically that only reason can answer them ... [T]hose parts of mathematics which have been systematically and completely developed ... show an amazing degree of beauty and perfection. In those fields, by entirely unsuspected laws and procedures ... means are provided not only for solving all relevant problems, but also solving them in a most beautiful and perfectly feasible manner. This fact seems to justify what may be called "rationalistic optimism".
>
> (Wang [1974, pp. 324–326].)

The rational optimist thus rejects the second of Gödel's disjuncts. Invoking disjunctive syllogism, the rational optimist holds that mechanism to be false. So with this suggestion concerning human mathematical abilities and the above disjunction, Gödel teams up with Lucas and Penrose. But Gödel does not regard the rational optimism as itself established. He only says that it is not utterly irrational to hold it.

In a personal anecdote, Penrose [1996, §4.2] expressed sentiments similar to Gödel's concerning the brain power of humanity at large, but perhaps a bit less guarded:

> I had vaguely heard of Gödel's theorem prior [to the first year of graduate school], and had been a little unsettled by my impressions of it . . . I had been disturbed by the possibility that there might be true mathematical propositions that were in principle inaccessible to human reason. Upon learning the true form of Gödel's theorem . . . I was enormously gratified to hear that it asserted no such thing; for it established, instead, that the powers of human reason could not be limited to any accepted preassigned system of formalized rules.

On the other hand, perhaps the existence of absolutely undecidable propositions is not that implausible. Boolos [1995] wonders why "should there *not* be mathematical truths that cannot be given any proof that human minds can comprehend?" Once again, if **K** is recursively enumerable, then some arithmetic propositions are undecidable "by human reason" even in principle. The mechanist does a modus ponens, while Penrose and Gödel, via rational optimism, invoke modus tolens (or disjunctive syllogism).

## 8.3 Idealization

The problem here is that the exact content of the mechanistic thesis is usually left unspecified. The same goes for the content of Gödel's second disjunct concerning what human mathematicians can and cannot know. To belabor the obvious, the relevance of the incompleteness theorems to mechanism depends on what the mechanist claims. The raw thesis that the human mind is, or can be modeled as, a digital computer or Turing machine, is too vague to apply anything as sharp and delicate as the Gödel theorems. The mechanist claims that there can be a machine whose outputs are the same as those of a human or a group of humans. What sort of machine? What outputs? And what sort of human?

Concerning the "outputs" of these machines and these humans, we can stick to propositions that can be rendered in the language of first-order Peano arithmetic, or even, following Penrose [1996], to $\Pi_1$-sentences. The totality of such sentences that a given person asserts in their lifetime is finite. The same goes for the totality of sentences asserted by any finite collection of humans, such as the professional mathematicians who lived or will live before the sun goes cold. Moreover, the totalities in question are inconsistent. It only takes one mistaken calculation.

A mechanist might claim that there could be a machine whose output is one of these finite sets, or the truths among one of these sets, or the logical consequences thereof. But, of course, the incompleteness theorems are irrelevant to such a thesis. Every finite set is recursive, and the set of (first-order) logical consequences of any finite set is recursively enumerable.

What of the second disjunct in Gödel's conclusion, the possibility that there are (absolutely) unknowable arithmetic truths? There is a sense in which it is obvious that there are unknowable mathematical truths. Let $p$ and $q$ be prime numbers, each of which is greater than $10^{1,000,000}$. If their product $n = pq$ were written in standard arabic, decimal notation, it would have about two million digits. Consider the proposition, or sentence, $P$ that $n$ (written in standard arabic notation) has exactly two prime factors. By hypothesis, $P$ is true. Yet it is extremely unlikely that any human being can know $P$. That is, no one can have compelling evidence for $P$. Indeed, it is unlikely that anyone can even *understand* $P$ or even parse the sentence.

In fact, there are probably unknowable truths that can be feasibly stated. Consider the proposition $Q$ that the $10^{200,208}$-th prime greater than $10^{200,721}$ leaves a remainder of 5 when divided by 32. If $Q$ is true, then, probably, no human can come to know it. And if $Q$ is not true, then its negation is probably an unknowable truth. So, either way, there is a feasibly stated, unknowable truth.

Clearly, to get an interesting and substantial debate going, we have to idealize, in several different directions. Things get philosophically murky almost from the start. Penrose speaks of "unassailable" truths. Unassailable by whom? Just about anybody can assail just about anything, say if they do not understand a proof or cannot parse all of the steps (or do not bother to). Presumably, Penrose means that the propositions should have proofs that are so rationally compelling that no one *ought* to challenge them, or that no one has the right to challenge them, once they understand the demonstration. But what does this mean? Some proofs are so long and complicated that no human can understand them, in the relevant sense.

Clearly, we are not talking about mathematicians composed (at least in part) of flesh and blood. The so-called "medical" limitations are not relevant. Presumably, the mechanist and anti-mechanist are both talking about what an *ideal* human, or the community of ideal human mathematicians, *can prove* or know for certain. Lucas and Penrose both refer to human abilities "in principle." What does this mean?

We here broach some fundamental issues concerning the epistemology of mathematics. First, we have to figure out how it is that flesh and blood mathematicians come to know the mathematics that they know—in practice. Then we have to somehow extrapolate this epistemology to our ideal humans, to figure out how *they* come to know things. If we can get clear on this, we'll have a handle on what Penrose means by "unassailable." But this is a big "if."

We must idealize on the "machines" as well. Like humans, actual digital computers have fixed limits on memory and materials they can use when doing what they do. That is, actual computers are more like finite state machines. Moreover, computers are subject to hardware malfunctions and software bugs. I do not know if there could be a physical computer whose output is one of the finite sets mentioned previously. Could there be a piece of hardware that matches a human being in their arithmetic (or $\Pi_1$) productions, reproducing both veridical output and error? Or could there be a piece of hardware that matches the *true* arithmetic (or $\Pi_1$) sentences of a given human?

Well, I suppose we could just beg every question and define "computer" in such a way that a human being counts as one. Then we *do* know how to produce such a "computer," since we know how to have children, but perhaps we cannot "build" such a computer—depending on what one means by "build."

We also have good empirical confirmation that any given physical machine (not to mention any given human being) will crash or otherwise malfunction eventually. Good theoretical reason to think so as well, given the nature of entropy.

In any case, none of these considerations is relevant to any mechanistic claim that is supposed to be settled by Gödel's incompleteness theorems. On the machine side, the idealizations are familiar, and the going is smooth: we simply ignore finite limits. In particular, we assume that our machines never run out of memory, energy, time, or working materials. We need not assume that our ideal machines have access to an actual infinity of material, say an infinite hard drive, but we do assume that they will never stop computing just because they run out of material or memory or storage space. In effect, we assume that there is always more storage to be had—the traditional idea of potential infinity. We also assume that the machines run indefinitely without crashing, flawlessly following their assigned programs. In effect, we enforce a familiar distinction between hardware and software, and then ignore the hardware. We only consider what happens when programs run as written (if not as intended). In sum, we assume that the machines in question are abstract objects like *Turing machines*, with their fixed programs and unlimited tapes.

Similar idealizations are familiar throughout mathematics, and have been almost from the beginning. Consider, for example, four of Euclid's five postulates:

1. To draw a straight line from any point to any point.
2. To produce a finite straight line continuously in a straight line.
3. To describe a circle with any centre and radius.
5. That, if a straight line falling on two straight lines makes the interior angles on the same side less than two right angles, the two straight lines, if produced indefinitely, meet on that side on which the angles are less than the two right angles.

We are not to balk at, say, Postulate 1, just because there are, or might be, two points that are so far apart that no one could draw a line connecting them without their hand slipping or without running out of whatever materials one uses to draw lines.[2]

I do not intend to raise any doubts about the idealizations on the machine side of the divide. That would be to raise a skepticism about the bulk of contemporary and historical mathematics. I just take it as given that we understand what Turing machines are, and that things here are as definite as anything in mathematics or anywhere else, Wittgensteinian worries about rule-following notwithstanding.

Now, what about the humans? As noted, for any finite set *S*, there is a Turing machine that prints out the members of *S* and nothing else, and there is a Turing machine that prints out the logical consequences of S and nothing else. But again, that is not relevant to any mechanistic thesis that can be adjudicated, one way or the other, through the incompleteness theorems. The principals to the present debate (try to) make idealizing assumptions about humans analogous to those of Turing machines. They do not speak of the theorems a subject *does* produce, but the theorems that they *can* produce, invoking what they take to be a similar modality to the one we invoke when dealing with Turing machines, as well as in Euclidean geometry and in a myriad of other places.

So, for example, our mechanist accommodates theorems whose shortest proofs are so long that no human can establish them without falling asleep or otherwise losing concentration, or without using up every particle in the universe. In short, the envisioned creatures have unlimited lifetimes, unlimited attention spans, unlimited energy, and unlimited materials at their disposal, just like our Turing machines do. This takes care of the previous examples of unknowable truths, involving large prime numbers. Yet, it is assumed that our ideal mathematicians are like humans in every other respect—whatever that means.

The principals to the debate also assume that our ideal subjects do not make mistakes—a sort of *normative* idealization. To get around the human propensity to make mistakes, we consider the *correct* theorems that our ideal subject can produce. Implicitly, the standard move is to postulate something like an arithmetic competence/performance distinction in actual humans and then ignore problems with performance.

But this cannot just be taken for granted. Here is one place where Wittgensteinian rule-following considerations might become serious. Is it clear which machine Lucas or Penrose (for example) would become if they were to undergo the envisioned modifications and idealizations—having an unlimited lifetime, with no attention lapses, and unlimited material available—even if we restrict attention to their abilities concerning arithmetic sentences?

Kripke [1982] raises doubts about these matters, on behalf of the later Wittgenstein. So does Kreisel [1972, pp. 317–318], on his own behalf. The literature on this is long, vexed, and contentious, and there is little need to rehash it here in any detail. For the briefest of reminders, when dealing with natural organisms, can we sharply distinguish a breakdown or limitation in memory recall, for example, from an error in the "software"? Is there a clear distinction between human "hardware" and human arithmetic "software"—or what would be human software if memory limitations (etc.) were waived? The principals to the present debates presuppose that there is.

The normativity idealization is not only that there are things like human arithmetic "routines" and "procedures," but that these are free of arithmetic bugs. The procedures themselves are infallible. Errors that actual humans make are attributed to lack of attention or memory failure. We flesh and blood types do not and, indeed cannot, always follow the procedures completely. Again, Penrose is concerned only with what he calls "unassailable" truths.

The idealizations in question—that is, ignoring finite limits—make sense in the cases of Euclidean geometry, the theory of computability, and the other mathematical endeavors, such as proof theory, because, in each case, we have a fixed, specified number of definite routines, and we consider the results of applying those fixed routines over and over. We know exactly what the routines are. In the case of geometry, it is compass and straightedge construction; in the case of a given Turing machine, it is the instructions coded in its machine table; and in the case of proof theory, it is the application of fixed rules of inference.

In these mathematical cases, we get sharp, rigorous, *negative* results only because we have a handle on the routines and techniques available. With this handle, we can get a rigorous conception of the results—and limits—of applying the specific routines over and over, with no finite limit imposed. We show that it is not possible to trisect an angle or square a circle with compass and straightedge alone; we show that a Gödel sentence cannot be derived in a certain formal system, and the like. To be sure, in all of these cases, it took hard and sustained mathematical and conceptual effort to get those rigorous handles, but without a clear and determinate account of what the routines in question are, the situation is hopelessly vague and under-specified.

So what are the fixed routines available to our idealized humans, who can supposedly out-Gödel any idealized machine? What is the analogue of compass and straightedge constructions or a Turing machine table, or a set of fixed axioms and rules of inference? Similarly, if we go with Gödel's second disjunct, and want to claim that certain propositions are unknowable, we need a determinate handle on what the specific knowing-techniques are? Are those techniques somehow fixed in advance? Or are they somehow indefinitely extensible, in a way that cannot be precisely delineated?

When it comes to the human activity of coming to know things, Lucas and Penrose do sometimes speak in terms of routines and procedures. They argue that the incompleteness theorems give idealized humans some tools they can deploy, to extend arithmetic knowledge. Lucas [1996, p. 105] once put the issue in pretty much these terms: "Having once got the hang of the Gödelian argument, the mind can adapt it appropriately to meet each and every variant claim that the mind is essentially some form of Turing machine." Lucas argues that the incompleteness theorem suggests a certain procedure that a human can "get the hang of" and wield against the mechanist. This "procedure" is demonstrably not effective, not Turing computable.

So what is this ability that idealized humans can supposedly deploy that is not available to any machine? It cannot be the ability to take any formal system or Turing machine and write down a Gödel sentence for it. *That* is an effective procedure, if ever there was one. Indeed, Judson Webb [1980] once argued that the effective nature of the incompleteness theorem actually *supports* mechanism. He claimed that the incompleteness theorems are the best thing that ever happened to mechanism.

Presumably, the special, inherently non-effective routine that is attributed to the idealized human mathematicians is the ability to *see that* the produced Gödel sentence is *true*. The mathematician uses their inherent, intuitive ability, and asserts the Gödel sentence, as a piece of unassailable arithmetic *knowledge*.

The supposed ability attributed to idealized humans thus seems to be some sort of intuitive insight. How far does this special insight go? If Lucas and Penrose are attributing the ability to tell whether a given first-order (or perhaps $\Pi_1$) sentence, in the language of arithmetic, is true, then, of course, our idealized humans go beyond any Turing machine. But the incompleteness theorems do not support *that* hypothesis, at least not by themselves. What Lucas and Penrose are claiming is that if these humans are presented with a candidate Turing machine $T$ which enumerates only knowable arithmetic sentences, then the idealized humans can *see that* the Gödel sentence $G_T$ for $T$ is true. But, as has been pointed out by opponents, many times, if $T$ has a minimal amount of arithmetic, it will produce the sentence stating that *if* the output of $T$ is consistent, then $G_T$. Presumably, Lucas and Penrose think that the idealized human can get beyond that, *knowing* that the output of $T$ *is* consistent, and thus that $G_T$ is true. But how?

## 8.4 Epistemology

We thus broach matters of epistemology. How do real, flesh and blood, human beings come to know arithmetic propositions? And how do our idealized counterparts do it? Of course, these are big issues, not to be settled here, and so I must be brief, resting content with a sketch.

Recall the "normative idealization" that the subjects proceed by absolutely certain, infallible methods, and that, in doing so, they produce all and only mathematically provable arithmetic sentences. The normative idealization, so articulated, is consonant with a longstanding epistemology for mathematics. The idea is that for mathematics at least, humans are capable of proceeding, and should proceed, by applying infallible methods. We begin with self-evident truths, called "axioms," and proceed by self-evidently valid steps, to all of our mathematical knowledge. In practice (or performance) we invariably fall short of this, due to slips of the pen, faulty memory, and the practical need to take shortcuts, but in some sense we are indeed capable of error-free mathematics, and, moreover, all of the mathematics we have, or at least all of the correct mathematics we have, could be reproduced, in principle, on the basis of rigorous deductions from self-evident axioms. It is absolutely certain. Lakatos calls this the "Euclidean" model for mathematical knowledge, but we should not attribute anything like this to Euclid. Call it the *foundationalist* model.

In the Gibbs lecture, Gödel [1951, p. 305] comes close to endorsing a foundationalist model:

> [The incompletability of mathematics] is encountered in its simplest form when the axiomatic method is applied, not to some hypothetico-deductive system such as geometry (where the mathematician can assert only the conditional truth of the theorems), but to mathematics proper, that is, to the body of those mathematical propositions which hold in an absolute sense, without any further hypothesis ... [T]he task of axiomatizing mathematics proper differs from the usual conception of axiomatics insofar as the axioms are not arbitrary, but must be correct mathematical propositions, and moreover, evident without proof.

Penrose [1994, Chapter 3] also seems to endorse something in the neighborhood of the foundationalist model, with his central notion of "unassailable knowledge." He admits that even ideal subjects may be subject to error, but he insists that errors are "correctable."

So the Lucas–Penrose anti-mechanist might be claiming that the productions of the ideal subjects, asserting the various Gödel sentences, are either "evident without proof," as Gödel put it, or else they follow from self-evident propositions by a self-evident chain of reasoning. So what is this self-evident insight, or this chain of reasoning, and why is it not available to the idealized machines?

Although the foundationalist model for epistemology has advocates today, it is under serious challenge, partly due to the actual development of mathematics. The standard way to establish new theorems about, say, the natural numbers, is to embed them in a much richer structure, and take advantage of the newer, more powerful expressive and deductive resources. The spectacular resolution of Fermat's theorem, via elliptical function theory, is not atypical in this respect. Since we do not know, in advance, just what rich theories we will need to prove future theorems about the natural numbers, we are not in a position to say what the "resources" are, in wondering if there are arithmetic sentences unknowable in principle. A Turing machine has to have a fixed alphabet, a single "language," and a fixed program for operating with that fixed language. Our real, let alone ideal, mathematicians do not have that. There is no fixed language, no fixed set of expressive resources, and there is no fixed set of axioms, once and for all, that we operate from.

Moreover, as the embedding structures get more and more complex and powerful—as our ability to postulate gets more and more bold—we eventually lose our "unassailable" confidence that the theories are consistent. In other words, the ideal of proceeding by self-evidently valid steps is compromised—if there ever was such an ideal.

To be sure, the overwhelming bulk of mathematics today can be recast in axiomatic set theory, and sometimes is, for foundational purposes. Set theory has a single language, in fact only one non-logical symbol, and it has a fixed axiomatization, ZFC. But it is highly contentious to think of those axioms as themselves self-evident. Do we really have unassailable knowledge (or unassailable rational insight) that every set has a unique powerset? Or of the axiom of choice? Or of replacement? Or, for that matter, of the axiom of infinity?

Moreover, is Zermelo–Fraenkel set theory sufficient for *all* unassailable mathematical knowledge? If so, the mechanist wins. But ZFC clearly isn't sufficient. Don't forget the Gödel sentence for ZFC. I presume we do know that. Do we require that any new axioms added to ZFC must themselves be self-evident? The bold large cardinal candidates explored by contemporary set theorists hardly meet that norm. And, I submit, there is no reason they should.

All of this tells against the foundationalist model. I would think that a more holistic epistemology would make sense for contemporary mathematics, if not for all of science. The idea is that successful theories support each other, and there is just no foundational bedrock, at least not in any epistemological sense (see Shapiro [2009]). From the holistic perspective, there is no difference in kind between a mathematical proof, an entrenched scientific thesis, and a well-confirmed working hypothesis. Those are more matters of degree. In principle, nothing is unassailable in principle.

Of course, this is not the place to settle broad questions concerning mathematical knowledge, either real or ideal. But perhaps we need to rethink the Lucas–Penrose argument, so as not to tie it to a contentious epistemology. Lucas himself [1996, p. 120] says that to "claim to know something is not to claim infallibility, but only to have adequate backing for what is asserted." If this comment applies to both the output of the mechanist's proposed system and its Gödel sentence, then Lucas himself rejects the foundationalist model, or at least he does not want to presuppose it.

So how does the Lucas–Penrose argument go? On a more holistic epistemology, it seems, we do not have the normative assumption that we, or our idealized counterparts, never make mistakes. We do sometimes propose inconsistent theories, and, of course, later retract or modify them. So it looks like we do not have the aforementioned set **K** of sentences knowable with absolute certainty. What do we focus on instead? What is the mechanistic thesis now?

Define **K**′ to be the collection of arithmetic sentences for which we can "have adequate backing," to use Lucas's phrase. If this is not a set, with sharp boundaries, we cannot apply the incompleteness theorem. So let us assume (or pretend) that it is.

Perhaps the interlocutor claims that his Turing machine enumerates **K**′ and Lucas proposes his Gödel sentence as a counterexample. Lucas argues that the mechanist must have adequate backing for the statement that their machine is consistent, and this same backing supports the Gödel sentence, even though the Gödel sentence is not produced by the Turing machine in question.

But do we have adequate backing for the claim that **K**′ is consistent? More importantly, do we *know* that it is consistent? We are in, or at least near, the familiar territory of the lottery paradox. Consider a lottery with 1,000,000 tickets. For each ticket *t*, we have adequate backing for the statement that *t* is not the winning ticket, even though the set of all such sentences is inconsistent with the rules of the lottery. The notion of "having adequate backing" is not monotonic. And, according to most epistemologists, "having adequate backing" and truth are not sufficient for knowledge.

There is no reason we cannot have adequate backing for each member of an inconsistent set of sentences. Surely, we do not have adequate backing for any explicit contradiction, once we realize that it is a contradiction—dialetheism notwithstanding once again. Thus, the set **K**′ might not be closed under deduction. We can have adequate backing for each of a large set of sentences without having adequate backing for every consequence of this set.

Admittedly, the situation is paradoxical. Can there be sentences $\Phi$ and $\Psi$ such that both $\Phi$ and $\Phi \rightarrow \Psi$ are in **K**′—we have adequate backing for both—and yet $\Psi$ is not in **K**′? All that follows from this consideration is that the notion of "having adequate backing" is vague, and we have a version of the sorites paradox on hand. Again, unless the set **K**′ is precise, we cannot apply the incompleteness theorem to it.

To apply the limitative theorems in the most straightforward manner, we need an epistemological notion that is both sharp *and* closed under deductive consequence, and which the mechanist can plausibly attribute to idealized subjects as the goal of mathematical activity. "Knowable with absolute certainty" is perhaps too strict a notion, not consonant with how mathematics is done, and "having adequate backing" is too weak and too vague. Is there something in between?

To make a stab at it, let us define a set $\Gamma$ of sentences to be *stably backed* in a given state of information if for each member $\Phi$ of $\Gamma$, $\Phi$ enjoys adequate backing in that state of information and $\Phi$ remains adequately backed no matter how many logical consequences of $\Gamma$ are added to the state of information. That is, $\Gamma$ is stably backed if no member of $\Gamma$ loses its backing as deductions are made on members of $\Gamma$. It follows that $\Gamma$ is stably backed only if $\Gamma$ is consistent. Define $\Gamma$ to be *maximally stably backed* if $\Gamma$ is stably backed and if there is no proper superset of $\Gamma$ that is stably backed. That is, if $\Gamma$ is maximally stably backed then no sentence outside of $\Gamma$ can get adequately backed without giving up some member of $\Gamma$. This notion is a Peircean limit, of sorts.

Suppose that someone comes up with a Turing machine and claims to have adequate backing for the statement that the set *S* of sentences produced by this machine is maximally stably backed. Assuming that *S* contains some arithmetic, Lucas wins the round. The mechanist has contradicted themselves. If someone has adequate reason to think *that S* is stably backed then they have adequate backing for $Con_S$ and if $S$ is consistent then it does not contain $Con_S$. So we either give up some member of $S$ or give up the statement that $S$ is *maximally* stably backed.

This conclusion does not threaten mechanism. How is the mechanist committed to the existence of a maximally stably backed set of arithmetic sentences? Even if there is such a set, why is the mechanist committed to the idea that humans are capable of producing such a set and that it is stable under Gödelian reflection? Why should the mechanist believe that if we idealize on things like limitations on lifespan and memory, we end up with procedures which enumerate a maximally stably backed set of arithmetic sentences?

Maybe there is some other important epistemological notion that can be put to work here. The burden is surely on Lucas and Penrose, and also on their mechanistic opponents, to say just what real and then idealized mathematical knowledge amounts to, and to show that the knowable arithmetic (or $\Pi_1$) sentences constitute a sharply defined set. Again, we have to be clear about what the game is before we can decide who wins it.

## 8.5 Ordinal Analysis

None of this is to say that the incompleteness theorems have nothing to teach us about arithmetic, or even its epistemology for some ideal mathematicians. There are some fascinating techniques for iterating the Gödel construction into the transfinite, and a number of important conceptual issues are broached along the way. What we do not get, so far as I can see, is any support for a mechanistic thesis, nor do we get support for a Lucas–Penrose–Gödel anti-mechanist perspective.

Solomon Feferman [1962] makes an important definition:

> By a *reflection principle* we understand a description of a procedure for adding to any set of axioms $A$ certain new axioms whose validity follows from the validity of the axioms $A$ and which formally express, within the language of $A$, evident consequences of the assumption that all of the theorems of $A$ are valid.

If one is somehow committed to the truth of every member of $A$, then one is intuitively also committed to the truth of the output(s) of a reflection principle, as applied to $A$. Nontrivial reflection principles thus sharply illuminate the indefinite extensibility of arithmetic truth—if not arithmetic knowledge.

The reflection principle in Turing's original [1939] is the Gödel sentence $G_A$ or, equivalently, the formal consistency $Con_A$. Feferman [1962] considers other reflection principles, the most powerful of which is a reflective ω-rule of sorts. Let $A^*$ consist of the axioms of $A$ together with every instance of

$$\forall x \mathbf{Pr}_A(\ulcorner \Phi(x) \urcorner) \rightarrow \forall x \Phi(x),$$

where $\mathbf{Pr}_A$ is a standard provability predicate for $A$, and $\mathbf{Pr}_A(\ulcorner \Phi(x) \urcorner)$ is a formula stating that $\mathbf{Pr}_A$ holds of the result of substituting the appropriate numeral for $x$ in $\Phi(x)$.

The program is to iterate a reflection principle $P$ through recursive ordinals. Let $A$ be a base theory, such as standard Peano arithmetic. For a first stab, let $A(0)$ be $A$ and, for each ordinal $\alpha$, if $A(\alpha)$ is defined, then let $A(\alpha + 1)$ be the theory consisting of $A(\alpha)$ together with the result of applying $P$ to $A(\alpha)$. In Turing's case, where $P$ is the consistency statement, $A(\alpha + 1)$ is $A(\alpha)$ together with $Con_{A(\alpha)}$. In Feferman's strongest case, $A(\alpha + 1)$ is $A(\alpha)^*$. If $\lambda$ is a limit ordinal and $A(\alpha)$ is defined for every $\alpha < \lambda$, then the theorems of $A(\lambda)$ are the union of the theorems of $A(\alpha)$ for every $\alpha < \lambda$.

For the two reflection principles mentioned previously, if $A(\alpha)$ is recursively enumerable, then so is $A(\alpha + 1)$. If $\lambda$ is a recursive limit ordinal and if the members of the members of $\{A(\alpha) \mid \alpha < \lambda\}$ can be enumerated by a single Turing machine, then $A(\lambda)$ is itself recursively enumerable.

One might be tempted to consider the (non-recursive) theory consisting of the union of

$$\{A(\alpha) \mid \alpha \text{ is a recursive ordinal}\}.$$

The problem is that, for any given $\alpha > \omega$, the theory $A(\alpha)$, as presented so far, is not well defined. In particular, neither of the reflection principles is extensional, as each is constructed from a provability predicate. If $B$ is a set of axioms, then a sentence $Con_B$, stating the consistency of $B$, depends not just on the members of $B$ (and their consequences), but also on how $B$ is given. For example, there are theories $B$ and $C$ that have the same theorems, but where $Con_B$ is not equivalent to $Con_C$ (see Feferman [1960]). Moreover, with the above notation, the theories $B^*$ and $C^*$ will be different, even extensionally.

The intensionality is especially pronounced at limit ordinals. If $\lambda$ is such a limit, then to apply the reflection principle at stage $\lambda + 1$, we need not just the theorems of $A(\lambda)$ but also a *description* of those theorems, and this depends on a description of the ordinal $\lambda$. If we have two different descriptions of $\lambda$, we can end up with two different theories at the next stage, $A(\lambda + 1)$.

The intensionality here serves as a reminder that neither actual humans, nor, presumably, the ideal humans we envision, deal with theories—sets of sentences—as such, but only with theories-under-a-description. Here the description matters. So to pursue our transfinite Gödelization, we must deal with *notations* for recursive ordinals.

Feferman invokes a common notation, which we adopt here. Recursive ordinals are "denoted" by natural numbers. The number 1 denotes the ordinal 0; if $n$ denotes an ordinal $\alpha$, then $2^n$ denotes its successor $\alpha + 1$; and if $e$ is the Gödel number of a Turing machine that enumerates numbers denoting an increasing sequence of ordinals, then $3 \cdot 5^e$ denotes the limit of that sequence. Let $O$ be the set of natural numbers that denote ordinals on this notation, and if $m \in O$ then let $|m|$ be the ordinal denoted by $m$. It turns out that the set $O$ is *not* recursively enumerable (see Rogers [1967, pp. 205–210]).

Let $P$ be a reflection principle. Let $R(1)$ be a standard enumeration of the theorems of our base theory $A$. We will hold both $A$ and this enumeration fixed. If $n \in O$, then let $R(2^n)$ be an enumeration of the theory consisting of the sentences enumerated by $R(n)$, together with the result of applying the reflection principle to that theory (under that description). If $e$ is the Gödel number of a Turing machine that enumerates numbers denoting an increasing sequence $S$ of ordinals, then let $R(3 \cdot 5^e)$ be a uniform enumeration of the union of the sets $R(s)$ for $s \in S$.

The idea is that for each $n \in O$, $R(n)$ is the theory $A(|n|)$—under that description. The notation makes the intensionality explicit, since the theorems of $R(n)$ depend not just on the *ordinal* $|n|$ but also on the *number* $n$ that denotes this ordinal. It is straightforward that if $n \in O$ and the base theory $A$ consists only of truths, then $R(n)$ also contains only truths.

In these terms, and with this notation, Turing's [1939] plan was to overcome incompleteness by using theories like $R(n)$, with $n$ ranging over $O$. His main theorem is that for the simple reflection principle $Con_A$, if $\Phi$ is a true $\Pi_1$-sentence, then there is an $n \in O$ (which can be found effectively from $\Phi$) such that $|n| = \omega + 1$ and $\Phi$ is among the theorems of $R(n)$. This astounding result is that there is a way to iterate the Gödel construction on theories, beginning with $A$, so that when we collect together the finite iterations and take one more Gödel sentence, $\Phi$ is decided.[3] There is thus a certain completeness for $\Pi_1$-sentences.

Feferman extended this result. With his stronger reflection principle $A^*$, he showed that for *any* true first-order sentence $\Phi$ in the language of arithmetic, there is a number $n \in O$, denoting a small countable ordinal, such that $\Phi$ is among the theorems of $R(n)$. That is, for any sentence $\Phi$, there is a way to iterate the reflection principle (not very far) and decide $\Phi$.

Turing [1939, §9] was aware that his result does not provide the wherewithal to *decide* the truth of any new arithmetic sentences, in any epistemologically relevant sense of "decide": "This completeness theorem . . . is of no value. Although it shows, for instance, that it is possible to prove Fermat's last theorem [with $R(n)$] (if it is true), yet the truth of the theorem would really be assumed by taking a certain" number as a member of $O$.

To switch the example, suppose that a mathematician wants to decide the truth value of the Goldbach conjecture. They calculate a number $n$, based on the sentence expressing the conjecture, and start enumerating the theorems of $R(n)$, looking for the Goldbach conjecture among the output. So far so good, since all this is effective. From Turing's result, the mathematician knows that if the Goldbach conjecture is true, then $n \in O$ and so every sentence in $R(n)$ is true. However, an examination of Turing's proof shows that if the Goldbach conjecture is false, then $n$ is not in $O$, and, even worse, $R(n)$ is inconsistent. Thus, the results of the enumeration can be believed only if the Goldbach conjecture is true. And the statement of the Goldbach conjecture will be among the results either way. Clearly, this procedure is of no help whatsoever in trying to determine the truth-value of the Goldbach conjecture.

Something similar holds for the Feferman result. For each sentence $\Phi$ we get a formal system $F$ that is sound, only provided that $\Phi$ is true; and if $\Phi$ *is* true, $F$ proves it. Nothing to celebrate here. We can get that much just by adding $\Phi$ as a new axiom to the base theory $A$. Feferman [1962, p. 262] concludes that "questions of completeness of sequences derived from progressions hinge . . . on more subtle questions on how paths through $O$ are obtained." He elaborates:

> Whenever [a mathematician] is given the information $d \in O$ he will be able to compute [$R(d)$] and prove theorems from [$R(d)$]; moreover, if he accepts the information that $d \in O$ he should find all these theorems acceptable. Unfortunately, . . . as he advances farther and farther out into the collection of systems [$R(d)$], he may not be able

> to gain the knowledge necessary to decide, of any given $d_0$, whether or not $d_0 \in O$. In other words, in order to proceed, he may have to appeal to an 'oracle'.
>
> (p. 279).

The key to wielding the Turing–Feferman results is the ability to decide membership in $O$, or to find effective notations for recursive ordinals generally. In these terms, the Lucas–Penrose contest to write and assert Gödel sentences becomes a contest to enumerate recursive ordinals. One might think that all Lucas or Penrose have to do is to iterate the procedure of adding Gödel sentences (or the Feferman reflection principle) far enough. The problem, however, is with the crucial notion of "far enough." At some point, we are no longer sure we are on the right road—we can't be sure that our number denotes an ordinal. We have a sort of transfinite sorites series. The run won't break down at a successor stage. But what of limit stages?

Clearly, no machine can iterate the Lucas–Penrose procedure, of adding Gödel sentences, through all and only the recursive ordinals. Can Lucas, or some idealized Lucas, do so? In [1996, p. 110], the flesh and blood Lucas envisions the shift from the Gödel sentences to the construction of ordinals (although he does not explicitly restrict the discussion to recursive ordinals). Lucas imagines a mechanist designing machine after machine in an effort to confound the Gödelizing. Lucas keeps winning, by producing Gödel sentences (again, in a purely mechanical way), and the mechanist keeps designing new machines to enumerate the new theories:

> Every now and again the mechanist loses patience, and incorporates in his machine a[n] ... operator designed to produce in one fell swoop all the Gödelian sentences the mentalist is trumping him with: this is in effect to produce a new limit ordinal. But such ordinals, although they have no predecessors, have successors just like any other ordinal, and the mind can out-Gödel them by producing the Gödelian sentence of the new version of the machine, and seeing it to be true, which the machine cannot.

Here, again, we have the human able to "see" things, that are supposedly not available to the machine.

Douglas Hofstadter [1979, p. 475] questioned Lucas's confidence, citing the Church–Kleene theorem "that we cannot program a machine to produce names for all the ordinal numbers," as Lucas put it. Of course, there is no machine that produces names for *all* the ordinals, because the collection of ordinals is a proper class. The Church–Kleene theorem is that there is no *recursive* enumeration of every *recursive* ordinal (in a way that allows us to determine the order type of each).

There are different anti-mechanist claims that might be made at this point. A strong one is that an appropriately idealized human *can* enumerate the members of $O$ (or some other effective notation for all recursive ordinals). This is a reasonably sharp thesis, assuming we can make sense of the idealizations (which, as stated previously, we can't—see §8.3). But what reason do we have to believe it?

Lucas argues that Hofstadter begs the question in the assumption that "since there is no mechanical way of naming all the ordinals, the mind cannot do it either." Lucas says that "this is precisely the point at issue." If this *is* the point at issue, then Lucas makes the strong claim that the mind *can* enumerate all of the recursive ordinals (i.e. the members of $O$)—if not all of the ordinals.

Clearly, the presumed ability to enumerate $O$ goes well beyond any Turing machine. So if Lucas could give some reason to think humans have this ability, he wins—hands down. Humans—suitably idealized—can do something that no machine can do, namely iterate the members of $O$. So far as I can tell, however, the limitative theorems do not give a reason to think any suitably idealized human has the ability to enumerate $O$, and so the limitative theorems do not support this strong anti-mechanist scenario. Or at least we have no argument that they do.

Lucas cites the authority of Gödel as "rejecting mechanism on account of our ability to think up fresh definitions for transfinite ordinals." Gödel did no such thing. First, his rejection of mechanism was tentative—he did not claim to have a compelling argument. Second, and more important here, Gödel's concerns about ordinals were with large cardinals and not with these puny recursive (and thus countable) things.

To defeat the mechanist, perhaps Lucas does not need the strong claim that some human can enumerate $O$. It is enough if for any given machine $M$ put forward by a mechanist, Lucas can "out-enumerate" *that* machine—if he could enumerate more members of $O$ than $M$ can. He puts the debate in these terms:

> All the difficulties are on the side of the mechanist trying to devise a machine that cannot be out-Gödelized. It is the mechanist who resorts to limit ordinals, and who may have problems devising new notations for them. The mind only needs to go on to the next one, which is always an easy, unproblematic step, and out-Gödelize whatever is the mechanist's latest offering.

It is not at all clear just what this weaker anti-mechanistic claim is. Suppose that Lucas claims that for any Turing machine $M$ that enumerates only members of $O$, there is some idealized human $h$ such that $h$ can enumerate more members than $O$ than $M$. This claim has the $\forall M \exists h$ form, and it allows that different Turing machines $M$ may get trumped by different idealized humans $h$. But this claim does not undermine mechanism at all. For any Turing machine $M$ that enumerates only members of $O$, there is another *Turing machine* that enumerates more members of $O$. This is an analogous $\forall M \exists M'$ theorem. Indeed, Lucas is correct that this trumping is always an "easy, unproblematic step." To get beyond a sequence enumerated by the Turing machine with Gödel number $e$, Lucas calculates $3 \cdot 5^e$; to go beyond $|n|$, to $|(|n| + 1)|$ he just calculates $2^n$. However, there is nothing non-mechanical going on in this calculation, just as there is nothing non-mechanical involved in writing out a Gödel sentence. The calculations are grade school drills (involving *very* large natural numbers).

So what exactly is this ability that Lucas claims on behalf of ideal humans? We are back where we were before we got fancy with ordinals. For any given machine $M$, Lucas cannot claim that the "easy, unproblematic step" results in knowledge of a new member of $O$ unless he *already knows* that $M$ produces only members of $O$. How does Lucas get this prior knowledge? As before, he shifts the burden to the mechanist. If the mechanist *proves* that

$M$ produces only members of $O$, then they have already managed to go beyond $M$, without Lucas's help. However, if the mechanist only claims some sort of quasi-empirical, or holistic evidence, on behalf of $M$, the Gödel–Turing–Feferman results do not apply. Lucas only knows that *if* $e$ is the Gödel number of a Turing machine that enumerates an increasing sequence of recursive ordinals, then $3 \cdot 5^e$ denotes an ordinal greater than any enumerated by that machine, and Lucas knows that *if* $n$ is a member of $O$, then $2^n$ denotes a greater ordinal. But these conditional statements are theorems of the base theory $A$, and so are "available" to all of the Turing machines.

How does Lucas get beyond the conditionals, to their consequents? Presumably, he claims some sort of rational insight, not available to any of the mechanist's Turing machines. He writes that in the ordinal-writing duel with the mechanist, "every now and again some new, creative step is called for, when we consider all the ordinal numbers hitherto named" ([1996, p. 111]). Lucas is correct that to defeat mechanism, it must be a *creative* step and not the application of an algorithm—not the mere construction of a Gödel sentence or the calculation of a new member of $O$. What is this creative step, and how do the incompleteness theorems indicate that we have it?

Lucas argues that Turing, "like Gödel, allows that the mind's ability to recognize new ordinals outruns the ability of any formal algorithm to do so." The key word here, I think, is "recognize." Lucas is not just attributing to his idealized self the ability to write out Gödel sentences or to print natural numbers, nor a mere ability to assert the conditional sentences. It is an *epistemic* ability that supposedly defeats the mechanist. The "creative step" is the presumed ability to *see that* every member of a given sequence of numbers is a member of $O$ (or denotes a recursive ordinal), or the ability to *see that* every theorem of a given formal theory is true. Similarly, Penrose speaks often of the human ability to "understand," which any algorithmic device (supposedly) lacks.

Once again, in light of the idealizations needed to even get started on this quest, it is hard to see how the incompleteness theorems give us any insight into what we, let alone our idealized counterparts, can "see" or "understand." Lucas's proposal seems to be a variant of the Gödel–Kreisel suggestion that humans, unlike machines, can traffic in abstract concepts. Here, the thesis is that humans are capable of epistemic states like "recognition" or "intuition." Unlike Lucas and Penrose, however, Gödel and Kreisel were both well aware that even if they are right about the non-mechanical nature of the human mind, and, in particular, the human ability to traffic in abstract objects and concepts, there is *still* a burden to show that humans can out-perform any machine. That is, even if we assume that humans do, and machines do not, have the ability to understand, manipulate, and work with abstract objects and concepts, how does it follow that, in enumerating truths about the natural numbers, a human, so endowed, can out perform any machine? How does the (idealized) human take advantage of his or her (supposed) intuitive abilities to get beyond anything a machine can do?

Again, the incompleteness theorems concern formal systems, algorithms, and Turing machines. To "apply" these theorems we need a sharp thesis. In this case, it is no longer clear what mechanistic thesis Lucas is refuting. What does it mean to say that an algorithm, a formal system, or a Turing machine "sees that" a number is a member of $O$, beyond printing out the relevant formula? In a footnote to [1939, §11], Turing wrote that the requirements on intuition are "very vague." Indeed. So, once again, we need some precise content to the supposed mechanistic theses before Lucas and Penrose can wield the Gödel theorem against them.

## Notes

1. See Milne [2007] for an illuminating account of what Gödel sentences "say."
2. I gather that Aristotle *would* have had a problem with Postulate 5, if not 2 and 3, because he held that geometric objects, like lines, correspond to physical objects, such as the edges of cube-shaped things, and Aristotle held that the universe is finite—a large sphere in fact (see White [1992, pp. 154–156]). Of course, this aspect of Aristotle's philosophy was not codified into Euclid's *Elements*
3. See Feferman [1988, §7] for readable sketches of Turing's results.

## References


[1] Boolos, G. Introductory note to Gödel [1951], in Gödel [1995], 290–304, 1995.

[2] Feferman, S. Arithmetization of mathematics in a general setting, *Fundamenta Mathematicae* **49**, 35–92, 1960.

[3] Feferman, S. Transfinite recursive progressions of axiomatic theories, *Journal of Symbolic Logic* **27**, 259–316, 1962.

[4] Feferman, S. Turing in the land of O(z), in R. Herken ed., *The Universal Turing Machine*, New York, Oxford University Press, pp. 113–147, 1988.

[5] Gödel, K. Some basic theorems on the foundations of mathematics and their implications, in Gödel [1995], pp. 304–323, 1951.

[6] Gödel, K. *Collected Works III*, Oxford, Oxford University Press, 1995.

[7] Hofstadter, D. *Gödel, Escher Bach*, New York, Basic Books, 1979.

[8] Kreisel, G. Which number theoretic problems can be solved in recursive progressions on $\Pi^1_1$ paths through *O*?, *Journal of Symbolic Logic* **37**, 311–334, 1972.

[9] Kripke, S. *Wittgenstein on Rules and Private Language*, Cambridge, Massachusetts, Harvard University Press, 1982.

[10] Lucas, J. R. Minds, machines, and Gödel, *Philosophy* **36**, 112–137, 1961.

[11] Lucas, J. R. Minds, machines, and Gödel: A retrospect, in P. J. R. Millican and A. Clark eds., *Machines and Thought: The Legacy of Alan Turing, Volume 1*, Oxford, Oxford University Press, 1996.

[12] Milne, Peter, On Gödel sentences and what they say, *Philosophia Mathematica (3)* **15**, 193–226, 2007.

[13] Penrose, R. *Shadows of the Mind: A Search for the Missing Science of Consciousness*, Oxford, Oxford University Press, 1994.

[14] Penrose, R. Beyond the doubting of a shadow: A reply to commentaries on *Shadows of the Mind*, *Psyche* **2**(23) (http://psyche, 1996.cs.monash.edu.au/v2/psyche-2-23-penrose.html), 1996.

[15] Priest, G. *In Contradiction: A Study of the Transconsistent*, second, revised edition, Oxford, Oxford University Press, 2006; first edition, Dordrecht, Martinus Nijhoff Publishers, 1987.

[16] Rogers, H. *Theory of Recursive Functions and Effective Computability*, New York, McGraw-Hill, 1967.

[17] Shapiro, S. Incompleteness, mechanism, and optimism, *Bulletin of Symbolic Logic* **4**, 273–302, 1998.

[18] Shapiro, S. We hold these truths to be self evident: But what do we mean by that?, *Review of Symbolic Logic* **2**, 175–207, 2009.

[19] Turing, A. Systems of logic based on ordinals, *Proceedings of the London Mathematical Society* **45**, 161–228, 1939.

[20] Wang, H. *From Mathematics to Philosophy*, London, Routledge and Kegan Paul, 1974.

[21] Webb, J. *Mechanism, Mentalism and Metamathematics: An Essay on Finitism*, Dordrecht, Holland, D. Reidel, 1980.

[22] White, Michael J. *The Continuous and the Discrete: Ancient Physical Theories from a Contemporary Perspective*, Oxford, Oxford University Press, 1992.

# PART III
## Absolute Undecidability

# 9

# Provability, Mechanism, and the Diagonal Problem

GRAHAM LEACH-KROUSE

If a deduction of the twin primes conjecture from the axioms of PA exists, then the twin primes conjecture is provable. If someone exhibits that deduction, then the twin primes conjecture is thereby proved. But, if there is a deduction of the twin primes conjecture from PA + $\phi$, where $\phi$ is some other famous open problem in number theory (or set theory, or geometry) then, as much as we might celebrate the ingenuity of such a deduction, the twin primes conjecture remains open. What's proved is merely the conditional "if $\phi$, then the twin primes conjecture holds." While propositions deduced from PA are thereby *proved* and admitted as full fledged pieces of mathematical knowledge, the results of deduction from PA + $\phi$, where $\phi$ is an open problem, are not; they're merely shown to follow, given $\phi$.[1]

This example is intended to draw your attention to a difference between proving a sentence only *relative* to some formal theory like PA + $\phi$ and proving it simpliciter, or in an absolute sense. We've noted that what's proved from PA is proved simpliciter. The same goes for ZFC and for certain extensions of ZFC.[2] Could it be that there is some formal system, or set of axioms, such that everything that is provable simpliciter is provable relative to those axioms?

There is a prominent tradition in the philosophy of logic which argues, on the basis of the incompleteness phenomena discovered in 1931 by Kurt Gödel, that, in fact, there is no such system. Hence, absolute provability is something other than provability in any fixed theory $T$ and this something else stands in need of further analysis. Two representatives of this tradition are Gödel himself and the logician Emil Post. Post also proved the existence of a variety of incompleteness and undecidability phenomena a decade before Gödel and, on the basis of those phenomena, argued that the notion of absolute provability was worth investigating.

Our primary goal in this chapter will be to understand and to make a start at evaluating Gödel and Post's conceptions of absolute provability. Our starting point will be Gödel's famous assertion from the 1951 Gibbs lecture [10], that either (a) there exist absolutely undecidable mathematical propositions, or (b) the process of human mathematical reasoning cannot be fully imitated by any machine.

*Gödel's Disjunction*. First Edition. Leon Horsten & Philip Welch (eds).

Interestingly, in 1921 Post independently arrived at an equivalent conclusion.[3] But while Gödel, at least in private, seems to have hoped to refute (a) and establish (b), Post, on the basis of a rejection of (b), apparently hoped to prove (a). This situation raises at least two interesting questions. First: why did Gödel take the side he did and Post the opposite? Second: Given such a dramatic divergence on an apparently basic question, is it clear that we even have a disagreement here? How do we know that Post and Gödel were talking about the same sense of absolute provability?

The first three sections of this chapter will be devoted to laying out some historical background and to articulating and defending an answer to the first of these two questions. In Section 9.1, I describe Post and Gödel's respective paths to incompleteness—the way in which they arrived at the results in logic which motivated their investigation of absolute provability. This will give us an opportunity to say something about their views on the relationship between formal systems and mathematical reasoning. In Sections 9.2 and 9.3, I describe Post and Gödel's reactions to their logical findings—Post's reasons for believing that there exists an absolutely undecidable proposition, and Gödel's reasons for ultimately rejecting this view. For the remainder of the chapter, I develop an answer to the second question, by articulating and investigating a condition of adequacy—what I call the *subgroundedness requirement*[4]—that Post and Gödel seem to have accepted on accounts of absolute provability. I claim that, in particular, it's acceptance of this condition of adequacy, rather than a shared conception of how mathematicians actually prove things, which anchors the core notion of absolute provability that Post and Gödel were both interested in, and which provides each of them with reasons for resisting the identification of provability with provability in a fixed theory $T$. In Section 9.5, I briefly describe some tentative steps that Post and Gödel took toward actually producing formal frameworks for the study of absolute provability. In Section 9.6, I explain a problem that confronts any attempt to produce a formally precise account of a notion of provability satisfying the subgroundedness requirement; I argue that both Post and Gödel's proposed frameworks are likely to be unable to overcome this challenge. Finally, in Section 9.7, I conclude by distinguishing some options for thinking about absolute provability and recommending one of these for further investigation.

## 9.1 Two Paths to Incompleteness

I claimed earlier that both Gödel and Post were impressed by the conflict between the idea, on the one hand, that the procedure of mathematical reasoning can be carried out in a mechanical or algorithmic way,[5] and on the other hand, the article of faith that all mathematical problems can be solved.[6] Both were, in fact, sufficiently impressed to urge the dilemma on their readers as an important philosophical insight. But we should be cautious about inferring too much from this surface agreement.

If we want to use Post and Gödel's thinking about mathematical reasoning as a guide to their conceptions of absolute provability, then we need to take care that we get their thinking right. So let's take a closer look at the context to their remarks. On the basis of what we'll find, I'll ultimately suggest that, in spite of the surface similarity of Post and Gödel's claims, their *reasons* for making the claims that they did—indeed, their understanding of what the claims came to—were quite different and turned on their idiosyncratic interpretations of the relationship between formal systems and mathematical thinking.

I'll give a narrative account of Post and Gödel's respective paths to the incompleteness theorems—their early influences and discoveries along the way, as well as what we know about the heuristics and analogies that led them to incompleteness. This will give me the opportunity to present some striking contrasts between Post and Gödel's metamathematical perspectives. It's these contrasts which I think are important for understanding their respective attitudes toward the mechanizability of mathematical reasoning and toward the existence of absolutely undecidable propositions and absolutely unsolvable problems.

Post's discovery was first. So let's begin with him.

### *9.1.1 Post's Path*

Post's first major result in logic was the "fundamental theorem" of his 1921 dissertation: that the propositional logic of Russell and Whitehead's *Principia* is sound and complete with respect to truth table semantics. The result is noteworthy, not so much for its great sophistication or difficulty, as for the way in which Post, first of all, singles out a fragment of the full logic of the Principia for special examination and, second of all, treats this fragment as a purely syntactical or combinatorial object—a set of meaningless strings and rules for their manipulation—to which ordinary mathematical reasoning can be applied.[7] It's this second idea that made Post's incompleteness theorem possible.

In the dissertation, Post credits this approach to C.I. Lewis and approvingly cites the section of Lewis' *A Survey of Symbolic Logic,* in which Lewis argues in favor of "the external view of mathematics," a purely syntactic interpretation of all mathematical activity. Lewis offers the following definition as an expression of his view:

> A mathematical system is any set of strings of recognizable marks in which some of the strings are taken initially and the remainder derived from these by operations performed according to rules which are independent of any meaning assigned to the marks.
>
> [27, p. 355].

and remarks that "Whatever the mathematician has in his mind when he develops a system, what he does is to set down certain marks and proceed to manipulate them in ways which are capable of the above description." [27, p. 355]. The perspective here is that the mathematician is something like an "operator," *working with* a formal system—the mathematician is responsible for implementing certain explicitly specified mechanizable rules in the actual physical activity of putting marks on paper or some other equivalent procedure.[8]

Lewis' influence can be seen in some of the ingenious generalizations and corollaries of the fundamental theorem that Post produced in the later section of the dissertation and in the following years. Two of these are particularly noteworthy for our purposes. The first is the observation that the fundamental theorem gives a decision procedure for derivability in the propositional fragment of the Principia (that is, a uniform procedure for determining, of an arbitrary sentence $\phi$, whether $\phi$ is provable).[9] The second is the generalization of the Principia's propositional logic to what Post would eventually call systems of canonical form A.

The idea behind a system of canonical form A is this. The propositional logic of the Principia can be compactly described as the smallest set $S$ of propositional sentences formed from infinitely many proposition letters $p_1, p_2, \ldots$ and the 1-ary and 2-ary operators $\neg, \vee$, such that (writing $\vdash \phi$ to indicate that $\phi \in S$, $\phi \vdash \psi$ to indicate that if $\phi \in S$ then $\psi \in S$, and abbreviating $p_1$ as $p$ and $p_2$ as $q$):

$$\begin{array}{ll}
\vdash \neg(p \vee p) \vee p & \vdash \neg[p \vee (q \vee r)] \vee [q \vee (p \vee r)] \\
\vdash \neg q \vee (p \vee q) & \vdash \neg[\neg q \vee r] \vee [\neg(p \vee q) \vee (p \vee r)] \\
\vdash \neg(p \vee q) \vee (q \vee p) & \\
\phi, \neg\phi \vee \psi \vdash \psi & \phi(p_\mu) \vdash \phi(\psi)
\end{array}$$

There are three distinguishable elements here: rules of inference stating closure conditions, a language $\{\neg, \vee\}$ and axioms. Post's canonical form $A$ systems allowed for completely arbitrary choices of all three, with the one restriction that the system be closed under a rule of substitution:

**Definition 1 canonical form A** *A propositional logic in infinitely many proposition letters $p_1, p_2, \ldots$, finitely many $m_i$-ary operators $f_1, \ldots, f_\mu$, with substitution $\phi(p_\mu) \vdash \phi(\psi)$ and finitely many axioms and rules of inference*

$$\begin{array}{ll}
\vdash h_1(\overline{p_{\ell_1}}) & g_{11}(\overline{\phi_{k_1}}), \ldots, g_{1\kappa_1}(\overline{\phi_{k_1}}) \vdash g_1(\overline{\phi_{k_1}}) \\
\vdots & \vdots \\
\vdash h_\lambda(\overline{p_{\ell_\lambda}}) & g_{\kappa 1}(\overline{\phi_{k_\kappa}}), \ldots, g_{\kappa\kappa_\kappa}(\overline{\phi_{k_\kappa}}) \vdash g_1(\overline{\phi_{k_\kappa}})
\end{array}$$

*is said to be of canonical form A.*

From a syntactical point of view this is a very natural generalization. Shortly after the completion of his dissertation, Post managed to produce a decision procedure for systems in canonical form A with languages consisting of only 1-ary operations.[10] This suggests the following natural question: *is there a decision procedure for every system in canonical form A?*[11]

It's this innocuous-seeming question that led Post to his encounter with incompleteness. Post's description of the process of discovery is contained in "Absolutely undecidable problems and relatively undecidable propositions—account of an anticipation" [35, p. 397] (henceforth just "the Anticipation"), a paper published in 1965 consisting mostly of annotated and corrected contents of notes that Post produced around 1921. This will be our primary source in what follows, though we'll also draw on some other articles by Post and secondary sources as well.

According to the Anticipation, Post began work on the decision problem for systems in canonical form $A$ by developing methods for reducing the decison problem for an arbitrary system in canonical form $A$ to a simpler problem. The main tool here was what's now known as 1-reduction.[12] A set of strings $X$ is 1-reducible to $Y$ if there exists some effective 1–1 correspondence $f$, transforming strings to strings in such a way that $f(\sigma) \in Y$ just in case $\sigma \in X$. If $X$ is 1-reducible to $Y$, then a decision procedure for $Y$ will naturally provide a decision procedure for $X$. To find out if $\sigma \in X$, produce $f(\sigma)$ and ask whether $f(\sigma) \in Y$.

The key reductions that Post managed to produce are these. Say that a set of strings $S$ is a system in normal form—or "a normal system," for brevity—if it is the closure of a single string $s$ under finitely many rules of the form (writing $s^\frown t$ for the result of concatenating $s$ and $t$)

$$s_1^\frown\sigma \vdash \sigma^\frown s_2,$$

that is, finitely many rules for string-transformation that say "remove the prefix $s_1$ and adjoin a suffix $s_2$."[13] Post showed, first, that every system in canonical form $A$ is 1-reducible to a system in normal form. And second, Post showed that even the full first-order fragment of the Principia's logic was 1-reducible to this remarkably simple sort of combinatorial setup. It was also clear to Post that systems in normal form ought to easily be reducible to systems of canonical form $A$ and that systems in canonical form $A$ were reducible to the logic of the Principia. So, a decision procedure for any one of these systems would yield a decision procedure for them all.

Hearing about this second reduction from a modern perspective, we may begin to experience a sinking feeling. The first-order fragment of the Principia is complete with respect to first-order validity. And—by Church's Theorem—first-order validity is algorithmically undecidable. So in fact, despite the apparent simplicity of normal systems, it is a corollary of Post's second reduction in the previous paragraph that their decision problem is algorithmically undecidable.

Post—after wrestling with the combinatorial structure of normal systems, which he had initially hoped would be tractable—came to recognize this undecidability in a striking and illuminating way. Assume that all normal systems draw the symbols that make up their strings from some fixed countable set $\{a_1, a_2, \dots\}$ of symbols. Imagine an ordering of the normal systems, with $N_e$ the $e^{th}$ such system.[14] Let $\bar{n}$ abbreviate the string $a_1 a_1 a_1 \dots$ consisting of $n$ inscriptions of the symbol $a_1$. There's then at least one set which *can't* be produced by a normal system: the set $P$ which contains $\bar{e}$, a sequence of $e$ inscriptions of $a_1$, just in case $N_e$ does not contain $\bar{e}$. For if $P$ were produced by a normal system $N_w$, we would have $\bar{w} \in P$ just in case $\bar{w} \notin P$. The situation is precisely analogous to what we encounter with Cantor's diagonal argument: the set $P$ is the anti-diagonal of an array containing all the normal systems.

Now consider the following hypothesis, which Post formulated after seeing decision problem after decision problem turn out to be reducible to a problem about normal systems:

> Post's thesis: Every set of strings producible by an effective procedure is the result of restricting some normal system $S$ to a subset of its vocabulary.[15]

If there were an effective decision procedure for normal systems, then there would be an effective procedure for producing $P$: for each $e$, check to see if $\bar{e} \in N_e$ and then write down

$\bar{e}$ if $\bar{e} \notin N_e$. But then, given Post's thesis, $P$ would be produced by some normal system, when it is not. Post's thesis—which is supported by the remarkable generality of the normal system concept—gives reason to expect that there is no solution to the decision problem for normal systems.

The argument also has a precise analog. In addition to our enumeration $N_e$ of normal systems, consider an ordering of the strings in such a way that $\rho_n$ is the $n^{th}$ string. Say that a *normal decision procedure* for the normal systems is a normal system $D$ which for each $e$, (a) contains a string $\bar{e}a_2\bar{n}$ iff $N_e$ contains $\rho_n$ and (b) contains $\bar{e}a_3\bar{n}$ iff $N_e$ does not. Then, on the basis of the rules of $D$, it's possible to define a normal system generating $P$.[16] Hence, there is no normal decision procedure for normal systems.

The barrier to the existence of a normal decision procedure isn't condition (a); Post was able to define a normal system $K$ consisting only of strings $\bar{x}a_2\bar{y}$ with the property that $\bar{e}a_2\bar{n} \in K$ iff $S_e$ contains $\rho_n$.[17] It's (b) that causes the difficulty. More specifically, since $K$ doesn't contain any strings $\bar{e}a_3\bar{n}$ and therefore satisfies the left to right of (b), it's the right to left direction of (b). $K$ functions as a reliable indicator, then, of when a normal system will and won't contain a given string. Its only shortcoming is that it is not total: though it never lies, sometimes it doesn't give an answer. Call a normal system which, like $K$, satisfies (a) and the left to right of (b) an *indicator system*.

So far what Post has uncovered is, in effect, the existence of algorithmically undecidable sets. There's one more key insight, which led him to something closer to Gödel's incompleteness theorem. The insight is this. Given an indicator system $S_w$, one can effectively find an indicator system $S_{w'} \supseteq S_w$ which includes more strings $\bar{x}a_3\bar{y}$ than $S_w$. Here's how.[18] Find a normal system $S_\ell$ enumerating each $\rho_n$ such that $\bar{n}a_3\bar{n} \in S_w$. If $\rho_\ell$ was in $S_\ell$, then by the left to right of $(a)$, $\bar{\ell}a_3\bar{\ell}$ would not be in $S_w$. But this would then contradict the construction of $S_\ell$. So $\rho_\ell$ is not in $S_\ell$. Hence, we may adjoin to $S_w$ rules generating the string $\bar{\ell}a_3\bar{\ell}$, thereby producing the normal system $S_{w'}$. Since $\rho_\ell$ is not in $S_\ell$, we need violate neither (a) nor the left to right of (b).

In more modern terminology, what Post discovered was that the sets $\{\bar{x}a_2\bar{y} \mid \rho_y \in S_x\}$ and $\{\bar{x}a_3\bar{y} \mid \rho_y \notin S_x\}$ are not just undecidable, but respectively "creative" and "productive."[19] The second set (relative to an appropriate coding) can be equipped with a total computable function $f$ such that for any recursively enumerable set $W_e$ contained in it, $f(e) \notin W_e$, but $f(e) \in \{\bar{x}a_3\bar{y} \mid \rho_y \notin S_x\}$. This is established, essentially, by the argument of the previous paragraph. The first of the two sets is recursively enumerable, but its complement is productive.[20]

Recall now Lewis' picture of mathematical activity, which Post inherited: what the mathematician does is work with antecedently specified mechanical rules, in order to produce strings of symbols called theorems. What Post discovered, at the close of the Anticipation, was that—if Lewis' description of the mathematician's procedure is correct—then the mathematician has unnecessarily hamstrung themselves. Given an indicator system meant to serve as a guide to normal system derivability,[21] a better indicator system, with a broader scope of application, can always effectively be constructed.

This is the reason for Post's striking remark at the end of the Anticipation, immediately after the proof of this theorem.

We are led to the following:

*A complete symbolic logic is impossible*

...better still, we may write

*The Logical Process is Essentially Creative*

> This conclusion... is not so much contrary to Russell's viewpoint (since he does not fully express himself) but to that of C. I. Lewis... It makes of the mathematician much more than a kind of clever being who can do quickly what a machine could do ultimately. We see that a machine would never give a complete logic; for once the machine is made *we* could prove a theorem it does not prove.[22]
>
> [35, p. 394].

### 9.1.2 Gödel's Path

Like Post's path to incompleteness, Gödel's path began with a decision problem: the problem posed by Hilbert toward the end of his 1928 Bologna address (and in the textbook [16], which is where Gödel probably encountered it). Hilbert described his question this way:

> One sets off [formulas of first-order logic] that cannot be refuted by an determinate specification of the variable predicates... Now the question arises whether all these formulas are provable from the rules of the predicate calculus, in other words whether the system of the usual logical rules is complete... In the domain of monadic predicate logic, a proof for the completeness of the rules can likewise be obtained from the method of the solution of the decision problem.
>
> [19, p. 233].

The question, to put it anachronistically, is whether the predicate calculus is complete for first-order validity. Hilbert had completeness results for propositional logic and for the monadic fragment of predicate logic, both going by way of decision procedures for these systems. It seems that he expected the completeness proof for the predicate calculus to follow the same pattern.

That is perhaps why Gödel beat him to it. In his 1929 dissertation, Gödel established that the first-order fragment of the logic of the Principia was complete, in the sense that each formula which could not be proved could be given a confuting assignment of variables. Gödel's result exhibits, on the one hand, his debt to Hilbert and, on the other, the ways in which he had already broken with the Hilbertian perspective.

The crucial fact about Gödel's argument, for our purposes, is this: it is entirely nonconstructive. Although his proof demonstrates the existence of a confuting assignment of variables, it gives no clue as to how one might find such an assignment effectively.[23] This is unavoidable, given the undecidability of first-order validity. But it makes certain aspects of Gödel's understanding of his own theorem, as presented in the unpublished introduction to his thesis, quite remarkable.

Two of Gödel's comments especially stand out. The first, regarding the formulation of the theorem, that, roughly, every consistent set of sentences has a model:

> The [a model exists for every irrefutable sentence] formulation [of the completeness theorem] seems also to be of some interest in itself, since the solution of this question represents in a certain sense a theoretical completion of the usual method for proving consistency (only, of course, for the special kind of axiom systems considered here), for it would give us a guarantee that in every case this method leads to its goal, that is, that one must either be able to produce a contradiction or prove the consistency by means of a model.
>
> [6, p. 61].

The second comment occurs in a discussion of the objection that his theorem, by making use of the excluded middle, involves some unacceptable circularity. In Gödel's view, this objection (which from a modern point of view seems like a non-sequitur) has some merit:

> For what is to be proved can, after all, be viewed as a kind of decidability (every expression of the restricted functional calculus either can be recognized as valid through finitely many inferences or its validity can be refuted by a counterexample). On the other hand, the principle of the excluded middle seems to express nothing other than the decidability of every problem.
>
> [6, p. 63].

In both of these quotations, we see Gödel implicitly conceiving of the infinitistic construction of a model for a satisfiable sentence as a kind of "non-constructive decision procedure."

This way of thinking about the completeness theorem directs our attention to a way in which Gödel's thinking about formal systems and about mathematical activity differs sharply from Post's Lewisian perspective. While for Post metamathematical reasoning is reasoning about what a mathematician can *do* with a formal system—what type of concrete transformations can be performed on symbols—for Gödel, metamathematical reasoning is often reasoning about mathematical thinking, with no assumption that this thinking consists of symbol manipulation.

While Post imagines a mathematician *working with* a formal system, Gödel imagines a mathematician whose thinking is (partly) *represented by* the workings of a formal system.[24] Above, Gödel imagines the mathematician finding a counterexample, or proving consistency by exhibiting a model—even if the counterexample or model can only be found by means of a search through an infinite tree. Gödel, of course, shouldn't be read as literally asserting that such a search is possible for flesh and blood mathematicians. The point is instead that Gödel has a different conception of what might be an intelligible task or problem—whether or not a concretely feasible one—than Post does.[25]

The key observation is this. For Gödel, the formal systems corresponding to mathematical practice[26] are representations of the mathematician's thinking, not just a tool with which the mathematician performs calculations. Given such an understanding, it makes sense to "jump levels"—to take any piece of mathematical reasoning and re-represent it in a formal system, or to take a formal system and ask about its intuitive meaning. This type of maneuver seems to have played a crucial role in Gödel's discovery of the incompleteness phenomenon.

In an article [42] (prepared in collaboration with Gödel), Wang describes Gödel's early work on incompleteness as inspired by the Hilbertian problem of proving the consistency of analysis. Gödel reports being puzzled by the idea that one should prove the consistency of analysis by finitistic means. Instead, he proposed to divide the question of consistency, proving the consistency of analysis by number theory and the consistency of number theory by finitistic number theory.

In the first part of the attempted argument—proving the consistency of analysis by number theoretic means—Gödel found himself making unavoidable appeals to the truth or falsity of sentences of number theory. These appeals, Gödel recognized, made it possible to reconstruct the Richard paradox and the liar paradox. Rather than abandoning this line of reasoning, however, Gödel realized that the barrier was that truth (for the language of number theory) cannot be defined in the language of number theory.[27] The idea that his informal number-theoretic reasoning about the consistency of analysis led to paradox, *because* he had assumed the expressibility of a concept not expressible in the language of number theory, requires the recognition that legitimate informal number theoretic reasoning can be recast as reasoning strictly in the formal language of number theory. And—to belabor the central point—this requires a conception of a formal system as a representation of, rather than as a tool for, mathematical reasoning.

From the undefinability of truth for the language of number theory in the language of number theory, plus the definability of provability in formal number theory, the incompleteness of number theory follows immediately.[28] Making this argument without using the notion of truth is considerably harder. And Gödel reports that he found the simpler argument first.[29]

So—to summarize the claims of the last two sections—Post and Gödel's paths to incompleteness were remarkably parallel: early work on decision problems gave way to impossibility results as they pushed the frontiers of investigation forward. But behind these parallels lay important differences in their conceptions of formal systems and metamathematics.

These differences are on display most dramatically in the opposed attitudes that Post and Gödel took toward the existence of absolutely undecidable propositions. Post ultimately took his experience with incompleteness and with the algorithmic undecidability of the decision problem for normal systems to be strong evidence for the existence of absolutely undecidable propositions. Gödel took the opposite view, arguing that the particular features of the incompleteness phenomena actually counted *against* the existence of absolutely undecidable propositions and against the mechanizability of mathematical reasoning. In Sections 9.2 and 9.3, we'll take a closer look at these responses.

## 9.2 Post's Response to Incompleteness: Absolutely Undecidable Propositions

We have briefly mentioned the heuristic role that Post's thesis—the claim that every effectively producible set of strings is produced by restricting the vocabulary of a normal system—played in Post's discovery of productive sets. In this section, we'll describe how that same thesis led Post to his belief in the existence of absolutely unsolvable problems and absolutely undecidable propositions.

We used the neutral term "effective procedure" in our earlier formulation of Post's thesis. Post's original formulation used what he called a "generated set": any set that can be "produced, created—in practice, written down"[35, n7] by a human agent employing some method. So by "effective procedures," Post means not just those procedures which might be carried out by a machine, but any procedure which can be carried out by a human agent.

Recall the set $P$: the set of strings $\bar{e}$ such that $\bar{e} \notin N_e$. The set $P$ is not a normal system. Hence, if $P$ could be produced by a human agent, $P$ would be a counterexample to Post's thesis. From a naive point of view, $P$ might appear to be easy to produce by the following method:

> for each $N_e$, check whether $\bar{e} \in N_e$ and if not, write it down

and this method did indeed first present itself to Post as a counterexample to his thesis. But it is only a counterexample if it can be carried out; and that requires the ability to decide, of each $N_e$, whether $\bar{e}$ is in it. That is, the procedure above is a counterexample only if the decision problem for normal systems is humanly solvable. Post was aware of the tension:

> Suppose that the [decision] problem were solved for the class of normal systems... This operation iterated would then actually generate the above defined set of a-sequences. That is, a solution of the [decision] problem for all normal systems would yield a counter-example disproving the correctness of our proposed generalization [i.e. Post's thesis].
>
> [35, p. 386].

So Post argues for this conditional: "if the decision problem for finite normal systems is solvable, then human beings can generate sets besides restrictions of normal systems"; or equivalently, this dilemma: "either the decision problem for finite normal systems is unsolvable, or human beings can generate sets besides restrictions of normal systems." He was unequivocal about which horn to grasp:

> an extended attempt to solve the simplified form of the [decision] problem "tag" led to ever increasing difficulties with all the complexities of number theory in the offing.[30] On the other hand nothing in the above argument directly weakens the reasoning that led us to [Post's thesis]. We therefore hold on to that generalization and conclude that the [decision] problem for the class of all normal system is unsolvable.
>
> [35].

However, Post regarded this claim as a hypothesis, as something yet to be demonstrated.[31] His proposed method of demonstration is surprising:

> for full generality [of proof] a complete analysis would have to be made of all the possible ways a human mind could set up finite processes for generating sequences. The beginnings of such an attempt will be found in our Appendix.
>
> [35, p. 387].

What such an analysis might possibly amount to isn't clear. But what is clear is that by asserting the unsolvability of the normal systems decision problem, Post meant to make a deep epistemological claim, very different from his combinatorial result that there's no normal decision procedure for normal systems.

### *9.2.1 From Unsolvability to Undecidability*

Post's claim—we might put it this way, as Post eventually came to—is that there is an absolutely unsolvable *problem*. By a *problem* Post means a problem in the classical sense: to perform some task, given some initial data. For example, to construct an equilateral triangle on a given line segment is a canonical example of a problem. The segment is the data and the task is to construct the triangle.[32] The solution of a problem requires the exhibition of a uniform method for performing the task on arbitrary data. So, an *unsolvable* problem is one for which no such uniform method can be exhibited.[33]

Post's idea, recall, was that the absolute unsolvability of a problem might be demonstrated by somehow running through all the humanly performable methods for solving problems—i.e. for "setting up finite processes"—since the solution of a problem is a finite process for transforming given data. However, at least as far as the unsolvability of the decision problem for normal systems was concerned, Post came to regard his approach as having been superseded by Turing and Church's work on Turing computability.

This comes out in writing in a number of ways. In a 1936 presentation of a formal model of computation,[34] Post contrasts "Gödel's theorem on the incompleteness of symbolic logics" with "Church's results concerning absolutely unsolvable problems" [32, p. 103].[35] By 1943, he states the unsolvability of the decision problem for normal systems as a "result" and glosses "unsolvable" in a footnote as "that is, absolutely unsolvable, to use a phrase due to Church"[33, p. 200].[36]

The most dramatic illustration of Post's turn away from absolute unsolvability, though, can be found in a diary entry from February, 1938, where he describes how others have beaten him to publication: "Gödel: has done completely what I planned to be a corollary of my finite process analysis... Church: Has an unsolvable (absolutely) problem. His sequence of definitions surpasses any I attempted... Turing: In large measure removed what was left in my point of view." [14, p. 82] Post concludes by taking stock of "what still remains untouched in my work." He lingers on two ideas from the appendix to the Anticipation, where he had attempted to work out how we might run through all the ways a human being could set up finite processes.

In a footnote to the Anticipation, Post describes the advantages of the approach of the appendix in the following way:

> In this connection we must emphasize the distinction between a formulation which includes an equivalent for every possible "finite process," and a description which will cover every possible method for setting up finite processes. It is was the latter goal that the Appendix of the present paper strained, the first, having been achieved in [Post's thesis]. While [Turing's argument for the Church–Turing thesis] may make the detailed development envisioned in the Appendix unnecessary for the analysis of

> process, though retaining an intrinsic interest as added description, it is doubtful if Turing considerations can replace such a development in the analysis of proof.
> [35, n. 6].

So, while Post believed that the Church–Turing thesis captured what an effective procedure *was* and that the analysis entailed the existence of an absolutely unsolvable problem, work remained to be done on the analysis of proof. And for that analysis, something like an analysis of how one might *set up* finite processes, rather than just of what a finite process was, was required.[37]

Post's objective in the analysis of proof was analogous to his earlier objective in defending Post's thesis. Describing current work, at the beginning of the Anticipation, he mentions: "We might add that since February 1938, we have given an occasional week to the continuation of this work, and largely in the spirit of the Appendix. Our goal, however, is now an analysis of proof, perhaps leading to an absolutely undecidable proposition, rather than an analysis of finite process." [35, n4]. This suggests two questions: (a) Why, if there was an absolutely unsolvable problem, was the existence of an absolutely undecidable proposition still open? (b) Why did Post expect that he would find such a proposition?

With regard to (a): the absolute unsolvability of, say, the normal systems decision problem *does* show that any theory $T$ for which we can find sentences $\phi_{n,e}$ such that $T \vdash \phi_{n,e}$ only if $\rho_n \in N_e$ and $T \vdash \neg\phi_{n,e}$ only if $\rho_n \notin N_e$, will be incomplete. But, given such a $T$, for each $n$ such that $n \notin N_e$, $T$ has an extension $T_{n,e}$ such that $T_n \vdash \neg\phi_{n,e}$. What's to prevent us from adopting or "setting up" each $T_{n,e}$ sequentially and thereby eventually proving—according to a Lewisian conception of proof—every proposition $\neg\phi_{n,e}$? Indeed, Post was in possession of a proof that, given a theory system $T$ like the one above, or an analogous indicator system, one can *effectively* find at least one of these theories $T_{n,e}$. What's to prevent you from finding them all?[38]

With regard to (b): Post's idea seems to have been to isolate some proper subset of what he called "finite processes" and what we now call r.e. sets as the ones that human beings could "set up." The fact that, given an indicator system, we can find a "better" such system surely seemed to him to be a clue that there might be a reasonable theory of how formal systems could be set up. But—this is the key point—Post's rejection of the idea that "provable" meant "provable in a fixed formal system" was not accompanied by any kind of anti-mechanism.

In spite of his rejection of Lewis' account of mathematical thinking (as mechanically working out the results of an antecedently accepted set of rules), Post firmly *accepted* the Church–Turing thesis, and accepted it on the reading that effective procedure means "procedure that can in principle be carried out *by a human being*." He seems only to have modified Lewis' picture of the mathematician's activity in this way: in addition to working out the results of antecedently accepted rules, the mathematician may also make use of effective means for the improvement of those rules—means which Post had proven to exist.[39] In light of the fact that the whole procedure of proving theorems was still given by a straightforward and presumably mathematically describable process, it seemed to Post very likely that the ability to prove would be just as limited in its reach as the ability to solve. As he puts it toward the end of the Anticipation:

> The unsolvability of the [decision] problem for all normal systems, and the essential incompleteness of all symbolic logics, are evidences of limitations in man's mathematical powers, creative though these be. They suggest that in the realms of proof, as in the realms of process, a problem may be posed whose difficulties we can never overcome; that is that we may be able to find a definite proposition which can never be proved or disproved.

### *9.2.2 Encounter, 1938*

In October 1938, Post and Gödel happened to become acquainted, at a regional meeting of the American Mathematical Society. Among other things, they discussed absolutely undecidable propositions. Post, in a journal entry two years later, describes the meeting in this way:

> In first meeting with Gödel. . . suggested to him abs. undec. prob. He said perhaps Cont. Hyp. I said this more like parallel axiom, i.e. would merely mean different theories of classes. Wanted rather abs. undec. arith. prop. where analyse all possible methods of proof & perhaps find a property of all such which would lead to abs. und. arith. prob. Gödel poo- poohed idea. Said (roughly) it was absurd

So, apparently, Post advanced roughly the procedure that he had previously tried to use to prove the existence of an absolutely unsolvable problem, now with the goal of demonstrating the existence of an absolutely undecidable proposition of arithmetic. Gödel, it seems, was not very receptive to this thought.

Post, on the day of the meeting, sent Gödel a letter, partly apologizing for being excessively enthusiastic at their first encounter. The day after, he sent a note emphasizing the undecidability of the normal systems decision problem by *any* formal system. Gödel's response was warm and collegial.

## 9.3 Gödel's Response to Incompleteness: Anti-Mechanism

Gödel gives some of his earliest statements about the mechanizability of mathematical reasoning in the unpublished lecture manuscript *Undecidable Diophantine Propositions* [13].[40] A reference in the manuscript to a paper by Kondo, published in December 1938, suggests that the manuscript was written shortly after Post and Gödel's first meeting. Unlike Post, Gödel regarded it as clear that the incompleteness phenomenon demonstrated that mathematical reasoning was not mechanizable. He is about as explicit as it is possible to be on this point: "Another way of putting the result is this: It is not possible to mechanize mathematical reasoning, i.e., it will never be possible to replace the mathematician by a machine, even if you confine yourself to number-theoretic problems."[13, p. 164].

In the lecture, Gödel credits Turing with having established beyond any doubt that the recursive functions are exactly the functions that can actually be computed. So Gödel accepted some version of the Church–Turing thesis. But, unlike Post, Gödel understood the thesis as stating that the recursive functions are exactly the *mechanically* computable functions,[41] *not* the functions computable by a humanly executable method.

Hence, Gödel's appraisal of an algorithmically unsolvable decision problem analogous to Post's, to find the truth value of diophantine propositions (more details on these in a moment): "It is true we have also found a problem which is absolutely unsolvable, namely, to find a mechanical procedure for deciding every [diophantine proposition]. But this is not a problem in the form of a question with an answer Yes or No, but rather something similar to squaring the circle with compass and ruler."[13, p. 175].

Gödel's suggestion is that the decision problem for diophantine propositions is only undecidable because the problem of finding a *mechanical* procedure restricts the types of possible solutions, just as the problem of squaring the circle *with* compass and straightedge restricts possible solutions.[42] As in his proof of the completeness theorem, in Gödel's view, effective solutions aren't the only intelligible solutions that can be offered to problems. Of course, this leaves open the question of *why* Gödel believed that it was possible to decide diophantine propositions in some non-mechanical way. Let's turn to that question.

To offer a little more detail: diophantine propositions are sentences of the form:

$$(\forall x_1, \ldots, x_n \in \mathbb{Z})(\exists y_1, \ldots y_m \in \mathbb{Z})p(x_1, \ldots, x_n, y_1, \ldots, y_m) = 0,$$

where $p$ is a diophantine polynomial.[43] It turns out that the question of whether a given normal system $N$ produces a certain string is equivalent to the question of whether a certain diophantine proposition $P_N$ is true. Hence the mechanical unsolvability of the decision problem for diophantine propositions: the decision problem for diophantine propositions is essentially the decision problem for normal systems, under another description.

But, it is worth noting one feature of how Gödel presents this problem, in contrast with how Post presents the finite normal systems problem. The difference between the two presentations is this: the diophantine propositions are meaningful, interpreted strings. They can be considered independently of any fixed formal system, or even language.[44] So it makes sense to consider proving them outside of the context of a formal system, by some means other than combinatorial manipulation. The fact that they can't be proved in a formal system, in Gödel's view, only tells us something about the imperfection of the formal system as a representation of mathematical reasoning.

Gödel's view is especially clear in his remarks at the beginning of the Diophantine Propositions lecture:

> If [we work in a formal system], then [the question of whether every proposition is decidable] becomes amenable to mathematical treatment and the answer turns out to be negative even in the domain of number theory. But it is clear that this negative answer may have two different meanings: (1) it may mean that the problem in its original formulation has a negative answer, or (2) it may mean that through the transition from evidence to formalism something was lost. It is easily seen that actually the second is the case.
>
> [13, p. 164].

In particular, Gödel claims, sentences like the consistency sentence for a formal system are just as evident—informally—as the axioms for the system. So, the fact that these aren't provable in a formal representation of mathematical reasoning just shows the shortcoming of that representation:

> the number-theoretic questions which are undecidable in a given formalism are always decidable by evident inferences not expressible in the given formalism. As to the evidence of these new inferences, they turn out to be exactly as evident as those of the given formalism.
>
> [13].

The reason for saying that the consistency sentence for a theory can be proven by inferences *exactly* as evident as those of the given theory is, presumably, that one has only as much reason to believe in one's axioms as one has reason to believe in their consistency—accepting axioms while rejecting the claim that they are consistent would involve a kind of irrationality.[45]

The nearest equivalent to Gödel's diophantine propositions in Post's set-up are the strings $\bar{x}a_3\bar{y}$, which an indicator system contains only if $\rho_x \notin N_y$. But these strings are system dependent; there's no sense in which one could be translated into an informal language, or into the language of arithmetic.[46] As a result, it's difficult to think of how you might "decide," say $\bar{75}a_3\bar{98}$ unless this means deducing it in an indicator system $N_w$ or some suitable improvement on $N_w$. The idea that something is "lost in formalization" isn't available from this point of view.

Since for Post, the mathematician's activity consists of performing combinatorial manipulations on formal systems, it wouldn't have appeared to him that there was any loss in focusing on languages like the indicator system language when studying absolute undecidability, or that propositions might have system-independent properties relevant to the decision of their truth values. For Gödel, on the other hand, since formal systems relevant to absolute provability are *representations* of genuine mathematical thinking, it's crucial—for thinking about absolute undecidability—that the formal systems under consideration actually "formalize" meaningful portions of mathematics. As a result, the focus on system-independent undecidable propositions and the possibility of a decision procedure based on non-syntactic properties of these propositions is much more natural from his perspective.

## 9.4 Subgroundedness

So, let's once again take stock. Post, we've argued, came to the question of absolute undecidability by way of questions about absolute unsolvability. While rejecting the Lewisian idea that the theorem-proving activity of a mathematician consists only of working out proofs by following a set of mechanical rules specified in advance he nevertheless held that Post's thesis was correct and that the only procedures that could be implemented by human beings were those that could be represented by a formal system. For him, the question of absolute provability was the question of *which* among all the possible formal systems were ones that human beings could "set up" as reliable sources of theorems—a question that became pressing with the realization that, for each indicator system, a "better," more inclusive indicator system could be effectively found. So Post retained part of Lewis' perspective, that mathematical activity consists of working with a formal system as a tool—he just added a further element, that mathematical activity may also involve *setting up* a formal system.

Gödel, on the other hand, came to the question of absolute provability by way of questions about what is lost in representing mathematical activity by means of a formal system. For example, one has only as much reason to believe in axioms as to believe in their consistency. If one commits oneself to the former, one is thereby committed to the latter. But, since the axioms will not prove their own consistency, it seems that the axioms don't include everything that can be legitimately asserted in the course of a proof. The question of what might be absolutely provable was then, for Gödel, the question of determining which propositions one might have the same reasons for believing as one has for believing the axioms of one's theory.

So, in some sense we have two questions of absolute provability, coming out of two incompatible pictures of the role that formal systems play in mathematical thinking. Were Post and Gödel, in their discussions of absolute provability and solvability, simply talking past each other?

There is a charitable interpretation we can give to what they said about absolute provability on which they were not. Post did conceive of absolute provability as something like provability in any formal system that can be set up in such a way as to be a reliable instrument for producing theorems. And Gödel did regard it as sufficient for absolute provability that a sentence be provable on the basis of axioms with the same type of evidence as axioms we already regard as sufficient for proof. But both conditions for absolute provability can be seen as expressions of a common requirement on an adequate conception of proof.

Say that one method $S$ of producing sentences is *subgrounded* to another $T$, if every reason or ground to believe that what $S$ produces is correct depends on some equally compelling reason to believe that what $T$ produces is correct.[47] Say that $S$ is *proof-consituting* if a sentence produced by $S$ is thereby proved.[48] The requirement is then this:

> **Subgroundedness requirement:** If $M$ is proof constituting, and $M$ is subgrounded to $M'$, then $M'$ is proof constituting.

Post's rejection of the Lewisian conception of proof—proof as merely the result of a mathematician working with a set of rules specified in advance—can be seen as the result of an application of this requirement, together with the recognition that, for a given indicator system, one can effectively construct a better indicator system. The construction of the improved system demonstrates that if one has grounds for accepting a given indicator system $N_e$—conceived of as a set of rules one accepts for mathematical reasoning—one *also* has equally compelling grounds for accepting the improved system $N_{f(e)}$. Hence, by the subgroundedness principle, $N_{f(e)}$ is proof-constituting and the Lewisian conception of proof is missing something: the provable sentences are not just those produced by $N_e$, but also those produced by extensions like $N_{f(e)}$, which we can set up on the basis of $N_e$.

Similarly, Gödel's argument for the absolute provability of the Gödel sentences appears to turn on an appeal to the subgroundedness principle. The grounds for believing in the theorems of a piece of axiomatized mathematics are equally grounds for believing in some method of proof that establishes the consistency of that piece of mathematics—since the consistency will be provable, according to Gödel, by inferences "exactly as evident" as those licensed by the axioms.

I should say a little more in support of the subgroundedness requirement, as a condition of adequacy on a conception of proof. So, it is worth pointing out that one finds something like it in, for example, Myhill's argument for the absolute provability of the consistency of a given proof-constituting formal system:

> We have logically compelling grounds to believe Gödel's undecidable statement [Con($S$)]. Since this statement asserts merely the consistency of $S$, if we have no grounds for believing [Con($S$)] we have no grounds either for believing anything established by the use of $S$. The fact that a system stronger than $S$ is needed in order to give a formal proof of the consistency of $S$ should not mislead us into thinking that the consistency of $S$ is 'less certain' than the theorems established by the use of $S$. It is possible to prove [Con($S$)] by methods which we must admit to be correct if we admit that the methods available in $S$ are correct.
>
> [29, p. 466].

Though Myhill might accept the following minor emendation. It's not quite right that rejecting Con($S$) means rejecting everything established by the use of $S$. We might have other grounds for believing what's established by $S$, for example, if it were established by some other system than $S$.[49] The real force of this argument is when we judge that the theorems of $S$ are *proved* by $S$, then Con($S$) must be regarded as provable as well—since it is "exactly as evident" or, if you like, "certain," as the theorems of $S$.

A second example is suggested by a point of Kreisel's, in [26, p. 168]. We formally express induction in Peano arithmetic as a scheme, saying for each formula in the language $\{0, S, +, \times\}$ that if $\phi(0)$ and for every $n$, $\phi(n)$ implies $\phi(n+1)$, then $(\forall n)\phi(n)$. But we don't accept this scheme because of some special insight about the formulas in the language $\{0, S, +, \times\}$.[50] It's rather because we accept the scheme for any property whatsoever, whether definable in the language of arithmetic or not: if 0 has the property and the property is closed under $+1$, then every number has the property. So, any reason we have for accepting the scheme depends on the second-order axiom. At least without denying this claim of subgroundedness, there would be something unintelligible about regarding the theorems of PA proved while at the same time, upon expanding the language of PA with a new predicate, expressing a property not definable in terms of $+, \times$,[51] refusing to accept as proved the results of applying induction to these new axioms.

So, as a result of differing conceptions of the relationship between mathematical thinking and formal systems, Post and Gödel differed on the nature of the grounds for belief in sentences like the Gödel sentence $G_T$ of a theory $T$, or in the reliability of something like the extended indicator system $S'_w$ of a given indicator system $S_w$. But they shared the view that the grounds for acceptance of $T$, or for $S_w$, depend on, or are identical to, grounds for accepting $T + G_T$, or $S_{w'}$. As a result Post and Gödel were lead by the subgroundedness requirement to conceptions of absolute provability on which the products of $S_{w'}$, or of $T + G_T$, were proved, just as much as were the products of $S_w$ and $T$.

## 9.5 Studying Absolute Provability

In the previous section, I described how Post came to believe that there existed an absolutely undecidable proposition and why Gödel rejected arguments that there ought to exist such a proposition. While, at the time of writing the diophantine propositions paper, Gödel apparently considered it likely that there did exist an absolutely undecidable proposition (just not one of the form given by the incompleteness theorem), later in his career he would come to reject this possibility, in favor of what he called "rationalistic optimism," the view that every proposition of mathematics is either provable or refutable in an absolute sense.

Neither Post nor Gödel, however, was content to leave their attitudes about absolute undecidability as articles of faith; both—at least for a time—regarded the existence of absolutely undecidable propositions as a research problem in logic, one that could be made mathematically precise and either proven or refuted. To actually prove theorems about absolute provability, a precise definition of the notion is required.

In this section, I'll sketch Post and Gödel's respective approaches to finding such a definition. Neither attempt at definition succeeded.[52] I'll argue that this is, at least in part, the result of a basic conceptual difficulty—what I'll call *the diagonalization problem*—with the idea that one can describe a mathematically precise notion of absolute provability satisfying the subgroundedness requirement.

### *9.5.1 Post's Approach*

For Post, recall, formal systems are tools employed in the course of mathematical reasoning. The reason for rejecting the identification of provability with provability in a fixed formal system is the fact that, given a formal system which one takes to be an accurate indicator of, say, derivability in normal systems, it's possible to effectively construct a more inclusive indicator system.

The problem of finding an absolute notion of provability then comes down to determining which formal systems can be set up on the basis of the formal systems that we actually use in mathematical reasoning. The hope was, as we've seen, to produce an absolutely undecidable proposition by exhaustively enumerating all of the possibilities for setting up formal systems.

Post seems to have hoped to do this by means of a kind of phenomenological investigation of mathematical thinking. His writings on this part of his program are fragmentary and obscure. Unpacking them would be work for a different article than this. However, at least one technical idea is suggested by our preceding discussion and by Post's other remarks.

We know that for each indicator system $N_e$, an indicator system $N_{f(e)}$ can effectively be found extending $N_e$, which gives more information about the behavior of normal systems. Post held that the second system was just as proof-constituting as the first. So, a minimal condition for a notion of proof which includes every proof-constituting formal system which can be "set up," is that it include the results of working with $N_{f^n(e)}$, the result of arbitrarily many iterations of the process of setting up $N_{f(e)}$ on the basis of $N_e$.

Is this enough? It seems to depend on whether this is the limit of our ability to set up formal systems. As it turns out, arbitrarily large finite numbers of iterations isn't the limit—we can continue into the transfinite. A competent recursion theorist can, without much difficulty, describe a normal system $N_a$ (or equivalently, a recursively enumerable set) that enumerates precisely the contents of $\bigcup_{n\in\omega} N_{f^n(e)}$. We then have $N_{f(a)}$ and so on. So the process goes onward. The question is (a) how far does it go? and (b) *which* systems $N_a$ at limit stages are subgrounded to the original system $N_e$, in such a way that their outputs are proof-constituting?[53]

The general outline here—given a presentation $N_e$ of a formal system in terms of the rules for generating it, finding an improved formal system $N_{f(e)}$ in an effective way, and iterating this process—leads to an *ordinal logic*, of the kind studied by Turing and later by Feferman and Kreisel. Post, in particular, seems to have considered the type of ordinal logic where one forms the improved formal system $T_{f(e)}$ by adding to a theory $T_e$ containing arithmetic, the Gödel sentence for $T_e$.[54] A great deal is known about the results of such iterations as they are continued into the transfinite; we'll return to this topic in the next section.

### *9.5.2 Gödel's Approach*

For Gödel, formal systems were the result of an attempt to represent or set down the rules of some prior process of mathematical reasoning. The reason for rejecting the identification of absolute provability with provability in a fixed formal system, is that the grounds for accepting the set of axioms that we incorporate into a formal system will equally be grounds for accepting, for example, a sentence expressing the consistency of those axioms. So, if the former is proof-constituting, so is the latter. Since the consistency sentence isn't proved by the axioms, the axioms only partly represent the full range of provable statements available.

The problem of defining absolute provability then comes down to attempting to define, not necessarily a formal system, but a set of sentences which are provable on the basis of axioms which are evident on the same grounds as axioms already known to be proof-constituting.

Gödel's considered opinion about the proper definition of absolute provability seems to have evolved over his lifetime.[55] I'll take up just one stage in that evolution here, the approach tentatively proposed in the 1946 Princeton bicentenial lecture on problems in mathematics. This view seems to have roots going back to Gödel's early work on incompleteness,[56] and in any case, strikes me as particularly illustrative.[57]

The basic idea is straightforward enough. The development of axiomatizations for Cantorian set theory suggests that the axioms of ZFC aren't especially privileged. They're true and we have grounds for believing in them on the basis of the standard "cumulative hierarchy" description of the universe of sets. But, the cumulative hierarchy description—in Gödel's view—also gives us evidence for axioms going beyond just the axioms of ZFC. In particular, it gives us evidence for axioms of infinity: sentences which can't be proved in ZFC, expressing the existence of extremely large infinite sets. The canonical expression of this attitude about axioms of infinity can be found in Gödel's paper on the continuum problem:

> the axioms of set theory by no means form a system closed in itself, but, quite on the contrary, the very concept of set on which they are based suggests their extension by new axioms which assert the existence of still further iterations of the operation "set of". These axioms can be formulated also as propositions asserting the existence of very great cardinal numbers (i.e., of sets having these cardinal numbers). The simplest of these strong "axioms of infinity" asserts the existence of inaccessible numbers (in the weaker or stronger sense) ... These axioms show clearly, not only that the axiomatic system of set theory as used today is incomplete, but also that it can be supplemented without arbitrariness by new axioms *which only unfold the content of the concept of set* explained above.
>
> [9, p. 261, my emphasis].

But variations on the same theme can be found throughout Gödel's corpus. This aspect of Gödel's thought has received a great deal of attention, so I won't linger too much on it here. Instead, let's move directly to Gödel's proposal, on this basis, for a notion of absolute provability:

> Let us consider, e.g., the concept of demonstrability. It is well known that, in whichever way you make it precise by means of a formalism, the contemplation of this very formalism gives rise to new axioms which are exactly as evident and justified as those with which you started, and that this process of extension can be iterated into the transfinite... In set theory, e.g., the successive extensions can most conveniently be represented by stronger and stronger axioms of infinity. It is certainly impossible to give a combinational and decidable characterization of what an axiom of infinity is; but there might exist, e.g., a characterization of the following sort: An axiom of infinity is a proposition which has a certain (decidable) formal structure and which in addition is true. Such a concept of demonstrability might have the required closure property, i.e., the following could be true: Any proof for a set-theoretic theorem in the next higher system above set theory (i.e., any proof involving the concept of truth which I just used) is replaceable by a proof from such an axiom of infinity. It is not impossible that for such a concept of demonstrability some completeness theorem would hold which would say that every proposition expressible in set theory is decidable from the present axioms plus some true assertion about the largeness of the universe of all sets.
>
> [8, pp. 152–3].

So Gödel's proposal, in 1946, is that a sentence is absolutely provable if it is provable from true large cardinal axioms. From the suggestion of the last sentence, it is clear that this isn't just a heuristic or a slogan. It's a proposal for a precise definition of absolute provability, admitting formalization and metamathematical investigation. Moreover, the proposal is precisely what we would expect, on the basis of the subgroundedness requirement, if the grounds for believing in the standard axioms for set theory is a grasp of the concept of set (or to put it a different way, a clear picture of the cumulative hierarchy), and if axioms of infinity are suggested by that concept in the same way that the accepted axioms for set theory are suggested by that concept.

## 9.6 Diagonalization Problem

At this point, I've argued that, as a result of different conceptions of the relationship between mathematical thinking and formal systems, Post and Gödel's accounts of absolute provability lead us to interestingly different notions of absolute provability. The sense in which they're both accounts of absolute provability—both attempts at describing the same thing—is not that they, for example, abstract from some shared conception of the actual activity of a mathematical thinker or from empirically ascertainable facts about how mathematicians actually proceed. Rather, what unifies Post and Gödel's accounts as accounts of absolute provability is the adherence to something like the subgroundedness requirement.

In this section, I'll argue that the ambitions that Post and Gödel had for metatheoretic study of the notion of absolute provability are in serious tension with their adherence to the subgroundedness requirement.

Here's the central difficulty.

> THE DIAGONALIZATION PROBLEM: The following properties cannot all be satisfied by a consistent set of sentences $P$, a predicate $\mathcal{P}$, a language $\mathcal{L}$ containing $\mathcal{P}$, the language of arithmetic, and a quotation operation $\ulcorner \bullet \urcorner$ mapping the formulas of $\mathcal{L}$ into the terms of $\mathcal{L}$.
>
> (1) $P$ is closed under first-order implication[58]
> (2) For some sentence $G_{\mathcal{P}}$ of $\mathcal{L}$, the biconditional $G_{\mathcal{P}} \leftrightarrow \neg\mathcal{P}(\ulcorner G_{\mathcal{P}} \urcorner)$ is in $P$
> (3) for each $\phi$, $\psi$ in $\mathcal{L}$,
>   (a) $\mathcal{P}(\ulcorner \phi \urcorner) \rightarrow \mathcal{P}(\ulcorner \mathcal{P}(\ulcorner \phi \urcorner) \urcorner) \in P$
>   (b) If $\phi \in P$, then $\mathcal{P}(\ulcorner \phi \urcorner) \in P$
>   (c) $\mathcal{P}(\ulcorner \phi \rightarrow \psi \ulcorner) \wedge \mathcal{P}(\ulcorner \phi \urcorner) \rightarrow \mathcal{P}(\ulcorner \psi \urcorner) \in P$
> (4) $\neg\mathcal{P}(\ulcorner \bot \urcorner) \in P$

If $P$ is taken to be the set of absolutely provable sentences, then (1) seems unavoidable. (2) is an immediate consequence of (1) plus the absolute provability of a small amount of arithmetic—enough to prove the diagonal lemma, so that for each formula $\phi$ of one free variable in $\mathcal{L}$, a sentence $\rho$ can be found such that the biconditional $\phi(\ulcorner \rho \urcorner) \leftrightarrow \rho$ is provable. If $\mathcal{P}$ is a definition of absolute unprovability, then (3c) just expresses that (1) is absolutely provable. (3b) expresses that what's absolutely provable is provably so, and (3a) expresses that what (2b) expresses is absolutely provable. So for a set of absolutely provable sentences $P$ and a predicate intended to represent $P$, about which certain minimal facts are absolutely provable, the fact that $P$ is consistent will not be absolutely provable. The argument is more or less a simplified version of one part of the standard proof of Gödel's second theorem, so I save it for a footnote.[59] The point, though, is that one aspect of the incompleteness phenomenon—the unprovability of consistency—has very little to do with the fact that the theorems of first-order foundational theories are effectively enumerable. Rather, it follows from the fact that the notion of provability under consideration is represented by a predicate and that predicate meets constraints (1)–(3).

As a result, escaping from the incompleteness phenomenon is not as simple as adopting a notion of provability which is not effectively enumerable. This is a problem for Post and Gödel, because of the close relationship between incompleteness—particularly, the unprovability of consistency—and the subgroundedness requirement. But this puts the point pretty abstractly. For concreteness, let's consider how Post and Gödel's proposed definitions actually get into trouble.

Recall Post's idea that one might simply run through all the ways of setting up formal systems and arrive at a satisfactory definition of absolute provability by, roughly, considering the union of these. One kind of "setting up" of a new system that Post apparently considered (he mentions this fact in [34]) was "Gödelization," adding the Gödel sentence of a given theory $T$ to $T$ and in that way arriving at a more inclusive theory $T'$.

If this works, why not just add "all" the Gödel sentences? The problem is this: even if you adjoin every Gödel sentence of every consistent extension of your theory to your theory (say, $\mathsf{PA}$), the result is simply a certain theory—call it $\mathsf{PA}_{\Pi_1}$.[60] This theory is no longer effectively presented, but it is still definable in the language of $\mathsf{PA}$. As a result, we can raise questions about its consistency. When a sentence is provable from $\Pi_1$ truths, the fact that it is so provable is also provable from $\Pi_1$ truths.[61] The other conditions of the diagonalization problem are straightforwardly met. So $\mathrm{Con}_{\Pi_1}$ won't be provable from $\mathsf{PA}_{\Pi_1}$. If $\mathrm{Con}_{\Pi_1}$ can nonetheless be recognized from the fact that each Gödel sentence is true, then $\mathsf{PA}_{\Pi_1}$ doesn't satisfy the subgroundedness requirement. If it can't be recognized, then it is hard to understand how we can be confident that this definition—provability from Gödel sentences of consistent theories—expresses absolute provability.

This is a contrived example, but incompleteness phenomena related to the diagonalization problem do affect "progression of theories" notions analogous to Post's as they occur in the wild. For example, soon after Turing presented his early results on ordinal logics, Kleene published a straightforward diagonal argument that Turing's ordinal logics were incomplete [22, p. 68].[62] More sophisticated progressions of theories can be shown to have the property that, very roughly, every true sentence of arithmetic is provable somewhere in the progression. But this is possible only because it is *not* possible to define the relevant progression in the language of arithmetic.[63] Once one actually has a definition of the progression, one automatically has an incompleteness result. Since the definition of the set of sentences proved in the progression is $\Pi^1_1$, it's possible, in a trivial way, to produce a $\Sigma^1_1$ sentence—approximately "I am not proved in this progression"—which cannot be proved in the progression, if the progression proves only true sentences.[64]

What about Gödel idea, defining absolute provability as provability from a true large cardinal axiom? The details are different, but again, the diagonalization problem presents a serious difficulty. The trouble is with explaining what is meant by "large cardinal axiom."

To take an example here, consider Koellner's suggestion [25, p. 180] that a large cardinal axiom is a statement $(\exists x)\phi(x)$, where $\phi(x)$ is $\Sigma_2$ and certain other facts obtain.[65] The fact that there's a bound here on the syntactic complexity of the axiom (it's $\Sigma_2$) means that the set of sentences provable from true axioms of this form is actually definable in the language of ZFC (since $\Sigma_2$-truth for the language of ZFC is definable in ZFC). The resultant notion of large cardinal provability then, once again, straightforwardly satisfies all but the last condition of the diagonalization problem and therefore does not establish its own consistency.

Hence, if we have grounds to believe in the correctness of this notion of large cardinal provability, then we have grounds to believe in the correctness of some extension of it as well: the result of adjoining the relevant consistency sentence to the notion. One might hope to avoid this type of problem by giving a definition of large cardinal axiom which allows sentences of unbounded syntactic complexity. But then, if we incorporate this definition into a definition of large cardinal provability (provability from *true* large cardinal axioms), it appears that we will need to extend our language with a notion of truth for set theoretic sentences; and—while it is difficult to make a decisive case against a hypothetical proposal like this—it seems that we will face the diagonalization problem at this higher level.[66] We might try, again, to evade the difficulty by devising a "non-compositional" definition of "true large cardinal axiom" in which the notion of truth doesn't figure.[67] But then, if we do have a definition in the language of set theory, it's hard to see how we'll avoid re-encountering the diagonalization problem.

## 9.7 Conclusions

I take the lesson of the diagonalization problem to be this. In attempting to study absolute provability, we have a problem not just of clarity, but of coherence. Saying, in a precise way what it is for a sentence to be absolutely provable—as Gödel and Post attempted to do—is difficult not just because the concept involved is complicated and requires a lot of analysis, or because evidence for the correctness of an analysis is hard to come by. Nor is the sole source of difficulty that we don't presently have much knowledge about how mathematical thinking actually works, or about how the future development of the practice of set theory will play out.

It is also difficult because the subgroundedness requirement is in tension with the diagonalization problem. But the subgroundendess requirement is one of the central motivations for rejecting a definition of absolute provability as just provability in some fixed formal theory, and the study of the notion of absolute provability in the hands of Post and Gödel suggests that the subgroundedness criterion is a condition of adequacy on a definition of absolute provability.

What does this tell us about the possibility of a systematic study of absolute provability? It seems like we're left with a few options. One way of sorting this out is to look at a set of incompatible propositions that are getting us into trouble.

Such a list would look something like this:

A. There exists a formally precise definition of absolute provability.
B. The grounds for accepting the sentences produced by a method $P$ depend on grounds for accepting the outputs of some method $P'$ which produces the consistency sentence of $P$.
C. The subgroundedness requirement.
D. Absolute provability satisfies the condition (3) of the diagonalization problem.

This gives us a division of our options for thinking about absolute provability. We might deny (B) or (C), consigning all that troublesome epistemology to the flames. Call this option positivism. But without those two conditions, it's not clear why we should be skeptical of the suggestion that absolute provability isn't just provability in some fixed theory $T$. So the positivist needs to either find a new explanation for the history of skepticism, or take the skepticism to be unjustified (which would be a non-trivial mutilation of practice).

We could reject (D). Call this option *deus ex machinism*. The difficulty is that most reasonable notions of provability *do* satisfy the conditions in (3). Arguably, any good notion of provability ought to, since these conditions just express properties that proof intuitively ought to have. The machinist needs then to find some reasonable notion of absolute provability which does not satisfy (1)–(3) and which does satisfy the subgroundedness requirement. This is easier said than done (hence the name for this view).[68] But one can't rule it out at the outset.

The last option is to resist (A). One might hold that there exists a meaningful notion of absolute provability, but that we are—in principle—incapable of finding a definition of it. Call this view *quietism*. There are some conceptual difficulties here too.[69] But there isn't obviously the same type of mutilation of practice as positivism involves, and the view amounts to more than just the kind of promissory note that mechanism gives us. So, perhaps this option is worth exploring.[70]

## Notes

1. And this would be so even if we happened to know that $\phi$ was true; whether there exists an open problem to which we know the answer (though we haven't found a proof) is a question I won't take up here. But one can certainly have a lot of rationally compelling evidence for a conjecture (say, $P \neq NP$) and believe it, without having proven it. In this case—if the conjecture is correct—it is at least hard to see why we shouldn't be said to know the answer.
2. If the twin primes conjecture was proved in ZFC + a proper class of inaccessible cardinals, it would still be proved.
3. We'll take this point up in more detail in Section 9.2.
4. Roughly, that if the grounds for a method of proof equally support some extension of those methods, then the the extension of the methods is also proof constituting.
5. Here and below, by "formal system" I'll mean any effective procedure for generating a set of sentences. First-order deduction, according to a finite set of effective inference rules, from a recursively enumerable set of axioms, is one example of a formal system.

6. This is Hilbert's famous axiom of solvability—Hilbert's canonical statement can be found in [18]. An explicit discussion of the axiom by Gödel can be found in [11].
7. Both ideas are far ahead of their time. For a description of some of the potential pitfalls on the way to a completely syntactic notion of proof and a detailed comparison of Post's thesis with concurrent work by Hilbert and Bernays, see [44].
8. At least, in the ideal case. Lewis recognizes that the actual practice of mathematics is messier than this. But the *possibility* of proceeding in this way is for Lewis, a norm for rigorous mathematics: "The important consideration is the fact that the operations of any abstract and really rigorous mathematical system are capable of formulation without any reference to truth or meanings" [27, pp. 359–60].
9. Though Post didn't officially give this result as a corollary, perhaps because of the difficulty, at the time, of giving a suitably precise definition of a decision procedure.
10. The result is described in [31].
11. Note, though, that the concept of a decision procedure was, for Post in 1921, not precisely defined. He knew one when he saw it (the method of truth tables being an obvious example). But he had limited means for proving any type of generalization about decision procedures, including, crucially, the non-existence of a decision procedure for a given problem.
12. Though textbook presentations of 1-reducibility generally use numbers in place of strings and computable functions $\mathbb{N} \to \mathbb{N}$, rather than Post's "effective" procedures.
13. This formulation is rather abstract, but normal systems are, in practice, extremely simple—the sort of thing you could implement with a pile of colored blocks and an unusually patient eight year old. An example would be the result, say, of closing $\{s\}$, for $s$ a string of ones and zeros under the rule "remove the first three symbols. If the first is 1, add 1101 to the end. If the first is 0, add 00 to the end"; for $s = 100$, the result is $\{10, 1011, 11011, 111011, 0111011, 10110, 101101, 1011101, 11011101, 111011011, \ldots\}$.
14. We can set up such a correspondence effectively—in such a way that $S_e$ can be recovered from its index and vice versa.
15. Post's precise statement is:

   > Every generated set of sequences" on a given set of letters $a_1, a_2, \ldots, a_\mu$ is a restriction of the set of assertions of a system in normal form with primitive letters $a_1, a_2, \ldots, a_\mu, a'_1, a'_2, \ldots, a'_\mu$, i.e. the subset consisting of those assertions of the normal system involving only the letters $a_1, a_2, ..., a_\mu$.
   >
   > [35, p. 385].

   As it turns out, every recursively enumerable set is of this form. So this is roughly equivalent to the Chruch–Turing thesis.
16. Post states an equivalent result as a theorem in the Anticipation.
17. Contemporary recursion theorists use $K$ to denote an analogous set, with Turing-computable enumerations replacing normal systems.
18. Post left the details of these constructions to the reader. Since normal systems are all effectively enumerable, it is clear (by Post's thesis or the Church–Turing thesis) that the constructions can be done.
19. The term "creative" is due to Post. The term "productive" is apparently due to Dekker [37, p. 84], but is suggested by remarks of Post [35, p. 397].
20. One has similar results as a corollary of the first incompleteness theorem. Indeed, one way of understanding part of the incompleteness phenomenon is simply as the fact that the set of true sentences is productive; given any recursively enumerable set of axioms, inducing a recursively

enumerable set of theorems $T$, one can effectively find a true sentence $G_T$ which is not one of the theorems $T$.

21. Or, indirectly, to any problem that can be solved on that basis—e.g. first-order validity, or derivability in ZFC.
22. Post's enthusiasm for this apparently anti-mechanistic conclusion soon faded. In the appendix to the Anticipation, which consists of journal entries from around 1921, he writes:

    > The conclusion that man is not a machine is invalid. All we can say is that man cannot construct a machine which can do all the thinking he can. To illustrate this point we may note that a kind of machine-man could be constructed who would prove a similar theorem for his mental acts.
    >
    > [35, p. 399].

23. This is because, at a key juncture, Gödel makes use of a version of Kőnig's lemma (not by that name—there's reason to believe that Gödel was not aware of Kőnig's presentation [40, p. 451]). He infers, from the existence of a certain infinite, finitely branching tree, the existence of an infinite branch. But given such a tree, one might search forever without being able to find a path through.
24. Gödel's primary influence in thinking about formalization seems to have been Hilbert's program, in which formal systems are not tools for doing mathematics, but representations of mathematical activity, constructed with specific metatheoretic investigations in mind (see for example [17, pp. 380–1].
25. Gödel eventually came to be well aware of this aspect of his own thinking and of its methodological advantages over a syntactical conception of the subject matter of metamathematics. See for example the discussion of the completeness theorem in [43, p. 78] and of the discovery of the constructible hierarchy at [43, pp. 250–1].
26. That is, foundational theories like ZFC and PA.
27. For reasons presumably related to Tarski's theorem. Gödel would later use Tarski's theorem to explain the first incompleteness theorem [7, pp. 362–3].
28. Since if provability in number theory were complete and sound, its formal definition would be a definition of truth.
29. cf. Wang [42, p. 654].
30. The problem of "tag" is a more or less special case of the decision problem for normal systems. See [28] for a detailed study and a proof that even this special case is algorithmically undecidable.
31. Indeed, his hesitancy on this point (and the chilly reception of some other work) is apparently the reason he delayed publication of his results. See [39, §4.2].
32. That if triangles have equal sides enclosing equal angles, then they also have equal bases, is a canonical example, by way of contrast, of a theorem. The distinction is marked in, for example, "the Elements," by the different concluding sentences of the solutions of problems: "which is what was required to do" and the proofs of theorems: "which is what was required to show." For discussion, see [15, p. 126]. This is also how Church—Post's source for the phrase "absolutely unsolvable"—seems to have thought of problems in the opening passages of [1].
33. This analogy with problems in the classical sense is underlined by Post, in the opening footnote to the Anticipation:

    > The writer cannot overemphasize the fundamental importance to matics of the existence of absolutely unsolvable combinatory problems. True, with a specific criterion of solvability under consideration, say recursiveness ... the unsolvability in question,

as in the case of the famous problems of antiquity, becomes merely unsolvability by a given set of instruments. Indeed, the corresponding proofs for combinatory problems are almost trivial in comparison with the classic unsolvability proofs. The fundamental new thing is that for the combinatory problems the given set of instruments is in effect the only humanly possible set.

[35, p. 340].

34. The model was similar to Turing's, but discovered independently of [38].
35. Though in the same paper, he describes the Church–Turing thesis as a hypothesis, still in need of further confirmation, he qualifies his qualification in a footnote, adding: "Actually the work already done by Church and others carries this identification considerably beyond the working hypothesis stage. But to mask this identification under a definition hides the fact that a fundamental discovery in the limitations of the mathematicizing power of Homo Sapiens has been made and blinds us to the need of its continual verification."[32, p. 105].
36. He also takes the Church–Turing thesis to entail the existence of absolutely unsolvable problems in [34].
37. This type of distinction—between the demonstrable existence of an absolutely unsolvable problem and the open problem of finding an absolutely undecidable proposition—also occurs in Kleene's writings. See [24, p. 341].
38. To make a more technical point (and to fall back on contemporary terminology), let $f$ be a computable function mapping indices of indicator systems $N_e$ to indices of "improved" indicator systems $N_{f(e)} \supseteq N_e$, such that there's some $\bar{x}a_3\bar{y} \in N_{f(e)} - N_e$. At least in 1921, when Post first began to consider this sort of question, it would have been epistemically possible for Post that the union of an $\omega$-sequence of normal systems $N_e, N_{f(e)}, N_{f(f(e))} \ldots$ was not itself generated by any normal system; so it remained open that the series settled every proposition about normal systems membership, even if the individual systems did not.
39. Post seems to have used precisely this picture as an objection in discussion after a talk by Curry. In a journal, Post recalls the objection as follows: "my criticism in meeting of Curry's talk i.e. Curry argued many formal systems as conseq. of Gödel's theorem,' I suggested conseq. more drastic, i.e. setting up formal syst. as much math. as working in it & more important of two; that all math. activity one. [14, p. 83].
40. The main technical result of the lecture is that the truth of diophantine propositions, in the sense defined below, is recursively undecidable.
41. By which Gödel meant the functions that could literally be computed by a machine: Gödel describes Turing's achievement this way: "He has shown that the functions defined in this way are exactly those for which you can construct a machine With a finite number of parts which will do the following thing. If you write down any number $n_1, \ldots, n_r$ on a slip of paper and put the slip into the machine and turn the crank, then after a finite number of turns the machine will stop and the value of the function for the argument $n_1, \ldots, n_r$ will be printed on the paper."[13, p. 168]. References to human psychology are completely absent.
42. It is interesting to compare this claim to Post's claim in footnote 33.
43. That is, a polynomial with integer coefficients.
44. Independently of language in what sense? Well, Gödel does not need to specify, during the lecture, whether he is working in PA or in ZFC; but in those two settings, the diophantine propositions would be represented by radically difference sentences. Thus, the propositions aren't tied to any particular language.
45. Two qualifications here. First qualification: believing in one's axioms shouldn't mean just believing in *each* of one's axioms—it would be perfectly rational to do this while believing that the

axioms taken as a whole are likely inconsistent (it would be analogous to believing of each sentence in a book that it is true, while believing that the book must contain at least one falsehood). Rather, believing in one's axioms should mean whatever attitude it is that one takes toward them, sufficient to license a commitment to whatever they logically entail. The point is that, according to Gödel, *this* attitude also commits you to some propositions that your axioms *do not* logically entail, like the consistency sentence. Second qualification: I of course assume here that the background logic is classical—paraconsistent logics are beyond the scope of this discussion.

46. Of course, one could take them to mean what they indicate—perhaps that $\rho_x \notin N_y$. But indication is a one–many relation. They could equally informatively be used to indicate that a certain Turing machine halts, that a certain diophantine proposition is true, and so on... (indeed, if all that "$s$ indicates $\phi$" comes to is that the indicator system produces $s$ only if $\phi$, then each string of the indicator system incidentally also indicates either CH or ¬CH (whichever is true)).
47. I'll leave "reason", "depends", and "equally compelling" unanalyzed here, though there is more to be said about what types of reasons are relevant in this context.
48. Not every sentence produced by a formal system is proved—one can't prove the inconsistency of PA by deducing it from PA + ¬Con(PA).
49. For example, suppose $S$ was ZFC + ¬Con($T$), where $T$ is a theory which is, for all we know, provably consistent in ZFC. Then we have no reason to believe Con($S$). But, if $T$ is actually inconsistent, then the theorems of ZFC + ¬Con($T$) are just the theorems of ZFC, in which case we do have reason to believe them.
50. A view opposed to Kreisel's can be found in [20].
51. One such property would be "the Gödel code of a sentence being true in the language of PA."
52. At least in producing the definition originally intended. Gödel's program was extremely fruitful in motivating twentieth-century developments in set theory related to large cardinals. See [21, §7].
53. The second question is important because, although the exact strings enumerated by $N_a$ are determined by the strings enumerated by the preceding systems $N_{f(e)}$, the method of enumeration $a$ is not thereby determined—one has some latitude for choice here. And whether a method of producing sentences actually proves what it produces is not just a matter of which sentences are produced, as the example of footnote 49 shows.
54. Post describes studying a candidate absolute proof procedure "involving, besides the elementary, only mathematical induction and "Gödelization.[34], where Gödelization" presumably means something like the adjunction of a Gödel sentence. This interpretation of this passage, along with the connection to ordinal logics, also appears in [39, §4.2].
55. For a survey of Gödel's attitudes, see [41] or [25].
56. That is, the suggestions of [5, p. 181] and [12, p. 48] that incompleteness could be eliminated by admitting higher types into one's theory.
57. Though, there are some indications that Gödel would not have accepted the 1946 definition as a definition of absolute provability without qualification. Some years later, he told Wang that

> the bankruptcy view only applies to general concepts such as proof and concept. But it does not apply to certain approximations where we do have something to lean back on. In particular, the concept of set is an absolute concept [that is not bankrupt], and provable in set theory by axioms of infinity is a limited concept of proof [which is not bankrupt].
>
> [43, p. 270].

which suggests that he held that the 1946 concept (glance ahead for details) was just an approximation to absolute provability. And even in 1946, he qualifies a suggestion that absolute provability might be definable, by adding "or, at least, that, certain very closely related concepts may be definable in an absolute sense."[8, p. 151]. More explicitly, he adds at the end that

> You may have noticed that, in both examples I gave, the concepts arrived at or envisaged were not absolute in the strictest sense, but only with respect to a certain system of things, namely the sets... i.e., although there exist proofs and definitions not falling under these concepts, these definitions and proofs give, or are to give, nothing new within the domain of sets"
>
> [8, pp. 152–3].

So we may need to take some of the other claims of the 1946 address with a grain of salt.

58. In particular, it contains first-order validities and is closed under modus ponens.
59. For supposed conditions (1)–(3) being met. First note that if $G_{\mathcal{P}} \in P$, then by (3b), we have that $\mathcal{P}(G_{\mathcal{P}}) \in P$ and, by (2), that $\neg\mathcal{P}(\ulcorner G_{\mathcal{P}} \urcorner) \in P$, violating the consistency of $P$.

    Now, by (2), (3b), (c), we have $\mathcal{P}(G_{\mathcal{P}}) \rightarrow \mathcal{P}(\ulcorner \neg\mathcal{P}(\ulcorner G_{\mathcal{P}} \urcorner)\urcorner) \in P$. But by (3*a*), we have $\mathcal{P}(G_{\mathcal{P}}) \rightarrow \mathcal{P}(\ulcorner \mathcal{P}(\ulcorner G_{\mathcal{P}} \urcorner)\urcorner) \in P$. Hence, by some straightforward applications of (1), (3*b*), and (3*c*), we have $\mathcal{P}(G_{\mathcal{P}}) \rightarrow \mathcal{P}(\ulcorner \bot \urcorner) \in P$. Contra-posing (by (1)), $\neg\mathcal{P}(\ulcorner \bot \urcorner) \rightarrow \neg\mathcal{P}(G_{\mathcal{P}})$. So, $\neg\mathcal{P}(\ulcorner \bot \urcorner) \rightarrow G_{\mathcal{P}} \in P$. Since we have just argued that $G_{\mathcal{P}} \notin P$, we must conclude that $\neg\mathcal{P}(\ulcorner \bot \urcorner)$ isn't either.
60. PA plus the $\Pi_1$ fragment of true arithmetic, a set of sentences definable in the language of arithmetic. Since, for $T$ a consistent extension of PA, Con$(T)$, is equivalent to the $\Pi_1$ local reflection scheme for $T$ and every true $\Pi_1$ sentence is deducible in some consistent extension of PA, adding the Gödel sentences above yields $\mathsf{PA}_{\Pi_1}$.
61. A sentence will be provable in $\mathsf{PA}_{\Pi_1}$ if there exists some proof $p$ of $\phi \rightarrow \psi$ and $\phi$ is a $\Pi_1$-truth. Say that a proof $p$ of $\phi \rightarrow \psi$ where $\phi$ is a $\Pi_1$ truth is a $\Pi_1$-proof of $\psi$. Since $\Pi_1$ truth is $\Pi_1$ definable, the relation of being a $\Pi_1$ proof of $\psi$ will be $\Pi_1$-definable. So if there is such a proof, say $p$ $\mathsf{PA}_{\Pi_1}$ will prove that $p$ is one and infer that $\psi$ is a theorem of $\mathsf{PA}_{\Pi_1}$.
62. Though Kleene's argument actually contains an error—he believed at the time of writing that the system of ordinal notations that Turing used was $\Pi^0_2$-definable, so that the theorems of Turing's progressions would be $\Sigma^0_3$ and fail to prove some $\Pi^0_3$ sentences saying "I am not provable." The correction can be found in [23]. For the progression which just adds a Gödel sentence, however, the contrived example of the previous paragraph indicates an argument for $\Pi^0_2$-incompleteness. For a progression more like Turing's original $\Lambda_P$, which added a $\Pi^0_2$ local reflection principle, the same argument works—the progression turns out to be axiomatized by the set of all true $\Pi^0_1$ sentences [3, p. 130]. This turns out to refute Turing's original conjecture that such a progression would prove all true $\Pi^0_2$ sentences.
63. In particular, one needs to appeal to Kleene's set $\mathcal{O}$ of ordinal notations, which is $\Pi^1_1$-complete. See [2] for the original presentation of some of the main results on transfinite progressions, or [4] for a readable overview.
64. This observation is due to Feferman. As for the full set of properties (1)–(3) occurring in the diagonalization problem—the complication involved in verifying these for the various possible ordinal logics is formidable. so I put that discussion to the side for now. Given the $\Pi^1_1$-completeness of certain of Feferman's logics, it seems that these conditions will be met in many cases.
65. That ZFC proves (1) that everything satisfying $\phi$ is strongly inaccessible, and (2) that for all $\kappa$ and $\mathbb{P} \in V_\kappa$ if $\phi[\kappa]$ then $\mathbb{P} \Vdash \phi[\check{\kappa}]$.
66. Gödel may have anticipated this objection and restricted claims about the adequacy of his definition for that reason. See Footnote 56.

67. For example, if there are no large cardinals, then "true large cardinal axiom" is easy to define: a sentence such that $1 = 0$.
68. The theorems of many known strong logics—e.g. $\omega$-logic, $\beta$-logic, and Woodin's $\Omega$-logic—all satisfy the conditions (1)–(3). What about Rosser provability? This does avoid satisfying (3) of the diagonalization problem; I would suggest that it does not satisfy the subgroundedness requirement, but that would require more discussion than we have space for here.
69. One challenge, which seems rather serious to me, is this. Even if we can't find a coherent definition of the notion, why can't we just enrich the vocabulary of our theories with a primitive predicate for absolutely provable sentences? Well, then, if we are not going to be positivists, we ought to admit conditions (1)–(3) in the diagonalization problem as well, and soon we'll be in the same fix we were in originally. So it is not just that we can't find a definition—we can't even admit a predicate for absolute provability (though we might study it as a modal operator, as in [36]). So the notion, on this view, is ineffable in some difficult to understand way. Making sense of that ineffability is an open problem for the quietist.
70. In defense of quietism, it's worth pointing out that neither Post nor Gödel was ultimately satisfied with their proposed definitions, even as schematic characterizations.

    Post, in the 1954 abstract mentioned, argues that he has been tackling the proof of existence of absolutely undecidable propositions in the wrong order; prior to a definition of absolute provability, a definition of absolute definability is required. His work on that last notion fills two notebooks. In the course of the study, Post seems to have uncovered essentially all of the basic properties of the hereditarily ordinal definable sets. For details, see [30]. But—to my knowledge—he never constructed a new and more satisfactory definition of absolute provability on this basis.

    Gödel, in late conversations with Wang on the subject of absolute provability, had this to say:

    > What is evident about AP [absolute provability] leads to contradictions which are not much different from Russell's paradox. Intuitionism is inconsistent if one adds AP to it. AP may be an idea [in the Kantian sense]: but as soon as one can state and prove things in a systematic way, we no longer have an idea [but have then a concept].
    >
    > [43, p. 208].

    He added that "It is not satisfactory to concede [before further investigation] that AP ... is an idea."; but the possibility was at least seriously open, in his view, that there might simply be no way of formally defining absolute provability. Gödel also told Wang that "Absolute demonstrability and definability are not concepts but inexhaustible [Kantian] Ideas. We can never describe an Idea in words exhaustively or completely clearly. But we also perceive it, more and more clearly."[43, p. 208]. So—perhaps there was a shift of terminology, or perhaps Gödel went back and forth on whether it was possible to define provability.

# References

[1] Alonzo Church. An unsolvable problem of elementary number theory. *American Journal of Mathematics*, **58**(2):345–363, 1936.

[2] Solomon Feferman. Transfinite recursive progressions of axiomatic theories. *Journal of Symbolic Logic*, **27**(3):259–316, 1962.

[3] Solomon Feferman. Turing in the land of O(z). In *The Universal Turing Machine: a Half-Century Survey*, pp. 103–134. Springer, New York, 1995.
[4] Torkel Franzén. Transfinite progressions: a second look at completeness. *Bulletin of Symbolic Logic*, **10**(3):367–389, 2004.
[5] Kurt Gödel. On formally undecidable propositions of principia mathematica and related systems I. In Solomon Feferman, ed., *Kurt Gödel, Collected Works, Volume I: Publications 1929–1936*, pp. 144–195. Oxford University Press, New York, 1986.
[6] Kurt Gödel. On the completeness of the calculus of logic. In Solomon Feferman, ed., *Kurt Gödel, Collected Works, Volume I: Publications 1929–1936*, pp. 61–101. Oxford University Press, New York, 1986.
[7] Kurt Gödel. On undecidable propositions of formal mathematical systems. In Solomon Feferman, ed., *Kurt Gödel, Collected Works, Volume I: Publications 1929–1936*, pp. 346–371. Oxford University Press, New York, 1986.
[8] Kurt Gödel. Remarks before the Princeton bicentennial conference on problems in mathematics. In Solomon Feferman, ed., *Kurt Gödel, Collected Works, Volume II: Publications 1938–1974*, pp. 150–153. Oxford University Press, New York, 1990.
[9] Kurt Gödel. What is Cantor's continuum problem? In Solomon Feferman, ed., *Kurt Gödel, Collected Works, Volume II: Publications 1938–1974*, pp. 254–270. Oxford University Press, New York, 1990.
[10] Kurt Gödel. Some basic theorems on the foundations of mathematics and their implications. In Solomon Feferman, ed., *Kurt Gödel, Collected Works, Volume III: Unpublished Essays and Lectures*, pp. 290–323. Oxford University Press, New York, 1995.
[11] Kurt Gödel. The modern development of the foundations of mathematics in light of philosophy. In Solomon Feferman, ed., *Kurt Gödel, Collected Works, Volume III: Unpublished Essays and Lectures*, pp. 374–387. Oxford University Press, New York, 1995.
[12] Kurt Gödel. The present situation in the foundations of mathematics. In Solomon Feferman, ed., *Kurt Gödel, Collected Works, Volume III: Unpublished Essays and Lectures*, pp. 45–53. Oxford University Press, New York, 1995.
[13] Kurt Gödel. Undecidable diophantine propositions. In Solomon Feferman, ed., *Kurt Gödel, Collected Works, Volume III: Unpublished Essays and Lectures*, pp. 165–175. Oxford University Press, New York, 1995.
[14] Ivor Grattan-Guinness. The manuscripts of Emil L. Post. *History and Philosophy of Logic*, **11**(1):77–83, 1990.
[15] Thomas Little Heath, Johan Ludvig Heiberg, and Euclid. *The Thirteen Books of Euclid's Elements*, volume 1. Cambridge University Press, London, 1908.
[16] D Hilbert and W Ackermann. *Grundzage der theoretischen Logik*. Springer, Berlin, 1928.
[17] David Hilbert. On the infinite. In Jean Van Heijenoort, ed., *From Frege to Gödel: A Source Book in Mathematical Logic, 1879–1931*, pp. 367–392. Harvard University Press, Cambridge, 1967.
[18] David Hilbert. Mathematical problems. In William Ewald, ed., *From Kant to Hilbert: A Source Book in the Foundations of Mathematics*, pp. 1096–1105. Oxford University Press, New York, 1996.
[19] David Hilbert. Problems of the grounding of mathematics. In Paolo Mancosu, ed., *From Brouwer to Hilbert: The Debate on the Foundations of Mathematics in the 1920s*, pp. 227–233. Oxford University Press, New York, 1998.
[20] Daniel Isaacson. Some considerations on arithmetical truth and the $\omega$-rule. In Michael Detlefsen, ed., *Proof, Logic and Formalization*, pp. 39–59. Routledge, New York, 1992.
[21] Akihiro Kanamori. Gödel and set theory. *Bulletin of Symbolic Logic*, **13**(2):153–188, 2007.
[22] Stephen Cole Kleene. Recursive predicates and quantifiers. *Transactions of the American Mathematical Society*, **53**(1):41–73, 1943.

[23] Stephen Cole Kleene. On the forms of the predicates in the theory of constructive ordinals (second paper). *American Journal of Mathematics*, **77**(3):405–428, 1955.
[24] Stephen Cole Kleene. Introductory note to 1934. In Solomon Feferman, ed., *Kurt Gödel, Collected Works, Volume I: Publications 1929–1936*, pp. 338–345. Oxford University Press, New York, 1986.
[25] Peter Koellner. On the question of absolute undecidability. *Philosophia Mathematica*, **14**(2):1–36, 2006.
[26] Georg Kreisel. Informal rigour and completeness proofs. In Imre Lakatos, ed., *Problems in the Philosophy of Mathematics*, pp. 138–171. North-Holland Amsterdam, 1967.
[27] Clarence Irving Lewis. *A Survey of Symbolic Logic*. University of California Press, Berkley, 1918.
[28] Marvin L Minsky. Recursive unsolvability of Post's problem of "tag" and other topics in theory of Turing machines. *Annals of Mathematics*, **74**(3):437–455, 1961.
[29] John R Myhill. Some remarks on the notion of proof. *Journal of Philosophy*, **57**(14):461–471, 1960.
[30] John R Myhill and Dana S Scott. Ordinal definability. In *Axiomatic Set Theory*, pp. 1–8. The American Mathematical Society, Providence, 2011.
[31] Emil Leon Post. On a simple class of deductive systems. *Bulletin of the American Mathematical Society*, **27**:396–397, 1921.
[32] Emil Leon Post. Finite combinatory processes-formulation 1. *Journal of Symbolic Logic*, **1**(3):103–105, 1936.
[33] Emil Leon Post. Formal reductions of the general combinatorial decision problem. *American Journal of Mathematics*, **65**(2):197–215, 1943.
[34] Emil Leon Post. Solvability, definability, provability; history of an error. *Bulletin of the American Mathematical Society*, **53**:245–246, 1953.
[35] Emil Leon Post. Absolutely unsolvable problems and relatively undecidable propositions. In Martin Davis, ed., *The Undecidable*, pp. 340–406. Dover Publications, Mineola, 2004.
[36] William N Reinhardt. Absolute versions of incompleteness theorems. *Noûs*, **19**(3):317–346, 1985.
[37] Hartley Rodgers. *Theory of Recursive Functions and Effective Computability*. McGraw-Hill Book Company, New York, 1967.
[38] Alan M Turing. On computable numbers, with an application to the Entscheidungsproblem. *Proceedings of the London Mathematical Society*, **42**(2):230–265, 1936.
[39] Alasdair Urquhart. Emil Post. In Dov M. Gabbay and John Woods, eds., *Logic From Russell to Church*, pp. 617–666. Elsevier, New York, 2009.
[40] Mark van Atten and Juliette Kennedy. Gödel's logic. In *Logic From Russell to Church*, pp. 449–509. Elsevier, New York, 2009.
[41] Mark van Atten and Juliette Kennedy. Gödel's modernism: on set-theoretic incompleteness, revisited. In Sten Lindström, Erik Palmgren, Krister Segerberg, and Viggo Stoltenberg-Hansen, eds., *Logicism, Intuitionism, and Formalism*, pp. 303–355. Springer, New York, 2009.
[42] Hao Wang. Some facts about Kurt Gödel. *Journal of Symbolic Logic*, **46**(3):653–659, 1981.
[43] Hao Wang. *A Logical Journey:* From Godel To Philosophy. MIT Press, Cambridge, 1996.
[44] Richard Zach. Completeness before Post: Bernays, Hilbert, and the development of propositional logic. *Bulletin of Symbolic Logic*, **5**(3):331–366, 1999.
[45] Alfred North Whitehead and Bertrand Russell. Principia Mathematica Vol. 1. *Cambridge University Press, Cambridge*, 1910.

# 10

# Absolute Provability and Safe Knowledge of Axioms

TIMOTHY WILLIAMSON

## 10.1 Absolute Provability

When mathematicians discuss proofs, they rarely have a particular formal system in mind. Indeed, they are typically not thinking of formal systems at all, although they might accept the suggestion that a genuine proof can in principle be reconstructed in an appropriate formal system. The picture is more like this. At any given time, the mathematical community has a body of knowledge, including both theorems and methods of proof. Mathematicians expand mathematical knowledge by recursively applying it to itself, adding new theorems and sometimes new derived methods of proof. Of course, present mathematical knowledge itself grew out of a smaller body of past mathematical knowledge by the same process. Since present mathematical knowledge is presumably finite, if one traces the process back far enough, one eventually reaches 'first principles' of some sort that did *not* become mathematical knowledge in that way. Some principles of logic and axioms of set theory are good candidates for such first principles. They became human mathematical knowledge within the last few thousand years, in some cases within the last century or two, but in some other way. Most mathematicians are hazy about what such other ways might be. Let us apply the term 'normal mathematical process' to all those ways in which our mathematical knowledge can grow. Normal mathematical processes include both the recursive self-application of pre-existing mathematical knowledge and the means, whatever they were, by which first principles of logic and mathematics originally became mathematical knowledge. By contrast, appeals to mere inductive support or alleged divine testimony presumably do not count as normal mathematical processes.

A mathematical hypothesis is *absolutely provable* if and only if it can in principle be known by a normal mathematical process. When mathematicians discuss provability, they typically have in mind something like absolute provability. No assumption is made either way as to whether a single formal system generates all and only the absolutely provable hypotheses.

*Gödel's Disjunction*. First Edition. Leon Horsten & Philip Welch (eds).

Since, necessarily, only truths are known, absolute provability entails possible truth. Therefore, since mathematical hypotheses are entirely non-contingent and so true if possibly true, for them absolute provability entails actual truth (where 'necessarily' and 'possibly' express metaphysical rather than epistemic modalities).

A mathematical hypothesis is *absolutely decidable* if and only if either it or its negation is absolutely provable; otherwise it is *absolutely undecidable*. Suppose that every mathematical hypothesis is either true or false; in the latter case, it has a true negation. Then every mathematical hypothesis is absolutely decidable if and only if every true mathematical hypothesis is absolutely provable; in other words, truth coincides with absolute provability for mathematical hypotheses. If every mathematical hypothesis is absolutely decidable, then for Gödelian reasons no single formal system generates all and only the absolutely provable hypotheses.

Are there absolutely undecidable mathematical hypotheses? Before attempting to answer that question, we must further clarify its terms. Section 10.2 contributes to that process.

## 10.2 Propositions and Normal Mathematical Processes

The first clarification is mainly a matter of book-keeping. On the natural reading, a mathematical hypothesis is a *proposition*. A proposition is not a sentence, although it can be expressed by a sentence, indeed, by different sentences in different languages. Goldbach's conjecture is the same proposition whether expressed in English, German, or an interpreted formal language. By contrast, the theorems of a formal system are standardly equated with the formulas themselves, not with the propositions they may express.

Unfortunately, some prominent theories of propositions are so coarse-grained as to make the propositions expressed literally by mathematical formulas useless for epistemological purposes. More specifically, if one follows Robert Stalnaker in treating a proposition as the set of (metaphysically) possible worlds at which it is true, then all true mathematical formulas literally express the same proposition, the set of all possible worlds, since all true mathematical formulas literally express necessary truths. It is therefore trivial that if one true mathematical proposition is absolutely provable, they all are. Indeed, if you already know one true mathematical proposition (that $2 + 2 = 4$, for example), you thereby already know them all. Stalnaker suggests that what mathematicians really learn are in effect new contingent truths about which mathematical formulas we use to express the one necessary truth; but his view faces grave internal problems, and the conception of the content of mathematical knowledge as contingent and metalinguistic is in any case grossly implausible.[1] A subtler alternative within the framework of coarse-grained propositions is to interpret mathematicians as coming to know the same old necessary truth under the guise or mode of presentation of a new formula. However, many philosophers of language react to the problem by appealing to finer-grained theories of propositions, although they face their own difficulties.[2] For present purposes, we may put the problem to one side, by treating the bearers of absolute provability as interpreted formulas rather than propositions, despite the unnaturalness of making formulas the objects of knowledge-that.

A more urgent need for clarification concerns the scope of the phrase 'normal mathematical process'. If our interest is in making mathematical progress over the next few decades, for example in deciding the continuum hypothesis (CH) and other issues undecidable within our currently accepted formal systems, then what matter are the mathematical processes feasible for humans physically and psychologically not too different from ourselves, in social and cultural circumstances not too different from our own. For practical purposes, it is presumably pointless to speculate about proofs or disproofs of CH graspable by creatures with brains much larger than ours (but still finite), unless we treat computers or humans-plus-computers as such creatures. For more theoretical purposes, however, it is rather arbitrary to postulate a specific finite upper bound on the intellectual capacity of the mathematicians with respect to whom 'normal mathematical process' is defined. Any such bound is likely to be incompatible with the closure of normal mathematical processes under natural operations. For example, we cannot assume that provability within such a bound is closed under modus ponens, since even if a proof of $A \rightarrow B$ and a proof of $A$ both fall below a finite complexity threshold, the result of combining them into a proof of $B$ may fail to do so. Such an accidentally contoured standard of provability is unlikely to be theoretically very rewarding. Consequently, the phrase 'normal mathematical process' will here be understood without any specific finite bound or restriction to the physical, psychological, social, and cultural limitations characteristic of humans, now and perhaps even in the distant future. However, we will understand 'normal mathematical process' as restricted to the capacities of mathematicians whose operations are finite in some appropriate sense. Checking Goldbach's conjecture for every natural number one after another, the check on $n + 1$ of course taking half the time of the check on $n$, does not count.

## 10.3 The Epistemic Status of Axioms

One normal mathematical process, even if comparatively uncommon, is adopting a new axiom. If set theorists finally resolve CH, that is how they will do it. Of course, just arbitrarily assigning some formula the status of an axiom does not count as a normal mathematical process, because doing so fails to make the formula part of mathematical *knowledge*. In particular, we cannot resolve CH simply by tossing a coin and adding CH as an axiom to ZFC if it comes up heads, ~CH if it comes up tails. We want to *know* whether CH holds, not merely to have a true or false belief one way or the other (even if we could get ourselves to believe the new axiom). Thus the question arises: when does acceptance of an axiom constitute mathematical knowledge?

In answering the question, we must not allow ourselves to be distracted by the use of the word 'axiom' to describe clauses in the definitions of various classes of mathematical structure, such as groups, rings, fields, and topological spaces. Axioms in that sense do not even express independent propositions, and the question of knowledge does not arise, although we can of course ask whether someone is familiar with the definition. Rather, in the present and more traditional sense, an axiom is a principle relied on without further proof. The usual phrase is just 'without proof', but a standard definition of a proof makes the sequence whose only member is an axiom (in the relevant sense) constitute a proof of that very axiom, and we may accept that more elegant way of thinking. The practice of proof in mathematics requires some axioms in that sense, some first principles even if which principles have that status is sometimes in dispute.

If some form of logicism is correct, on a liberal conception of logic, then mathematical proof can make do with purely logical first principles. Even if mathematics requires some non-logical first principles, it may be possible to resolve outstanding mathematical problems such as CH by adding new first principles, all of which are purely logical: for instance, principles of second-order logic (which is incomplete). For present purposes, we can simply treat logical reasoning as a normal mathematical process, and be neutral as to whether the first principles at issue are purely logical or distinctively mathematical.

To start a line of thought, let $A$ be a true interpreted mathematical formula that cannot be proved with just the resources of current human mathematics. When presented with $A$, human mathematicians are simply agnostic.

We may assume here that mathematical truths are not contingent: neither the existence nor the non-existence of absolutely unprovable mathematical truths puts that into question. Thus $A$ is a (metaphysically) necessary truth: in the relevant respect, things could not have been otherwise. Obviously, that does not mean that the formula $A$, if individuated typographically, could not have had a quite different meaning and expressed a falsehood. Such counterfactual interpretations are always possible, but irrelevant. For example, '0 = 0' would have expressed a falsehood if we had used '=' to mean *is less than*. What matters is that $A$ as *actually* interpreted expresses a necessary truth. If we introduce the operator $\Box$ to mean 'It is necessary that', we can put the point by saying that $\Box A$ is true.

For epistemological purposes, the upshot is that error is impossible in believing $A$ when it expresses what it actually expresses.[3] But that does not mean that whoever believes $A$ on its actual interpretation thereby knows $A$. For instance, if I am gullible enough to believe whatever my guru tells me, and he decides on the toss of a coin to tell me $A$ rather than $\sim A$, I do not thereby know $A$, let alone have specifically *mathematical* knowledge of $A$. I could far too easily have come to believe the falsehood $\sim A$ in a relevantly similar way.

However, there could in principle be mathematicians, perhaps non-human ones, who believe $A$ and could *not* easily have come to believe $\sim A$ or any other falsehood in a relevantly similar way. As a by-product of the evolutionary history of their kind, their brains have come to be wired so as strongly to predispose them to accept as obvious any formula which they interpret as we actually interpret $A$, when the formula is presented to them. As is sometimes said, they find such formulas 'primitively compelling'.[4] For simplicity, we may suppose that the specific feature of their brains that causes them to find a formula with $A$'s meaning primitively compelling does not cause them to find any other formulas primitively compelling. Their brains could not at all easily have been wired differently enough for them to lack that feature. These creatures could not at all easily have had a false belief in $A$ on its actual interpretation, nor have we any reason to suppose that they could easily have had a false belief in $A$ on some other interpretation. Do the creatures know $A$, once it is presented to them on its actual interpretation?

One might be tempted to deny that the creatures know $A$, on the grounds that their belief in $A$ is not appropriately connected with its truth. But what sort of 'appropriate connection' is being demanded here? Doubtless, if the creatures are asked 'Why does *A* hold?', they can only stutter and say 'It's just obvious', or perhaps glibly produce some rigmarole that begs the question in favour of *A*. But isn't that the very reaction that normal people give when asked why their first principles hold, principles that they take themselves to know?

Let us elaborate the story. We may suppose that the creatures develop a form of mathematics that includes all the methods and results of current human mathematics, together with *A*, suitably interpreted, as an additional axiom. Thus their mathematics is strictly somewhat more powerful than ours. They use *A* to settle various previously open questions. Since both *A* and all of their other first principles are true, no inconsistency ever results. Indeed, *A* will have various consequences that the creatures can independently test on the basis of the *A*-free part of their mathematics; of course, those consequences always pass the test. The creatures also have a vague sense that *A* coheres with their other first principles into a unified picture of mathematical reality, although they are unable to substantiate that claim in any rigorous way. Do they lack anything with respect to *A* that we have with respect to our first principles of logic and mathematics?

Some philosophers think that the key to the epistemology of first principles of logic is that they are *analytic*, in the sense that a disposition to assent to them is essential to linguistic or conceptual competence with the logical constants that figure in them.[5] Clearly, *A* is not analytic in that sense, since by hypothesis we interpret *A* as the creatures do, but are not disposed to assent to it. But it is very doubtful that *any* principles of logic are analytic in that sense. Someone may acquire the relevant logical constants in the normal way, and have normal linguistic or conceptual competence with them, but come to reject the principle at issue on sophisticated (albeit mistaken) theoretical grounds, and even find the original disposition to assent to the principle morphing into a different disposition to assent to some more qualified principle, while still retaining normal competence with those logical constants, and using them within the bounds of everyday normality.[6] The point is even clearer for mathematical expressions. A well-trained set theorist may come to challenge one of the axioms of ZFC on theoretical grounds, and cease to find it primitively compelling, without incurring any credible charge of linguistic or conceptual incompetence with the word 'set' or the '$\in$' symbol for set membership. The case for the analyticity of first principles of mathematics in the relevant sense does not stand up. Thus *A* should not be disqualified from the status of a mathematical axiom on grounds of its non-analyticity. Nor is there any good reason to deny that the creatures use the constituent expressions of *A* with the same meanings as we do. Their mathematical practice subsumes ours; that they know more than we know does not imply a difference in reference. For example, if the relevant mathematical language is that of arithmetic, it would be bizarre to suggest that '+' refers to a different operation from the one it refers to in our mouths.

A more moderate constraint on mathematical axioms is that they should be known *a priori*. They are supposed to be known independently of experience, in some appropriate sense. The nature and significance of the distinction between *a priori* and *a posteriori* knowledge are much in dispute.[7] For present purposes, however, we may simply note that the creatures' knowledge of *A* no more depends on experience than does human knowledge of our mathematical axioms. Perhaps the experience of the creatures' distant ancestors played some role in the evolutionary history of the brain structures underlying the creatures' assent to *A*, but then the experience of our distant ancestors surely played some role in the evolutionary history of the brain structures underlying our assent to our mathematical axioms. The human capacity for mathematics did not come about by magic or divine inspiration; it is rooted in our capacity to recognize the shapes and compare the sizes of external objects, to divide and combine those objects in different ways, to sort them according to different principles, to permute them and count them, and so on. If our knowledge of our mathematical axioms is *a priori*, then so too is the creatures' knowledge of *A*.

At least provisionally, we should admit that the creatures' knowledge of *A* is no worse than our knowledge of our axioms. In current epistemological terms, their knowledge of *A* meets the condition of *safety*: they could not easily have been wrong in a relevantly similar case. Here the relevantly similar cases include cases in which the creatures are presented with sentences that are similar to, but still discriminably different from, *A*, and express different and false propositions; by hypothesis, the creatures refuse to accept such other sentences, although they may also refuse to accept their negations.[8] Thus *A* is fit to be a mathematical axiom for the creatures, if not for human mathematicians. Therefore *A* is absolutely provable, because the creatures can prove it in one line. It would be pointless to disqualify the proof for its triviality, for the story could just as well have had another true formula *B* instead of *A* playing the axiomatic role in the creatures' mathematics, where both the creatures and we can prove $B \rightarrow A$, but only with some difficulty.

The claim is not just that *A would* be absolutely provable *if* there were such creatures. The point is the stronger one that *A is* absolutely *provable* because there *could* in principle be such creatures. Absolute provability is not supposed to be contingent on what creatures happen to be around: it is not subject to the accidents of evolutionary history.

Since the foregoing argument made no special assumptions about the formula *A*, it generalizes: every true formula of mathematics is absolutely provable, and every false formula is absolutely refutable. Thus every bivalent mathematical formula is absolutely decidable. If every mathematical formula is either true or false, then every mathematical formula is absolutely decidable. At least for the language of first-order arithmetic, it is overwhelmingly plausible that every formula *is* either true or false, so in particular every formula of the language of first-order arithmetic is absolutely decidable.[9]

## 10.4 Gödel on the Human Mind

The absolute provability of all arithmetical truths in this sense obviously does not imply that human or humanoid minds can somehow do more than machines. It means that for every arithmetical truth *A*, it is possible for a finitely minded creature somewhat like us to prove *A* (in the absolute sense). But nothing has been said to exclude the hypothesis that for every arithmetical truth *A*, some implementation of a Turing machine can prove *A* (in the absolute sense). That is not just the trivial claim that for every arithmetical truth *A*, some implementation of a Turing machine can print out *A*. Rather, the implementation of the Turing machine is required to come to *know A* by a normal mathematical process, and so to be minded. Presumably, on a suitable understanding of the relevant terms, no possible implementation of a Turing machine can prove *all* arithmetical truths. But nothing has been said to support the claim that some possible finitely minded creature somewhat like us *can* prove all arithmetical truths. More generally, no relevant asymmetry has been proposed between finitely minded creatures somewhat like us and implementations of Turing machines.

The possibilities under consideration do not form a linear structure, such as an idealized time sequence in which knowledge grows cumulatively and every truth of the mathematical language is sooner or later known. Rather, they form a branching tree structure in which what is known at one point may neither include nor be included by what is known at another. If the points are time-indexed, it is not even required that along some or every possible history (an infinite path through the tree, taking in every time once) every truth is sooner or later known; nor is it required that knowledge grows cumulatively. We simply identify what is absolutely provable with what is known somewhere or other in the whole branching structure. The truths known at any given point are recursively axiomatizable; the truths known somewhere or other in the structure are not. The coincidence of absolute provability and truth for the mathematical language is even consistent with the recursive axiomatizability of the totality of truths ever known along any given possible history, so that not every mathematical truth ever gets known along any given possible history. For example, in some models every possible history reaches a point after which knowledge stops growing; but every mathematical truth is known in some possible history or other. However, no special reason has emerged for doubting the possibility of a history in which every mathematical truth is sooner or later known.

Someone might nevertheless speculate that some possible finitely minded creature might be able to prove all arithmetical truths, by having a brain structure that somehow encoded a non-recursive pattern recognition capacity for true formulas of arithmetic in an understood notation. But such a scenario goes far beyond the rather mundane extensions of human cognitive capacity on which the argument above relies. For present purposes, we need not indulge in such wild speculations.

We are in no danger of anti-mechanist conclusions even if we grant, for the sake of argument, that the possible finitely minded creatures somewhat like us are indeed possible *humans*, members of our species, perhaps at some later stage of evolution. Then we have that for every arithmetical truth *A*, it is possible for there to be a human who in the factive absolute sense can prove *A*. But we still do not have that it is possible for there to be a human who, for every arithmetical truth *A*, can in that sense prove *A*. There is still no asymmetry between humans and implementations of Turing machines.

In this respect, Gödel's generic talk of 'the human mind' in his Gibbs lecture is dangerously misleading: '*Either* [...] *the human mind* (*even within the realm of pure mathematics*) *infinitely surpasses the powers of any finite machine, or else there exist absolutely unsolvable diophantine problems*'.[10] For if the mathematical powers of 'the human mind' are understood as comprising all those mathematical powers it is possible for a human to have, then by parity of definition the mathematical powers of 'the finite machine' should comprise all those mathematical powers it is possible for a finite machine to have. Then, on our envisaged scenario, although Gödel's first disjunct is true, because within the realm of pure mathematics 'the human mind' infinitely surpasses the powers of any finite machine, it is equally true that within the realm of pure mathematics 'the finite machine' surpasses the powers of any finite machine. For just as any diophantine problem can in principle be solved by some human, although no human can solve them all, so any diophantine problem can in principle be solved by some finite machine, although no finite machine can solve them all. On the other hand, if the mathematical powers of 'the human mind' are understood as restricted to those mathematical powers that any first-rate human mathematician would have if granted infinite supplies of pencils, paper, and life without mental decay, then although Gödel's first disjunct is false on the envisaged scenario, the second disjunct is true only with respect to those limited powers, in a sense compatible with every diophantine problem's being solvable in principle by some possible human or other, typically with greater mathematical powers than those of 'the human mind' as just redefined. Gödel's use of the generic definite article obscures crucial quantificational structure.

Talk of the powers of 'the human mind' may work better within a conception on which all normal humans have the same intellectual competence, all differences coming from accidental limitations on performance. The argument in Section 10.3 did not suggest that the envisaged knowledge of axioms would be attainable within the limits of such normal human competence. But it is hard to see why evolutionary processes should be *in principle* incapable of giving our distant descendants non-accidentally greater intellectual powers than we have, just as we presumably have non-accidentally greater intellectual powers than our distant ancestors had. Whether those descendants count as humans, members of our species, is a zoological question of only moderate epistemological interest.[11]

## 10.5 Mathematical Certainty

Some readers may find the argument in Section 10.3 for the absolute provability of mathematical truths quite dissatisfying. It lacks both the mathematical difficulty and the epistemological depth one might expect an argument for such a conclusion to need. What raises those expectations? At least two factors are relevant.

First, the argument does not help *us* solve recalcitrant mathematical problems. We may be incapable of mathematical knowledge of the new axioms. We may be capable of knowing them only on the testimony of others, perhaps creatures of a different kind, and in any case we are virtually certain not to meet those creatures. To help *us* attain mathematical knowledge of new axioms, an argument surely would need to engage in detail with issues of great mathematical difficulty. But once we separate the practical question 'How can we know the new axioms?' from the theoretical question 'Can the new axioms be known?', we should not expect the latter to inherit all the difficulties of the former.

Second, the account in Section 10.3 of possible knowledge of the new axioms does not fit the traditional picture of the epistemology of mathematical proof. Gödel emphasizes that he is talking about knowledge by purely mathematical means, not just knowledge by any means of purely mathematical propositions. He contrasts 'mathematical certainty' with 'empirical certainty'.[12] In Section 10.3, the new axioms were envisaged as known by an ordinary standard of safety, not by some extraordinary standard of epistemic certainty unattainable elsewhere. In Gödel's own case, it is anyway not clear how much weight he put on 'mathematical certainty', since he did not require mathematical intuition to be infallible.

For anyone who objects to the previous argument that it does not involve mathematical certainty of the new axiom, the challenge is to explain the nature of this 'mathematical certainty' we are supposed actually to have for the current axioms, but lack for the new one in the hypothetical scenario. It is not the necessity of the axioms, because that is the same in the two cases. Nor is it their subjective certainty, the doxastic state of the agent reported by 'I am certain that P' rather than the epistemic status of the proposition reported by 'It is certain that P', for subjective certainty, unlike objective certainty, is not even factive; whereas if it is certain that P it follows that P, if someone is certain that P it does *not* follow that P. In any case, the subjective certainty may also be the same in the two cases; the new axiom is just as primitively compelling for the hypothetical agent as the current axioms actually are for us. Nor does the crucial difference consist in whether the knowledge is *a priori*, since we have seen it to be so in both cases by the usual standard. Epistemological analyticity is also not the difference, since we have seen it to be absent in both cases. Although one may feel tempted to start using phrases like 'mathematical intuition', 'rational insight', or 'self-evidence', without further explanation they merely obscure the issue, because it is so unclear that they add anything of epistemological significance to the distinctions already made. Even if they imply some distinctive extra phenomenology, we may legitimately suppose it to be present in the possible creatures who know the new axiom.

I am content to leave the matter with this challenge: if mathematical knowledge of axioms takes more than Section 10.3 assumed, exactly *what* more does it take?

## Acknowledgement

An early version of this material was presented at the 2012 Workshop on the Limits and Scope of Mathematical Knowledge, at the University of Bristol. Later versions were presented in talks at Tel Aviv University, the University of Michigan at Ann Arbor, and the University of Oxford. I thank the audiences for helpful discussion, and Leon Horsten for useful written comments.

## Notes

1. See Stalnaker, 1999. One problem for his view is that for any formula $A$ in standard mathematical notation, the biconditional $\mathrm{True}(\ulcorner A \urcorner) \leftrightarrow A$ will follow logically from known axioms of a standard compositional theory of truth for the mathematical language. Since on Stalnaker's view our knowledge is closed under such logical consequence, it implies that we already know the biconditional, so for us, the metalinguistic claim is equivalent to the original mathematical claim. Thus semantic ascent to the metalinguistic level only postpones the problem. I believe that Saul Kripke made a similar objection to Stalnaker's view.
2. For example, Russellian theories of structured propositions have trouble with Russellian paradoxes. Let <O, p> be the structured proposition that results from applying the propositional operator O to the structured proposition p. Let N be propositional negation. Define a propositional operator R such that <R, <O, p>> is equivalent to <N, <O, <O, p>>> for any such operator O and proposition p (let <R, q> be equivalent to q if q is not of the form <O, p>). But

then <R, <R, p>> is equivalent to <N, <R, <R, p>>>, its own negation. Although measures can be taken to block such paradoxes, they significantly complicate the theory of propositions. Since banning all iterations of propositional operators would be much too restrictive, one needs paradox-free ways of defining the effect of an operator O on a structured proposition p of which O may itself be a constituent. The problem does not arise for more coarse-grained theories, since when an operator O is applied to an unstructured proposition p to yield an unstructured proposition O(p), there is no general way of uniquely recovering O and p from O(p), so the analogue of R is manifestly ill-defined.

3. Of course, an indexical sentence such as 'I am hungry' can express a different proposition with a different truth-value in a different context without changing its linguistic meaning (its character, in the sense of Kaplan, 1989). But mathematical formulas do not normally exhibit such indexicality. Arguably, many sentences containing a vague word such as 'heap' could easily have had slightly different linguistic meanings, and expressed different propositions with different truth-values, if the vague word had been used slightly differently (Williamson, 1994). But that phenomenon too is not relevant to normal mathematical ignorance, for example when the sentence at issue is in the language of first-order arithmetic.
4. See Peacocke, 1992, although in this chapter the phrase is used without commitment to Peacocke's theory of concepts.
5. See Boghossian, 2003.
6. See Williamson, 2007 and, for a recent exchange, Boghossian, 2012 and Williamson, 2012.
7. For an argument that the distinction between *a priori* and *a posteriori* does not correspond to a deep epistemological difference, see Williamson, 2013.
8. See Williamson, 2000. Williamson, 2009 provides apparatus for extending the safety conception from sentences to arguments.
9. The thought experiment sketched above is similar in spirit to the thought experiment about the mathoids, developed for a different but related purpose in Berry, 2013. The two papers were written independently of each other; their conclusions are different, but compatible. Berry considers in detail various attempts to find an epistemological asymmetry between the mathoids and us, and argues convincingly that they all fail. Analogues of her considerations are relevant to the present case too.
10. Gödel, 1951, p. 310. For good discussion, see Shapiro, 1998 and Feferman, 2006.
11. Perhaps Gödel individuated human minds more flexibly than he did machines, treating the former as able to preserve their identity over time through changes that would not preserve the identity of a machine. But such an asymmetry is not motivated by, and does not depend on, his incompleteness theorems.
12. Gödel, 1951, p. 309.

# References

[1] Almog, Joseph, Perry, John, and Wettstein, Howard (eds.) (1989). *Themes from Kaplan.* Oxford: Oxford University Press.

[2] Berry, Sharon (2013). Default reasonableness and the mathoids, *Synthese* **190**, 3695 3713.

[3] Boghossian, Paul (2003). Blind reasoning, *Aristotelian Society* **66**, 225–248.

[4] Boghossian, Paul (2012). Inferentialism and the epistemology of logic: reflections on Casalegno and Williamson, *Dialectica* **66**, 221–236.

[5] Casullo, Albert and Thurow, Joshua (eds.) (2013). *The A Priori in Philosophy*. Oxford: Oxford University Press.

[6] Feferman, Solomon (2006). Are there absolutely unsolvable problems? Gödel's dichotomy, *Philosophia Mathematica* **14**, 134–152.
[7] Gödel, Kurt (1951). Some basic theorems on the foundations of mathematics and their implications. In Gödel (1995), pp. 304–323.
[8] Gödel, Kurt (1995). *Collected Works, Vol. III. Unpublished Essays and Lectures*. S. Feferman, et al. (eds.). New York: Oxford University Press.
[9] Kaplan, David (1989). Demonstratives. In Almog, Perry, and Wettstein (1989), pp. 481–563.
[10] Peacocke, Christopher (1992). *A Study of Concepts*. Cambridge, MA: MIT Press.
[11] Shapiro, Stewart (1998). Incompleteness, mechanism, and optimism, *Bulletin of Symbolic Logic* **4**, 273–302.
[12] Stalnaker, Robert (1999). *Context and Content*. Oxford: Oxford University Press.
[13] Williamson, Timothy (1994). *Vagueness*. London: Routledge.
[14] Williamson, Timothy (2000). *Knowledge and its Limits*. Oxford: Oxford University Press.
[15] Williamson, Timothy (2007). *The Philosophy of Philosophy*. Oxford: Blackwell.
[16] Williamson, Timothy (2009). Probability and danger, *Amherst Lectures in Philosophy* **4**, 1–35. http://www.amherstlecture.org/williamson2009/
[17] Williamson, Timothy (2012). Boghossian and Casalegno on understanding and inference, *Dialectica* **66**, 237–247.
[18] Williamson, Timothy (2013). How deep is the distinction between a priori and a posteriori knowledge? In Casullo and Thurow (2013), pp. 291–312.

# 11

# Epistemic Church's Thesis and Absolute Undecidability

MARIANNA ANTONUTTI MARFORI AND LEON HORSTEN

## 11.1 Introduction

On December 26, 1951, Gödel delivered the 25th J. W. Gibbs Lecture at a Meeting of the American Mathematical Association at Brown University. In the lecture, he formulated a disjunctive thesis concerning the limits of mathematical reasoning and the possibility of the existence of mathematical truths that are inaccessible to the human mind. This thesis, known as *Gödel's disjunction*, is introduced as a direct consequence of the incompleteness theorems [Gödel 1951, p. 310]:

> Either ... the human mind (even within the realm of pure mathematics) infinitely surpasses the powers of any finite machine, or else there exist absolutely unsolvable diophantine problems [henceforth, *absolutely undecidable mathematical sentences*] ... (where the case that both terms of the disjunction are true is not excluded, so that there are, strictly speaking, three alternatives).

That is, either the output of the human mathematical mind exceeds the output of a Turing machine (called the *anti-mechanist thesis*) or there are true mathematical sentences that are undecidable "not just within some particular axiomatic system, but by any mathematical proof the human mind can conceive." The latter are called *absolutely undecidable mathematical sentences*, i.e. mathematical sentences that cannot be either absolutely proved or refuted [Gödel 1951].

According to Gödel, the fact that the disjunctive thesis above holds is a "mathematically established fact" of great philosophical interest which follows from the incompleteness theorems, and as such, it is "entirely independent from the standpoint taken toward the foundation of mathematics" [Gödel 1951]. Indeed, most commentators agree that Gödel's arguments for this disjunctive thesis are compelling.

Since Gödel's disjunction was first formulated in 1951, much effort has gone into finding equally compelling arguments for or against either of the disjuncts. In particular, attempts

*Gödel's Disjunction*. First Edition. Leon Horsten & Philip Welch (eds).

were made to establish the first disjunct by arguing on *a priori* grounds that the capacities of the human mathematical mind exceed the output of any Turing machine as a consequence of the incompleteness theorems (see chiefly [Lucas 1961], [Penrose 1989], and [Penrose 1994]). These attempts have sparked a lively debate, but there is wide consensus that they have so far turned out to be inconclusive. Fewer efforts have been concentrated on shedding light on the second disjunct, and, at present, no conclusive argument that decides either of the disjuncts has been found.[1] In this chapter, we focus on the second disjunct.

We make use of the framework of *Epistemic Arithmetic* proposed by Shapiro in the mid 1980s. In this framework, an absolute or informal notion of provability is taken as primitive and axiomatically investigated. Since this framework can also express classical (and constructive) mathematical propositions, it constitutes a good setting for the investigation of the concept of absolute undecidability. In this framework, a variant of the Church–Turing thesis can be formulated: this variant has been labelled *ECT* ("Epistemic Church's Thesis") in the literature. While there are strong reasons to think that this variant is not a very faithful approximation of the content of the original Church–Turing thesis, we will show that an analogue of Gödel's disjunction can be established which states that either *ECT* fails, or there are absolutely undecidable propositions (or both).

In analogy with Gödel's disjunction, this raises the question of what the truth value of each of the disjuncts is. While it can be shown that *ECT* implies the existence of absolutely undecidable non-contingent propositions (of low arithmetical complexity), we will see that it seems hard to argue convincingly that *ECT* is true. Consequently, it is not easy to see how the analogue of Gödel's disjunction can be used to show that there are absolutely undecidable propositions. Thus we conclude that the truth value of the disjuncts is not easy to ascertain.

## 11.2 Absolute Undecidability

### *11.2.1 Absolute Undecidability in Epistemic Arithmetic*

Following suggestions in [Myhill 1960], the notion of informal or absolute provability, and more generally the notion of *a priori* knowability of mathematical propositions, can be investigated in an axiomatic way. In particular, the investigation can be carried out within the framework of epistemic arithmetic developed in [Shapiro 1985b]. This is the framework that shall be adopted throughout the chapter. Since its background is a formal theory of *arithmetic*, the propositions that contain set-theoretic concepts fall outside the scope of the present chapter.

The formal framework of epistemic arithmetic can be described as follows. The formal language $\mathcal{L}_{EA}$ consists of the first-order language of arithmetic plus an intensional propositional operator $\Box$; the arithmetical vocabulary receives its intended interpretation, and the operator $\Box$ is interpreted as *a priori* knowability. The axiomatic theory *EA* that is proposed by Shapiro as describing the laws of *a priori* knowability consists of the axioms of Peano arithmetic plus the laws of **S4** modal logic. It should be noted that the modal logic

**S4** contains the necessitation rule and the axiom $\Box\phi \rightarrow \Box\Box\phi$ (the so-called "4 axiom"), so that $\Box$ is indeed an iterable notion.

The *absolute undecidability* of a sentence $\phi$ can then be expressed as

$$\neg\Box\phi \wedge \neg\Box\neg\phi.$$

This is the notion of absolute undecidability that will figure in the new disjunctive thesis that will be investigated in Section 11.3.

One may wonder whether there are *arithmetical* sentences that are absolutely undecidable in this sense. This is a version of Gödel's question whether there are absolutely undecidable *mathematical* propositions. But one may also ask in this framework whether there are sentences $\phi \in \mathcal{L}_{EA}$ that are undecidable. This question is of course closely related to Gödel's question. But it is not identical to it, for $\mathcal{L}_{EA}$ not only contains mathematical concepts, but also contains the concept of *a priori* knowability, which is not a mathematical concept. Both of these questions about absolute undecidability will be discussed in this chapter.

### *11.2.2 Other Concepts of Absolute Undecidability*

The notion of absolute undecidability in which we are interested here concerns statements that are non-contingent, may contain the concept of absolute or informal provability, and have a determinate truth value.

#### 11.2.2.1 Fitch's Undecidables

In [Fitch 1963], Fitch has argued for the following claim:

**Thesis 1 (Fitch)** *If there are unknown truths, then there are unknowable truths.*

The kinds of truths that Fitch adduces as witnesses of the consequent of Gödel's disjunction are propositions of the form "*p* and it is not known that *p*"; Fitch's argument of course does not provide a concrete witness.

Some believe that Fitch's argument is sound and that furthermore, since the antecedent of its conclusion is true, we must accept that there are unknowable truths. Others think that Fitch's argument is unsound.[2] In any case, Gödel would probably not have been satisfied with Fitch's cases of absolutely undecidable propositions as a way of sharpening Gödel's disjunction (if Fitch's propositions do indeed qualify as absolutely undecidable), for they are *contingent* propositions. And they also fall outside the scope of our analogue of Gödel's disjunction because, using the notion of a priori knowability and arithmetical notions, only non-contingent propositions can be formed in $\mathcal{L}_{EA}$.

#### 11.2.2.2 Formally Undecidable Arithmetical Statements

As said earlier, in his discussion of his disjunctive thesis, Gödel seems to have had *mathematical* undecidable statements in mind. Feferman and Solovay have produced instances of arithmetical sentences of which they claimed that it is unlikely that they will ever be decided [Feferman and Solovay 1990, Remark 3, p. 292]. However, it is far from clear that the reasons for thinking that such statements will presumably never be decided go as far as establishing that they are *in principle* humanly undecidable, i.e. absolutely undecidable.

Similarly, Boolos has shown that there are infinitely many what he calls *extremely unprovable* arithmetical sentences [Boolos 1982]. These are (true) arithmetical sentences that are not only undecidable in Peano arithmetic, but that are such that Peano arithmetic can only prove them to have properties characterisable in terms of "provability in Peano arithmetic" that *every* arithmetical sentence can be proved (in Peano arithmetic) to have. But, again, Boolos' considerations do not establish that some such sentences cannot be proved *in principle* rather than just in Peano arithmetic. The uncertainty in this area is caused by the fact that we do not have a sufficiently strong grasp on what the right idealisations involved in the notion of absolute undecidability are.[3]

Indeed, it seems that the only argument that we have for establishing of a given arithmetical sentence that it is absolutely unprovable relies essentially on the connection between absolute proof and truth. If we have a proof that $\phi$ is false (for $\phi$ arithmetical), then we have *a priori* knowledge that it cannot be provable. Our tentative claim is that this line of argument is the *only* way in which an arithmetical sentence can be shown to be absolutely unprovable. In other words:[4]

**Thesis 2** *Provable unprovability of an arithmetical fact supervenes on the provability of the negation of that arithmetical fact.*

The content of this principle can be expressed in the framework of *EA*:

**Axiom 1** $\Box\neg\Box\phi \rightarrow \Box\neg\phi$, *for $\phi$ any sentence of the language of arithmetic.*

It is easy to see that:

**Proposition 1** *If axiom 1 is true, then there are no provably absolutely undecidable arithmetical sentences.*

The statement that there are no provably absolutely undecidable arithmetical sentences is a restricted version of the **S4.1** axiom (also known as *McKinsey's axiom*) of modal logic.

So, in a nutshell, it seems to us unlikely that we can find a conclusive *a priori* argument that there are absolutely undecidable *arithmetical* sentences. Or, in other words, we think that the second disjunct of Gödel's disjunction *for arithmetical sentences* cannot be established by *a priori* means.

This, however, completely leaves open the question of whether it can be established that there are sentences *of the language of epistemic arithmetic*—i.e. sentences that involve the notion of absolute provability itself—that are absolutely undecidable. Indeed, we cannot appeal to the generalisation of axiom 1 to argue that there are no provably absolutely undecidables in $\mathcal{L}_{EA}$. For if the principle $\Box\neg\Box\phi \rightarrow \Box\neg\phi$ is true for every $\phi \in \mathcal{L}_{EA}$, then absolute provability (demonstrably) coincides with truth. It is clear that for any $\phi$, $\Box\neg\Box(\phi \wedge \neg\Box\phi)$. The generalisation of axiom 1 then allows us to conclude $\Box(\phi \rightarrow \Box\phi)$

[Horsten 1997]. In the following sections we will return to the question whether it might be provable in principle that there are absolutely undecidable sentences in $\mathcal{L}_{EA}$.

#### 11.2.2.3 Truth-Indeterminate Undecidables

Another question is whether there are absolutely undecidable sentences that, in addition to the arithmetical vocabulary, contain a primitive notion of provability. Reinhardt has observed that absolutely undecidable sentences can be shown to exist if we have a provably sound *absolute provability predicate* [Reinhardt 1986]:

**Proposition 2** *Suppose that $P(x)$ is any formula with $x$ free, and let $S \supseteq EA$ be such that $S \vdash \Box(P(\ulcorner\phi\urcorner) \to \phi)$. Then there is a sentence $G_S$ such that $S \vdash \Box G_S \wedge \Box\neg P(\ulcorner G_S \urcorner)$.*

If $P(x)$ is an absolute provability predicate satisfying the condition of the proposition, then $G_S$ is an absolutely undecidable sentence.[5]

The sentence $G_S$ is produced by diagonalisation (the fixed-point lemma). Intuitively, $G_S$ is a sentence which says of itself that it is absolutely unprovable. It is not purely arithmetical, due to the predicate $P$, which figures in the instance of the diagonal lemma that is used to produce $G_S$.

In contrast to Fitch's propositions, $G_S$ is not contingent: if it is true (or false), then it is so necessarily. So Reinhardt's proposition might be taken to be more relevant to the second disjunct of Gödel's disjunction. However, $G_S$ is the so-called "knower sentence", see [Anderson 1983]. It is a paradoxical sentence: intuitively, it lacks a truth value just like the liar sentence does, and for similar reasons. And if $G_S$ lacks a truth value, then it is not even a candidate for being proven or refuted, so it does not seem very relevant to the assessment of Gödel's disjunction.

There is a rich literature about purely mathematical sentences—mostly set-theoretical—that *may* be truth-indeterminate.[6] The continuum hypothesis is perhaps the most famous candidate for this category. Since such propositions are purely mathematical, and at least not trivially truth-indeterminate, they are more relevant to the question of the second disjunct of Gödel's disjunction as understood by Gödel than is the Knower sentence. However, there are at present no arguably absolutely undecidable mathematical sentences that are *uncontroversially* truth-determinate. Indeed, the question of whether there are truth-indeterminate set-theoretic propositions turns on deep and unresolved foundational questions. In any case, as mentioned earlier, set-theoretic statements fall outside the scope of the present chapter.

## 11.3 A New Disjunction

In this section, we introduce a variant of Gödel's disjunction in the language of $EA$ that will be investigated in the remainder of the chapter. In order to do so, §11.3.1 will present the principle that has been called in the literature *Epistemic Church's Thesis*, and §11.3.2 will explore the connection between such a principle and Gödel's disjunction. Finally, §11.4 will test $ECT$ in a class of models which embed certain idealisations.

### *11.3.1 Epistemic Church's Thesis*

Since functions are infinite abstract objects, human subjects—even in the idealised sense—do not have epistemic access to functions independently of the interpreted linguistic expressions that denote them (call these *function presentations*). According to a thesis proposed in [Shapiro 1985b], a function presentation $F$ is *calculable* if and only if there is an algorithm $A$ such that it is *a priori* knowable that $A$ represents $F$, where calculability is a "pragmatic counterpart" of the notion of computability and directly involves the human ability of computing a function [Shapiro 1985b, pp. 42–43].

This leads Shapiro to define *calculability* as follows:

**Thesis 3** *A function $f$ is calculable if and only if, recognisably, for every number $m$ given in canonical notation, a canonically given number $n$ exists such that the statement $f(m) = n$ is absolutely provable.*

Using Shapiro's notion of calculability, we can express in $\mathcal{L}_{EA}$ a variant of the Church–Turing thesis [Shapiro 1985b, p. 31]:

**Thesis 4 (*ECT*)**

$$\Box\forall x\exists y\Box\phi(x,y) \rightarrow \exists e\,[e \textit{ is a Turing machine} \wedge \forall x : \phi(x, e(x))],$$ [7]

for $\phi$ ranging over formulae of the language of epistemic arithmetic.[8]

This principle is called *Epistemic Church's Thesis* (*ECT*) in the literature. Note that in order for the antecedent to ensure that $\phi(x,y)$ expresses a *function*, a choice principle is implicit in *ECT*. However, the choice principle can be eliminated by prefixing the functionality of $\phi(x,y)$ as a condition on *ECT*, so that it assumes the form

$$\phi(x,y) \text{ is functional } \rightarrow ECT.$$

Shapiro considers *ECT* "a weaker version of *CT* [in the standard formalisation] which is closer to Church's thesis [than the intuitionistic version of the latter]" [Shapiro 1985b, p. 31].[9] That *ECT* is closer to the Church–Turing thesis than the intuitionistic variant of Church's thesis is due to the fact that, as in the Church–Turing thesis, the existential quantifier in the consequent of *ECT* is *classical*. Thus, *ECT* does not require that any specific Turing machine can be *shown* to compute the effectively computable function described in the antecedent.

Nonetheless, there are strong reasons to be sceptical that *ECT* approximates the content of the Church–Turing thesis in *EA*. The antecedent of *ECT* does not involve the informal notion of algorithm, so it is implausible to consider the antecedent of *ECT* as expressing that $\phi(x,y)$ is algorithmically computable. Indeed, there is no way to *directly* express or quantify over algorithms in the language of *EA* [Shapiro 1985b, pp. 41–43]. Another reason why *ECT* does not capture the content of the Church–Turing thesis is that the *converse* of the Church–Turing thesis is obviously true, whereas the converse of *ECT* is not obviously true [Black 2000, § 2].[10]

### 11.3.2 *ECT and Absolute Undecidability*

The truth of *ECT* (or even *ECT* restricted to *arithmetical* relations) would have a significant consequence, for it entails that there are absolutely undecidable propositions in the language of *EA*:

**Proposition 3** *If ECT is true, then there are absolutely undecidable propositions expressible in* $\mathcal{L}_{EA}$.

***Proof*** Suppose that there were no such absolutely undecidable propositions. Then for such sentences, a priori demonstrability would coincide with truth. All such occurrences of *ECT* could therefore be erased without changing the truth values of these sentences. But for any non-computable functional predicate $\phi(x, y)$, the corresponding instance of *ECT* would be false. So, contrapositively, if *ECT* is true, then there must be absolutely undecidable propositions. ■

In other words, we have arrived at a new disjunctive thesis (henceforth, *ND*) that is somewhat reminiscent of Gödel's disjunction:

**Thesis 5 (*ND*)** *Either ECT fails, or there are absolutely undecidable statements (expressible in* $\mathcal{L}_{EA}$*), or both.*

The antecedent of *ECT* does not express the notion that the human mathematical mind is, or is not, a Turing machine. Therefore, the content of the new disjunctive thesis *ND* is not the same as that of Gödel's disjunction. In fact, it is consistent for *ECT* to hold and for the following to fail:

$$\exists e \forall x \in \mathcal{L}_{EA} : T(\ulcorner \Box x \urcorner) \leftrightarrow \text{“}x \text{ is enumerated by the Turing machine } e\text{”},$$

where $T$ is a truth predicate for $\mathcal{L}_{EA}$.

Proposition 3 can be sharpened by weakening its assumption and strengthening its conclusion:

**Theorem 1** *If ECT restricted to* $\Pi_1$ *arithmetical relations* $\phi(x, y)$ *holds, then there are absolutely undecidable* $\Pi_3$ *sentences of* $\mathcal{L}_{EA}$.

***Proof*** We prove the contrapositive, that is: If there are no $\Pi_3$ absolutely undecidable sentences of $\mathcal{L}_{EA}$, then *ECT* restricted to $\Pi_1$ arithmetical relations is false.

Suppose that there are no absolutely undecidable $\Pi_3$ sentences in $\mathcal{L}_{EA}$:

$$\Box\Psi \leftrightarrow \Psi \text{ for all } \Pi_3 \text{ sentences } \Psi \in \mathcal{L}_{EA}.$$

Choose a Turing-uncomputable total functional $\Pi_1$ arithmetical relation $\phi(x, y)$; from elementary recursion theory we know that such $\phi(x, y)$ exist.

Then, $\forall x \exists y \phi(x, y)$. But then we also have that $\forall x \exists y \Box\phi(x, y)$. The reason is that $\Pi_1 \subseteq \Pi_3$, so for every $m$ and $n$, $\phi(m, n)$, being a $\Pi_1$ statement, entails $\Box\phi(m, n)$. However, $\forall x \exists y \Box\phi(x, y)$ is now a $\Pi_3$ statement of $\mathcal{L}_{EA}$, so again from our assumption it follows that $\Box\forall x \exists y \Box\phi(x, y)$.

Therefore, for the chosen $\phi(x,y)$, the antecedent of *ECT* is true whereas its consequent is false. Therefore, for the chosen $\phi(x,y)$, *ECT* is false. ■

From the existence of absolutely undecidable propositions, certain consequences can be derived.

Recall McKinsey's axiom (the **S4.1** axiom) for propositional modal logic, which states (roughly) that there are no *provably* absolutely undecidable statements:

**Axiom 2 (McKinsey)** $\neg\Box(\neg\Box\phi \wedge \neg\Box\neg\phi)$.

**Corollary 1** *From the absolute provability of ECT restricted to $\Pi_1$ relations, it follows that McKinsey's axiom for absolute provability fails.*

***Proof*** The proof of Theorem 3 produces a concrete absolutely undecidable sentence from instances of *ECT*. If these instances can be "necessitated," then the proof yields a provably undecidable sentence, which contradicts McKinsey's axiom. ■

In the antecedent of this corollary, *ECT* is taken as an axiom scheme, the instances of which fall within the scope of the necessitation axiom.[11] So the corollary says that *if ECT* can be established by *a priori* means, then even though we probably cannot establish that there are arithmetical absolutely undecidable sentences (Section 11.2.2.2), we will have a proof of the existence of absolutely undecidable sentences in the language of *EA*. Or, in other words, in this situation, the **S4.1** axiom may well hold for arithmetical sentences, but not for all sentences of $\mathcal{L}_{EA}$.

**Corollary 2** *If ECT holds, then the converse of ECT fails.*

***Proof*** Suppose that *ECT* holds; then there are absolutely undecidable propositions. Let $\psi$ be an absolutely undecidable proposition. Define $\phi(x,y)$ as

$$[(x = x \wedge y = 1) \leftrightarrow \psi] \wedge [(x = x \wedge y = 0) \leftrightarrow \neg\psi].$$

Then $\phi(x,y)$ defines either the constant 1 function or the constant 0 function, whereby $\exists e\forall x\phi(x,e(x))$. But since $\psi$ is absolutely undecidable, we have that

$$\neg\Box\forall x\exists y\Box\phi(x,y).$$ ■

Earlier we observed that in contrast to the converse of the Church–Turing thesis, the converse of *ECT* is not obviously true. Now we see that the situation is in fact worse: if *ECT* is true, then its converse is plainly false. This brings in sharper relief the fact that *ECT* does not have the same content as the Church–Turing thesis.

Note that the proof of Corollary 2 also entails that there are coextensive functional relations $\phi(x,y)$ and $\psi(x,y)$ expressible in $\mathcal{L}_{EA}$ such that $\Box\forall x\exists y\Box\phi(x,y)$ is true, whereas $\Box\forall x\exists y\Box\psi(x,y)$ is false. That is, it also follows as a corollary that Shapiro's notion of calculability is an intensional notion:

**Corollary 3** *If ECT holds, then the antecedent of ECT is intensional.*

***Proof*** Suppose that *ECT* holds; then there are absolutely undecidable propositions. Let $\psi$ be an absolutely undecidable true proposition. Let $g$ be defined and calculable. Then we can define a function $f$ in terms of $g$ and $\psi$:

$$\begin{cases} \forall x f(x) := g(x) \text{ if } \psi \\ \forall x f(x) := g(x) + 1 \text{ if } \neg\psi. \end{cases}$$

Then we have that $f$ is co-extensional with $g$:

$$\forall x f(x) = g(x)$$

because $\psi$ is true. Thus, $f$ is *not* calculable because in order to compute $f$ we have to know whether $\psi$ is true. ■

Notice that $f$ is not *provably* coextensive with $g$, because if we could prove that $\forall x : f(x) = g(x)$, then we would be able to prove that $\psi$ is true.

## 11.4 Models for *ECT*

In this section, we will discuss the status of *ECT*. In §11.4.1, we will briefly survey some mathematical facts about *ECT* and we will discuss whether it is possible to have a conclusive *a priori* argument for the truth of *ECT*. In §11.4.2 and §11.4.3 we will evaluate *ECT* in models for $\mathcal{L}_{EA}$.

### *11.4.1 Is ECT True?*

Some key metamathematical facts are known about *ECT*. The principle *ECT* was shown to be consistent with *EA* in [Flagg 1985], so it must have models. The proof of the consistency of *ECT* with *EA* was simplified in [Goodman 1986], and then further in [Horsten 1997]. In [Halbach and Horsten 2000] it was shown that from instances of *ECT* as *hypothesis*, no arithmetical theorems can be proved that are not already theorems of Peano arithmetic. As far as we are aware, the question of whether when *ECT* is added as an *axiom* to *EA* this yields an arithmetically conservative extension of Peano arithmetic is still open. It is also known that *EA* + *ECT* has the disjunction property and the numerical existence property [Halbach and Horsten 2000].

The disjunctive principle *ND* that was argued for in §11.3.2 bears witness to the fact that the truth of *ECT* has significant philosophical consequences. *If ECT* is true, then there are non-contingent absolutely undecidable propositions that are perfectly truth-determinate. However, the most important question that is still open is whether *ECT is* a true principle (despite the fact that it is *not* a faithful formalisation of the Church–Turing thesis).

In the discussion about the Church–Turing thesis, a distinction is often made between *quasi-empirical* (*a posteriori*) and conceptual (*a priori*) evidence for the thesis. For a long time it was thought that the Church–Turing thesis cannot be proved, and that all our evidence for its truth is quasi-empirical. In particular, it has been argued that all reasonable attempts that have been proposed to capture the notion of algorithmically computable function have turned out to be extensionally provably equivalent, and that therefore we have strong inductive grounds for thinking that we have mathematically captured the notion of algorithm.

In recent decades, however, most scholars have come to believe that it is rather Turing's [1936] conceptual analysis of the notion of algorithm that gives us conclusive *a priori* evidence for the Church–Turing thesis.[12] Similar questions to those about evidence for the Church–Turing thesis can be asked about our evidence for *ECT*. More specifically, one can ask whether a convincing *a priori* argument for *ECT* can be found, or whether only weaker (and perhaps quasi-empirical, *a posteriori*) forms of evidence can be found for it, or whether we could construct putative counterexamples to *ECT*.

One might try to use the Church–Turing thesis to argue that *ECT* is true. In [Horsten 1998, §4, p. 15] the following argument was formulated. Suppose that the following thesis holds:

**Thesis 6** *The only way in which a statement of the form* $\forall x \exists y \Box \phi(x, y)$ *can be proved is by giving an algorithm for computing* $\phi$.

It follows from the Church–Turing thesis that the function expressed by $\phi(x, y)$ is Turing-computable. Given Turing's *a priori* evidence for the Church–Turing thesis, it then follows that if we have good *a priori* evidence for Thesis 6, we will have an *a priori* argument for *ECT*.

Unfortunately, evidence for Thesis 6 is lacking. We have at present no way of excluding that for some functional relation $\phi(x, y)$, it is absolutely provable *in a non-constructive way* that $\forall x \exists y \Box \phi(x, y)$. Such a proof would not involve an algorithm for generating, for every $x$, a proof of $\phi(x, y)$ for some $y$. In sum, the argument of [Horsten 1998, §4] does not carry conviction,[13] and thus the prospects for having strong *a priori* arguments for ECT are not promising.

### *11.4.2 Simple Machines*

We can try to test *ECT* in some models that incorporate reasonable-looking idealisations on the notion of *a priori* knowability: it is possible that *ECT* will hold (or will fail to hold) in a variety of models that embody reasonable idealisations,[14] and if this is the case, then we would have some evidence in support of (or against) *ECT*. To this task we now turn.

Theories formulated in $\mathcal{L}_{EA}$ can themselves be regarded as models. Given a theory $S$, we define *truth in S* ($S \models \ldots$) as follows:

- The interpretation of the arithmetical vocabulary is standard.
- The interpretation of the classical Boolean connectives is as usual.
- $S \models \Box\phi \Leftrightarrow \phi \in S$.[15]

This will be called the *theory-as-model* perspective. We have argued that *EA* is sound for *a priori* knowability or absolute provability; on this grounds, it is reasonable to assume of theories-as-models $S$ that $EA \subseteq S$. More generally, we will be interested in theories that are *sound* in the following precise sense:

**Definition 1** *A theory S is called* sound *if* $S \models S$.

Using Shapiro's adaptation of the Kleene slash [Shapiro 1985b, p. 18], which we shall not rehearse here, it can easily be seen that *EA* is a sound theory:[16]

**Theorem 2** $EA \models EA$.

For obvious reasons, we will say that *EA* is the *minimal* model for *EA*.

*EA* also makes *ECT* true. In order to show this, we first recall that *EA* has the epistemic analogue of the *numerical existence property* (henceforth, *NEP*) [Shapiro 1985a, pp. 19–20]:

**Theorem 3 (NEP)** $EA \vdash \exists x \Box \phi(x) \Rightarrow$ *there is an* $n \in \mathbb{N}$ *such that* $EA \vdash \phi(\overline{n})$,

with $\phi(x)$ being a formula with only $x$ free. It should be noted that the antecedent does not simply express that it is *a priori* knowable that the extension of $\phi$ is not empty ($\Box(\exists x \phi(x))$), but it expresses the stronger statement that there is a *particular* number $x$ such that it is knowable that $x$ has the property $\phi$. So the theorem tells us that if *EA* proves the existence of a number $x$ of which it is *a priori* knowable that it has the property $\phi$, then there is a particular number $x$ such that *EA* proves that it is *a priori* knowable that $x$ satisfies the property $\phi$.

We can now show that *EA* makes *ECT* true:

**Theorem 4** $EA \models ECT$.

***Proof*** $EA \models \Box \forall x \exists y \Box \phi(x,y) \Rightarrow (T)$

$EA \vdash \forall x \exists y \Box \phi(x,y) \Rightarrow (NEP)$

$\forall m \exists n : EA \vdash \phi(\overline{m}, \overline{n})$.

Now let $e$ be the Turing machine that successively for each m finds the shortest *EA*-proof of $\phi(\overline{m}, \overline{n})$ for some minimal $n$, and outputs $n$. Then by the soundness of *EA* we have that

$$EA \models \exists e \forall x \exists z \exists v : T(e,x,z) \wedge U(z,v) \wedge \phi(x,v),$$

where $T$ is Kleene's $T$-predicate and $U$ is Kleene's $U$-function. ■

So *EA*, seen as a model, is also the simplest model for *ECT*. In [Halbach and Horsten 2000] it is shown that $EA + ECT \models NEP$. Generalising from this, we can see the following:

**Theorem 5** *For all r.e. sound* $S \supseteq EA$ *that have NEP,*

$$S \models ECT.$$

Hence, there are many simple models of *ECT* that assign a recursively enumerable extension to $\Box$. Note that because of the independence of *ECT* from *EA* (see [Flagg 1985] and [Carlson 2015]), we have that

$$EA \not\models \Box ECT.$$

Indeed, it is known that models that also make $\Box ECT$ true are necessarily somewhat complicated [Carlson 2015]. However, this need not unduly concern us here, because, as we have seen, it is anyway somewhat difficult to see how we can *a priori* come to know $ECT$.

### 11.4.3 More Realistic Models?

In this section, we will construct somewhat more *realistic* models for the behaviour of an idealised mathematical community. The aim is to construct simple models for $ECT$ that do not necessarily assign a recursively enumerable extension to the $\Box$ operator in order to allow for the possible non-systematicity of the cumulation of knowledge over time, and to test whether $ECT$ is true in a wide class of such models.

We start by defining possible worlds models for $\mathcal{L}_{EA}$. We base our models on a *branching time* framework. The informal idea behind this is as follows. A possible world, or possible space-time, might be seen as a linear sequence of moments at which new proofs are generated. These linear structures may be taken to partially overlap in such a way that the union of all the possible space-time moments form a tree (partial ordering) under the earlier-than relation.

**Definition 2** *A frame* $\mathcal{F} = \langle \mathcal{T}, < \rangle$ *is a partial ordering relation.*

The elements of $\mathcal{T}$ can be seen as possible moments in idealised branching time. Instead of writing $t_i < t_j$ for $t_i, t_j \in \mathcal{T}$, we will often simply write $i < j$.

**Definition 3** *A possible worlds model* $\mathcal{M}$ *is a structure of the form* $\langle \mathcal{F}, f \rangle$, *with* $\mathcal{F}$ *a frame and* $f : \mathcal{T} \mapsto \mathcal{P}(\mathcal{L}_{EA})$.

Informally, $f(t_i)$ specifies the theory that is known at moment $t_i$.

Even though we have defined truth in a theory-as-model, we have not yet defined truth in a possible worlds model. Let us do that now:

- The interpretation of the arithmetical vocabulary is standard.
- The interpretation of the classical Boolean connectives is as usual.
- $\mathcal{M} \models \Box\phi \Leftrightarrow \phi \in \bigcup_i f(t_i)$.

Then we immediately see:

**Proposition 4** $\mathcal{M} \models \Box\phi \Leftrightarrow \bigcup_i f(t_i) \models \Box\phi$.

This proposition connects the possible worlds perspective with the theory-as-model perspective.

To ensure that the appropriate degree of idealisation is satisfied, we require that possible worlds models meet the following conditions for all $i, j$:

I. (**Closure**) $f(t_i)$ is closed under logic and contains the theorems of $EA$.

II. (**Cumulativity**) $\phi \in f(t_i) \Rightarrow \forall k > i : \phi, \psi \in f(t_k)$.

III. **(Positive Introspection)** $\phi \in f(t_i) \Rightarrow \forall j > i : \Box\phi \in f(t_j)$.

IV. **(Soundness)** $\phi \in f(t_i) \Rightarrow f(t_i) \models \phi$ and $\bigcup_i f(t_i) \models \phi$.

V. **(Finiteness)** $f(t_i)$ is recursively enumerable.

The requirement of soundness (IV) entails that each $f(t_i)$ has the numerical existence property (*NEP*).

The requirements of closure (I) and cumulativity (II) are idealisations. It is assumed that the mathematical community deduces the logical consequences of what it knows, and it is assumed that the mathematical community has a perfect memory. The requirement of positive introspection (III) is a reflective property. If a subject knows $\phi$ *a priori*, then they can, in the following moment in time, reflect on their grounds for believing $\phi$ *a priori* and conclude (*a priori*) that they are strong enough to warrant *a priori* knowledge that $\phi$. In other words, they know *a priori* that they know $\phi$ *a priori*. The soundness requirement (IV) might be argued for from the definition of the concept of knowability: a subject can only come to know *a priori* at a certain moment in time that $\phi$ if $\phi$ is true, for it is analytic of the concept of knowledge that it entails the truth of what is known. Condition (V) is a finiteness requirement since any r.e. theory is finitely axiomatisable in a language extension. It is motivated by the fact that since the human mind (even the mind of an idealised mathematical community) is finite at every given moment in time, the content of what is *a priori* implicitly known (given closure under logical consequence) is contained in a Turing machine.

We now see that for such $\mathcal{M}$ the following holds:

**Theorem 6**

1. $\mathcal{M} \models$ *EA except for the K-axiom.*
2. $\mathcal{M} \models$ *ECT.*

***Proof***

(1.) In order to show that $\mathcal{M} \models EA$, we show that $\mathcal{M} \models 4, T$. (We already know that $\mathcal{M}$ models the necessitations of these principles, by the closure condition on the $f(t_i)$'s.) That $\mathcal{M} \models 4$ follows from positive introspection; the fact that $\mathcal{M} \models T$ follows from the soundness condition.

(2.) $\mathcal{M} \models \Box\forall x\exists y\Box\phi(x,y) \Rightarrow (T)$
$\exists i : f(t_i) \vdash \forall x\exists y\Box\phi(x,y) \Rightarrow (NEP)$
$\exists i\forall m\exists n : f(t_i) \vdash \phi(\overline{m},\overline{n})$.

Now let $e$ be the Turing machine that successively for each m finds the shortest $f(t_i)$-proof of $\phi(\overline{m},\overline{n})$ for some minimal $n$, and outputs $n$. Then by the soundness condition (see IV above) on $\mathcal{M}$ we have that

$$\mathcal{M} \models \exists e\forall x\exists z\exists v : T(e,x,z) \wedge U(z,v) \wedge \phi(x,v).$$ ■

Note that if the frame of $\mathcal{M}$ is a total ordering, then also the $K$-axiom holds (by cumulativity), and then $\mathcal{M} \models EA$.

Theorem 6 shows that there are many possible worlds models of *EA* that make *ECT* true. A particularly simple and intuitive subclass of possible worlds models consists of those in which the frame is an $\omega$-sequence; let us call these models *$\omega$-sequence models.*[17] This may be taken to be a particularly natural idealised scenario insofar as it depicts discrete linear time going on indefinitely, which seems the appropriate idealisation of the actual structure of time.

Even though it is assumed that each $f(t_i)$ is recursively enumerable, it is *not* assumed that the extension of $\Box$ in the model as a whole (i.e. $\bigcup_i f(t_i)$) is recursively enumerable: there are many intuitive branching time models according to which there is a non-recursively enumerable collection of *a priori* knowable sentences of $\mathcal{L}_{EA}$ in which *ECT* is nonetheless true. Indeed, requirement (V) is no restriction at all on the complexity of the content of what is *a priori* knowable. Let $\langle\phi_i\rangle_i$ be an enumeration of the set of true sentences of the language of arithmetic; then, if we let the extension of *a priori* knowledge at time $i$ be the logical closure of $\{\phi_1, \ldots, \phi_i\}$, the constraint is satisfied, while this entails that the extension of *a priori* knowability over time is the collection of all arithmetical truths.

The possible worlds models that we have discussed above seem to incorporate reasonable idealisations on *a priori* knowability. There is, however, a worry about the justification of one part of the soundness requirement (IV). The statement $\phi \in f(t_i) \Rightarrow \bigcup_i f(t_i) \models \phi$ is indeed supported by the fact that a subject can only come to know *a priori* at a certain moment in time that $\phi$, if $\phi$ is true *from a timeless perspective.* However, it is unclear how we can justify that $\phi \in f(t_i) \Rightarrow f(t_i) \models \phi$. The concern here goes back to the worries that we expressed about Thesis 6 in §11.4.1. Might it not be the case that at some stage the mathematical community has proved in a *non-constructive* manner that $\exists x \Box \phi(x)$, that is, that it has proved this existential statement without producing a witness? However, we have seen that the doubtful part of the soundness requirement is needed to ensure that each $f(i)$ has the numerical existence property, which is in turn used in the proof of Theorem 6. Thus, the class of models which we have proved to validate *ECT* is perhaps not as broad as one might wish.

A specific class of models that one can consider is the class of models in which the extension of absolute provability is given by a *systematic* transfinite progression of formal theories in the sense of [Feferman 1962]. Here subsequent systems are generated from earlier ones by adding uniform reflection principles to what has already been obtained. These systems were the focus of [Kreisel 1972], in which Kreisel tried to determine the truth value of principles in the vicinity of (but probably not quite identical to) *ECT*. We have reserved a detailed discussion of these models for another occasion; let it suffice to say here that in all such models, *ECT* holds (as does *EA*).[18]

## 11.5 Conclusion

Gödel's disjunction is generally taken to have been shown to be true. But until now we have no compelling evidence for or against any of its two disjuncts.

In this chapter, we have investigated a related disjunctive thesis according to which either Epistemic Church's Thesis (*ECT*) is false, or there are absolutely undecidable propositions expressible in the language of epistemic arithmetic.

It has emerged that this new disjunctive thesis is in the same boat as Gödel's disjunction. The new disjunction is unassailable, but we have no convincing philosophical arguments for or against each of its disjuncts taken individually. In particular, at present we have no convincing *a priori* argument for *ECT*. And in the absence of such an argument, it also seems difficult to find an *a priori* argument for the thesis that there are absolutely undecidable propositions. So it is not immediately obvious how the new disjunctive thesis can be seen as a stepping stone to *a priori* knowledge about the limits of the extension of the notion of *a priori* knowability.

We therefore went on to "test" *ECT* in models for the language of epistemic arithmetic. It turns out that in a wide class of such models, *ECT* holds. However, the significance of this finding is limited by the fact that it is built into these models that there are no non-constructive ways of proving statements of the form $\forall x \exists y \Box \phi(x, y)$; we have seen that it is difficult to argue convincingly for (or against) this assumption. In the end, we must therefore conclude that it would be premature to claim that there is quasi-empirical evidence for the truth (or for the falsity) of *ECT*. At present we therefore also lack evidence for the existence of absolutely undecidable propositions expressible in the language of epistemic arithmetic.

## Acknowledgments

Earlier versions of this article were presented at the *Plurals, Predicates and Paradoxes* seminar at Birkbeck College, University of London, at the *Logic Colloquium* in Manchester, at the *Midwest Philosophy of Mathematics Workshop* at the University of Notre Dame, at the *Philosophy of Mathematics Seminar* at the University of Oxford, and at the *Intensionality in Mathematics* conference at the University of Lund: thanks to the audiences for valuable reactions. We are grateful to Roy Cook, Jan Heylen, Harold Hodes, Peter Koellner, Richard Pettigrew, Philip Welch, and Tim Williamson for insightful comments.

## Notes

1. For an extended discussion of the implications and non-implications of Gödel's theorems, see [Franzen 2005].
2. For a discussion of the different viewpoints, see [Williamson 2000, chapter 12].
3. See [Antonutti Marfori and Horsten subm] and [Kreisel 1972].
4. For a discussion, see [Horsten 1997, p. 640].
5. Of course, a provability predicate with the properties assumed in Proposition 2 cannot be *proven* to exist in *EA*.
6. See e.g. the debate in *The Bulletin of Symbolic Logic* 6(4), 2000.
7. The notion of being a Turing machine can be formalised in the background language of arithmetic in the standard way in terms of Kleene's $T$-predicate and the $U$ function symbol.
8. This is not necessary, though; it suffices for the purposes of this paper that functional predicates range over formulas in the language of arithmetic, or even over a fragment of this language.
9. For a discussion of the intuitionistic version of Church's thesis, see [Troelstra and van Dalen 1988].

10. Note that the converse of the Church–Turing thesis entails that if *ECT* holds, then for every functional $\phi \in \mathcal{L}_{EA}$: if $\phi$ is calculable, then the graph of $\phi$ is algorithmically computable. (Thanks to Jan Heylen for pointing this out.)
11. If *ECT* is instead treated as a *hypothesis*, then the conclusion of this corollary does not follow. For a more extensive discussion of the treatment of *ECT* as an axiom scheme versus its treatment as a hypothesis, see [Halbach and Horsten 2000] and [Horsten 2006].
12. For an extended discussion, see [Sieg 1994] and [Soare 1996]. Actually, it was Gandy who first (in print) argued that Turing had given an *a priori* proof of the Church–Turing thesis: see [Gandy 1988].
13. For a discussion of this argument, see [Horsten 2006].
14. See [Antonutti Marfori 2010] and [Antonutti Marfori 2013] for discussions of what might be reasonable idealisations built into the notion of provability in principle. For a contrasting view, see [Williamson 2015] in this volume.
15. Since we are quantifying into the context of $\Box$, what will be said below must be relativised to assignments: see [Alexander 2013] and [Heylen 2013]. For ease of reading, we will ignore this complication in what follows.
16. Carlson denotes this by saying that *EA* is a *machine* [Carlson 2000].
17. In $\omega$-sequence models, the cumulativity condition follows from positive introspection and closure.
18. See [Antonutti Marfori and Horsten subm].

# References


[Alexander 2013] Alexander, S.A. A machine which knows its own code. *Studia Logica* **102**, pp. 567–576, 2014.

[Anderson 1983] Anderson, C.A. The paradox of the knower. *Journal of Philosophy* **80**, pp. 338–355, 1983.

[Antonutti Marfori 2010] Antonutti Marfori, M. Informal provability and mathematical rigour. *Studia Logica* **96**, pp. 261–272, 2010.

[Antonutti Marfori 2013] Antonutti Marfori, M. *Theories of Absolute Provability*. Ph.D. Thesis, University of Bristol, 2013.

[Antonutti Marfori and Horsten subm] Antonutti Marfori, M. and Horsten, L. Human effective computability. Submitted, 2015.

[Black 2000] Black, R. Proving Church's Thesis. *Philosophia Mathematica* **8**, pp. 244–258, 2000.

[Boolos 1982] Boolos, G. Extremely undecidable sentences. *Journal of Symbolic Logic* **47**, pp. 191–196, 1982.

[Carlson 2000] Carlson, T. Knowledge, machines, and the consistency of Reinhardt's strong mechanistic thesis. *Annals of Pure and Applied Logic* **105**: pp. 51–82, 2000.

[Carlson 2015] Carlson, T. Can a machine know that it is a machine? This volume.

[Enderton 2001] Enderton, H.B. *A Mathematical Introduction to Logic*, Second Edition. San Diego, California, Academic Press, 2001.

[Feferman 1962] Feferman, S. Transfinite recursive progressions of formal theories. *Journal of Symbolic Logic* **27**, pp. 259–316, 1962.

[Feferman and Solovay 1990] Feferman, S. and Solovay, R. Introductory note to 1972a. In: S. Feferman et al. (eds.) *Kurt Gödel. Collected Works. Volume II: Publications 1938–1974*, pp. 281–304, Oxford: Oxford University Press, 1990.

[Fitch 1963] Fitch, F. A logical analysis of some value concepts. *Journal of Symbolic Logic* **28**, pp. 135–142, 1963.

[Flagg 1985] Flagg, R. Church's Thesis is consistent with Epistemic Arithmetic. In [Shapiro 1985a, pp. 121–172].

[Folina 1998] Folina, J. Church's Thesis: prelude to a proof. *Philosophia Mathematica* **6**, pp. 302–323, 1998.

[Franzen 2005] Franzen, T. *Gödel's Theorem: An Incomplete Guide to its Use and Abuse*. A.K. Peters, Wellesley, MA, 2005.

[Gandy 1988] Gandy, R. The confluence of ideas in 1936. In: R. Herken (ed.) *The Universal Turing Machine: A Half Century Survey*, pp. 55–111, New York: Oxford University Press, 1988.

[Gödel 1951] Gödel, K. Some basic theorems on the foundations of mathematics and their implications. [1951] In: S. Feferman et al. (eds.) *Kurt Gödel. Collected Works. Volume III: Unpublished Essays and Lectures*, pp. 304–323, Oxford: Oxford University Press, 1995.

[Goodman 1986] Goodman, N., Flagg realizability in Epistemic Arithmetic. *Journal of Symbolic Logic* **51**, pp. 387–392, 1986.

[Halbach and Horsten 2000] Halbach, V. and Horsten, L. Two proof-theoretic remarks about *EA + ECT*. *Mathematical Logic Quarterly* **46**, pp. 461–465, 2000.

[Heylen 2013] Heylen, J. Modal-epistemic arithmetic and the problem of quantifying in. *Synthese* **190**, pp. 89–111, 2013.

[Horsten 1997] Horsten, L., Provability in principle and controversial constructivistic principles. *Journal of Philosophical Logic* **26**, pp. 635–660, 1997.

[Horsten 1998] Horsten, L. In defense of Epistemic Arithmetic. *Synthese* **116**, pp. 1–25, 1998.

[Horsten 2006] Horsten, L. Formalizing Church's Thesis. In: A. Olszewski et al. (eds.) *Church's Thesis after 70 years*. Heusenstamm: Ontos Verlag, 2006.

[Kreisel 1972] Kreisel, G. Which number theoretic problems can be solved in recursive progressions on $\Pi^1_1$-paths through **O**? *Journal of Symbolic Logic* **37**, pp. 311–334, 1972.

[Lucas 1961] Lucas, J.R. Minds, machines and Gödel. *Philosophy* **96**, pp. 112–127, 1961.

[Myhill 1960] Myhill, J. Some remarks on the notion of proof. *Journal of Philosophy* **57**, pp. 461–471, 1960.

[Penrose 1989] Penrose, R. *The Emperor's New Mind: Concerning Computers, Minds, and the Laws of Physics*. Oxford: Oxford University Press, 1989.

[Penrose 1994] Penrose, R. *Shadows of the Mind. A Search for the Missing Science of Consciousness.* Oxford: Oxford University Press, 1994.

[Reinhardt 1986] Reinhardt, W. Epistemic theories and the interpretation of Gödel's incompleteness theorems. *Journal of Philosophical Logic* **15**, pp. 427–474, 1986.

[Shapiro 1985a] Shapiro, S. *Intensional Mathematics.* Amsterdam: North-Holland, 1985.

[Shapiro 1985b] Shapiro, S. Epistemic and intuitionistic arithmetic. In: [Shapiro 1985a, pp. 11–46].

[Sieg 1994] Sieg, W. Mechanical procedures and mathematical experience. In A. George (ed.), *Mathematics and Mind*, New York: Oxford University Press, 1994.

[Soare 1996] Soare, R. I. Computability and recursion. *Bulletin of Symbolic Logic* **2**, pp. 284–321, 1996.

[Troelstra and van Dalen 1988] Troelstra, A. and van Dalen, D. *Constructivism in Mathematics. An Introduction. Volume 1* Amsterdam: North-Holland, 1988.

[Turing 1936] Turing, A.M. On computable numbers, with an application to the Entscheidungsproblem. *Proceedings of the London Mathematical Society* **42**, pp. 230–265, 1936.

[Turing 1939] Turing, A.M. Systems of logic defined by ordinals. *Proceedings of the London Mathematical Society* Ser. 2, **45**: pp. 161–228, 1939.

[Williamson 2000] Williamson, T. *Knowledge and its Limits.* Oxford: Oxford University Press, 2000.

[Williamson 2015] Williamson, T. Absolute provability and safe knowledge of axioms. This volume.

# INDEX

The manufacturer's authorised representative in the EU for product safety is Oxford University Press España S.A. of El Parque Empresarial San Fernando de Henares, Avenida de Castilla, 2 - 28830 Madrid (www.oup.es/en or product.safety@oup.com). OUP España S.A. also acts as importer into Spain of products made by the manufacturer.
Printed and bound by CPI Group (UK) Ltd, Croydon, CR0 4YY
14/09/2026
02216305-0006